ELSEVIER'S DICTIONARY OF AMPHIBIANS

ELSEVIER'S DICTIONARY OF AMPHIBIANS

in

Latin, English, German, French and Italian

compiled by

MURRAY WROBEL
London, United Kingdom

2004

ELSEVIER
Amsterdam – Boston – Heidelberg – London – New York – Oxford
Paris – San Diego – San Francisco – Singapore – Sydney – Tokyo

ELSEVIER B.V.	ELSEVIER Inc.	ELSEVIER Ltd	ELSEVIER Ltd
Sara Burgerhartstraat 25	525 B Street, Suite 1900	The Boulevard, Langford Lane	84 Theobalds Road
P.O. Box 211	San Diego, CA 92101-4495	Kidlington, Oxford OX5 1GB	London WC1X 8RR
1000 AE Amsterdam	USA	UK	UK
The Netherlands			

First edition 2004

Library of Congress Cataloging in Publication Data
A catalog record is available from the Library of Congress.

British Library Cataloguing in Publication Data
A catalogue record is available from the British Library.

ISBN: 0-444-51374-4

♾ The paper used in this publication meets the requirements of ANSI/NISO Z39.48-1992 (Permanence of Paper).
Printed in The Netherlands.

Preface

This dictionary has been compiled with the aim of giving an overview of the English, German, French and Italian names of amphibians. The Basic Table contains, in alphabetical order, the scientific names of orders, families species and some sub-species with, again in alphabetical order, their identified names in English, German, French and Italian, which are given in the singular for species and sub-species and in the plural for other terms. The relevant order and family are shown for each term and the taxonomy is that used by the majority of the sources consulted.

The criteria to determine which names are hyphenated or shown as two or more words are based on the frequency they are used by, and the relative importance I have attached to, the sources consulted.

My special thanks go to Katayoun Asadollahzadeh-Sarab who prepared this work for publication and to my wife Myra for her support and encouragement.

Murray Wrobel

Contents

Abbreviations

The following abbreviations denote a general, but not necessarily exclusive, use in the various areas.

(Ants)	-	One or more of the French-speaking islands of the West Indies
(ANZ)	-	Australia and New Zealand
(CSA)	-	Central and Southern Africa
(ISC)	-	Indian Subcontinent
(NA)	-	North America
(Qué)	-	Quebec
(SEA)	-	South East Asia
(VIP)	-	Virgin Islands and Puerto Rico
(WI)	-	One or more of the English-speaking islands of the West Indies

Basic Table

A

1 ***Acanthixalus***
Anura - Hyperoliidae
e African Wart Frogs

2 ***Acanthixalus spinosus***
Anura - Hyperoliidae
e African Wart Frog

3 ***Acris***
Anura - Hylidae
e Cricket Frogs
d Heuschreckenfrösche; Grillenfrösche; Grillenlaubfrösche; Westliche Grillenfrösche

4 ***Acris crepitans***
Anura - Hylidae
e Northern Cricket Frog; Cricket Frog (NA)
d Westlicher Grillenfrosch; Westlicher Heuschreckenfrosch; Westlicher Grillenlaubfrosch; Nördlicher Grillenfrosch; Grillenfrosch; Heuschreckenfrosch

5 ***Acris crepitans blanchardi***
Anura - Hylidae
e Blanchard's Cricket Frog
f Rainette grillon de Blanchard

6 ***Acris crepitans crepitans***
Anura - Hylidae
e Northern Cricket Frog; Eastern Cricket Frog

7 ***Acris crepitans paludicola***
Anura - Hylidae
e Coastal Cricket Frog

8 ***Acris gryllus***
Anura - Hylidae
e Southern Cricket Frog
d Südlicher Grillenfrosch; Südlicher Heuschreckenfrosch; Südlicher Grillenlaubfrosch; Grillenfrosch; Heuschreckenfrosch

9 ***Acris gryllus dorsalis***
Anura - Hylidae
e Florida Cricket Frog

10 ***Acris gryllus gryllus***
Anura - Hylidae
e Southern Cricket Frog; Coastal Plain Cricket Frog

11 ***Adelastes***
Anura - Microhylidae
e Neblina Frogs

12 ***Adelastes hylonomus***
Anura - Microhylidae
e Neblina Frog

13 ***Adelophryne***
Anura - Leptodactylidae
e Shield Froga

14 ***Adelophryne adiastola***
Anura - Leptodactylidae
e Yapima Shield Frog

15 ***Adelophryne gutturosa***
Anura - Leptodactylidae
e Guiana Shield Frog
f Adelophryne goitreux

16 ***Adelotus***
Anura - Myobatrachidae
e Tusked Frogs
d Australische Unkenfrösche

17 ***Adelotus brevis***
Anura - Myobatrachidae
e Tusked Frog
d Australischer Unkenfrosch

18 ***Adenomera***
Anura - Leptodactylidae
e Tropical Bullfrogs
f Adénomères

19 ***Adenomera andreae***
Anura - Leptodactylidae
e Lowland Tropical Bullfrog
f Adénomère de Müller

20 ***Adenomera bokermanni***
Anura - Leptodactylidae
e Bokermann's Tropical Bullfrog

21 ***Adenomera hylaledactyla***
Anura - Leptodactylidae
e Napo Tropical Bullfrog
f Adénomère de Cope

22 ***Adenomera lutzi***
Anura - Leptodactylidae
e Lutz'sTropical Bullfrog

23 ***Adenomera marmorata***
Anura - Leptodactylidae
e Marbled Tropical Bullfrog

24 ***Adenomera martinezi***
Anura - Leptodactylidae
e Martinez's Tropical Bullfrog

25 ***Afrana***
Anura - Ranidae
e River Frogs

26 ***Afrana angolensis***
Anura - Ranidae
e Angola Frog; Dusky-throated Frog (CSA); Common River Frog (CSA); Angola River Frog (CSA); Northern Rana; Common Rana (CSA)

27 ***Afrana dracomontana***
Anura - Ranidae
e Sani Pass Frog; Drakensberg River Frog (CSA); Drakensberg Frog (CSA)

28 ***Afrana fuscigula***
Anura - Ranidae
e Brown-throated Frog; Cape River Frog (CSA); Dark-throated River Frog (CSA); Dusky-throated River Frog (CSA); Marble-throat Frog; Cape Rana (CSA)

29 ***Afrana johnstoni***
Anura - Ranidae
e Tshiromo Frog; Johnston's Rana; Johnstone's River Frog (CSA)

30 ***Afrana vandijki***
Anura - Ranidae
e Van Dijke's River Frog

31 ***Afrixalus***
Anura - Hyperoliidae
e Banana Frogs; Leaf-folding Frogs (CSA); Spiny Reed Frogs (CSA); Cat's-eye Reed Frogs (CSA)
d Bananenfrösche; Stachelfrosch

32 ***Afrixalus aureus***
Anura - Hyperoliidae
e Golden Banana Frog; Golden Leaf-folding Frog (CSA)

33 ***Afrixalus brachycnemis***
Anura - Hyperoliidae
e Malawi Banana Frog; Golden Spiny Reed Frog (CSA); Short-legged Banana Frog (CSA); Eastern Short-legged Banana Frog; Short-limbed Banana Frog; Lesser Banana Frog (CSA); Golden Leaf-folding Frog (CSA); Golden Spiny Frog (CSA)

34 ***Afrixalus clarkei***
Anura - Hyperoliidae
e Clarke's Banana Frog

35 ***Afrixalus crotalus***
Anura - Hyperoliidae
e Zimbabwe Banana Frog; Snoring Leaf-folding Frog (CSA)

36 ***Afrixalus delicatus***
Anura - Hyperoliidae
e Pickersgill's Banana Frog; Deliccate Leaf-folding Frog (CSA)

37 ***Afrixalus dorsalis***
Anura - Hyperoliidae
e Brown Banana Frog; Cameroon Banana Frog
d Bananenfrosch
f Grenouille afrixale

38 ***Afrixalus ensaticala***
Anura - Hyperoliidae
e Bonga Banana Frog

39 ***Afrixalus equatorialis***
Anura - Hyperoliidae
e Zaire Banana Frog

40 ***Afrixalus fornasini***
Anura - Hyperoliidae
e Silver-banded Banana Frog; Greater Leaf-folding Frog (CSA); Brown-striped Spiny Reed Frog; Brown-and-white Spiny Reed Frog (CSA); Mozambique Banana Frog; Spiny Leaf-folding Frog (CSA); Fornasini's Banana Frog
d Bananenfrosch

41 ***Afrixalus fulvovittatus***
Anura - Hyperoliidae
e Banded Banana Frog; Lesser Banana Frog

42 ***Afrixalus knysnae***
Anura - Hyperoliidae
e Knysa Banana Frog; Knysna Leaf-folding Frog (CSA)

43 ***Afrixalus lacteus***
Anura - Hyperoliidae
e Cameroon Banana Frog

44 ***Afrixalus laevis***
Anura - Hyperoliidae
e Liberia Banana Frog

45 ***Afrixalus leucostictus***
Anura - Hyperoliidae
e Makese Banana Frog

46 ***Afrixalus lindholmi***
Anura - Hyperoliidae
e Lindholm's Banana Frog

47 ***Afrixalus nigeriensis***
Anura - Hyperoliidae
e Nigeria Banana Frog

48 ***Afrixalus orophilus***
Anura - Hyperoliidae
e Kivu Banana Frog

49 ***Afrixalus osorioi***
Anura - Hyperoliidae
e Angola Banana Frog

50 ***Afrixalus paradorsalis***
Anura - Hyperoliidae
e Foulassi Banana Frog

51 ***Afrixalus schneideri***
Anura - Hyperoliidae
e Schneider's Banana Frog

52 ***Afrixalus septentrionalis***
Anura - Hyperoliidae
e Kenya Banana Frog

53 ***Afrixalus spinifrons***
Anura - Hyperoliidae
e Natal Banana Frog; Natal Leaf-folding Frog (CSA); Golden Spiny Reed Frog (CSA)

54 ***Afrixalus stuhlmanni***
Anura - Hyperoliidae
e Stuhlmann's Banana Frog

55 ***Afrixalus sylvaticus***
Anura - Hyperoliidae
e Forest Banana Frog

56 ***Afrixalus ulugurensis***
Anura - Hyperoliidae
e Uluguru Banana Frog; Montane Banana Frog

57 ***Afrixalus vittiger***
Anura - Hyperoliidae
e Weidholz's Banana Frog

58 ***Afrixalus wittei***
Anura - Hyperoliidae
e De Witte's Banana Frog; De Witte's Leaping Frog

59 ***Afrocaecilia***
Gymnophiona - Caeciliaidae
e African Caecilians

60 ***Afrocaecilia changamwensis***
Gymnophiona - Caeciliaidae
e Changamwensis African Caecilian; Changamwe Lowand Caecilian

61 ***Afrocaecilia taitana***
Gymnophiona - Caeciliaidae
e Taita African Caecilian; Taita Mountains Caecilians

62 ***Afrocaecilia ulugurensis***
Gymnophiona - Caeciliaidae
e Uluguru African Caecilian; Uluguru Pink Caecilian

63 ***Agalychnis***
Anura - Hylidae
e Leaf Frogs
d Greiffrösche; Rotaugenlaubfrösche; Rotaugenfrösche

64 ***Agalychnis annae***
Anura - Hylidae
e Blue-sided Leaf Frog; Orange-eyed Tree Frog

65 ***Agalychnis calcarifer***
Anura - Hylidae
e Splendid Leaf Frog

66 ***Agalychnis callidryas***
Anura - Hylidae
e Red-eyed Tree Frog; Red-eyed Frog
d Rotaugenlaubfrosch; Rotaugenfrosch
f Rainette aux yeux rouges

67 ***Agalychnis craspedopus***
Anura - Hylidae
e Fringed Leaf Frog

68 ***Agalychnis litodryas***
Anura - Hylidae
e Pink-sided Leaf Frog

69 ***Agalychnis moreletii***
Anura - Hylidae
e Morelet's Leaf Frog; Morelet's Frog
d Makifrosch
f Rainette de Morelet

70 ***Agalychnis saltator***
Anura - Hylidae
e Misfit Leaf Frog
d Kleiner Rotaugenlaubfrosch

71 ***Agalychnis spurelli***
Anura - Hylidae
e Gliding Leaf Frog; Spurrell's Tree Frog

72 ***Aglyptodactylus***
Anura - Rhacophoridae
e Madagascar Jumping Frogs

73 ***Aglyptodactylus madagascariensis***
Anura - Rhacophoridae
e Madagascar Jumping Frog

74 ***Alexteroon***
Anura - Hyperoliidae
e Midwife Frogs; Midwife Toad

75 ***Alexteroon obstetricans***
Anura - Hyperoliidae
e Midwife Frog; Midwife Toad

76 ***Allophryne***
Anura - Allophrynidae
e Turkeit Hill Tree Frogs

77 ***Allophryne ruthveni***
Anura - Allophrynidae
e Turkeit Hill Tree Frog
f Allophryne de Ruthven

78 **Allophrynidae**
Anura
e Turkeit Hill Tree Frogs

79 ***Alsodes***
Anura - Leptodactylidae
e Spinychest Frogs

80 ***Alsodes barrioi***
Anura - Leptodactylidae
e Cabreria Spinychest Frog

81 ***Alsodes gorgola***
Anura - Leptodactylidae
e Toncheck Spinychest Frog

82 ***Alsodes laevis***
Anura - Leptodactylidae
e Portrero Spinychest Frog

83 ***Alsodes montanus***
Anura - Leptodactylidae
e Mountain Spinychest Frog

84 ***Alsodes monticola***
Anura - Leptodactylidae
e Island Spinychest Frog

85 ***Alsodes nodosus***
Anura - Leptodactylidae
e Black Spinychest Frog

86 ***Alsodes pehuenche***
Anura - Leptodactylidae
e Pehuenche Spinychest Frog

87 ***Alsodes tumultuosus***
Anura - Leptodactylidae
e La Parva Spinychest Frog

88 ***Alsodes vanzolinii***
Anura - Leptodactylidae
e Vanzolini's Spinychest Frog

89 ***Alsodes verrucosus***
Anura - Leptodactylidae
e Olive Spinychest Frog

90 ***Alsodes vittatus***
Anura - Leptodactylidae
e Malleco Spinychest Frog

91 ***Altiphrynoides***
Anura - Bufonidae
e Ethiopia Toads

92 ***Altiphrynoides malcolmi***
Anura - Bufonidae
e Ethiopia Toad; Malcolm's Ethiopia Toad

93 ***Altirana***
Anura - Ranidae
e High Himalaya Frogs

94 ***Altirana parkeri***
Anura - Ranidae
e High Himalaya Frog

95 ***Alytes***
Anura - Discoglossidae
e Midwife Toads
d Geburtshelferkröten
f Alytes; Crapauds accoucheurs; Alytes accoucheurs

96 ***Alytes cisternasii***
Anura - Discoglossidae
e Brown Midwife Toad; Iberian Midwife Toad; Spanish Midwife Toad
d Iberische Geburtshelferkröte; Spanische Geburtshelferkröte
f Alyte de Cisternas; Crapaud accoucheur d'Espagne; Accoucheur ibérique
i Alite iberico; Sapo partero de cisternas

97 ***Alytes dickhilleni***
Anura - Discoglossidae
e Southern Midwife Toad

98 ***Alytes muletensis***
Anura - Discoglossidae
e Majorca Midwife Toad; Mallorcan Midwife Toad; Majorcan Midwife Toad
d Mallorca-Geburtshelferkröte; Balearen-Geburtshelferkröte
f Alyte de Majorque

99 ***Alytes obstetricans***
Anura - Discoglossidae
e Olive Midwife Toad; Midwife Toad; Common Midwife Toad
d Geburtshelferkröte; Glockenfrosch; Feßler; Gemeine Geburtshelferkröte; Feßlerkröte
f Alyte accoucheur; Crapaud accoucheur; Alyte accoucheur commun

100 ***Ambystoma***
Caudata - Ambystomatidae
e Mole Salamanders; Ambystomids; Ambystomas
d Echte Queerzahnmolche
f Ambystomes

101 ***Ambystoma amblycephalum***
Caudata - Ambystomatidae
e Blunt-headed Salamander; Blunthead Salamander

102 ***Ambystoma andersoni***
Caudata - Ambystomatidae
e Anderson's Salamander; Alligator Newt
d Japanischer Krokodilmolch
f Salamandre d'Anderson; Salamandre alligator d'Anderson; Salamandre alligator

103 ***Ambystoma annulatum***
Caudata - Ambystomatidae
e Ringed Salamander
d Ringel-Querzahnmolch; Ringel-Querzahnsalamander; Geringelter Salamander
f Salamandre annelée; Ambystome annelé

104 ***Ambystoma barbouri***
Caudata - Ambystomatidae
e Streamside Salamander

105 ***Ambystoma bombypellum***
Caudata - Ambystomatidae
e Delicateskin Salamander

106 ***Ambystoma californiense***
Caudata - Ambystomatidae
e California Tiger Salamander

107 ***Ambystoma cingulatum***
Caudata - Ambystomatidae
e Flatwoods Salamander

d Genetzter Querzahnmolch; Genetzter Querzahnsalamander

108 ***Ambystoma cingulatum bishopi***
Caudata - Ambystomatidae
e Reticulated Flatwoods Salamander

109 ***Ambystoma cingulatum cingulatum***
Caudata - Ambystomatidae
e Frosted Flatwoods Salamander

110 ***Ambystoma dumerilii***
Caudata - Ambystomatidae
e Lake Patzcuaro Salamander
d Patzcuaro-Querzahnmolch; Patzcuaro-Querzahnsalamander

111 ***Ambystoma flavipiperatum***
Caudata - Ambystomatidae
e Yellow-peppered Salamander; Northwestern Salamander
f Salamandre foncée; Ambystome foncé

112 ***Ambystoma gracile***
Caudata - Ambystomatidae
e Northwestern Salamander

113 ***Ambystoma gracile decorticatum***
Caudata - Ambystomatidae
e British Columbia Salamander

114 ***Ambystoma gracile gracile***
Caudata - Ambystomatidae
e Brown Salamander

115 ***Ambystoma jeffersonianum***
Caudata - Ambystomatidae
e Jefferson Salamander; Jefferson's Salamander
d Jeffersons Querzahnmolch; Jeffersons Querzahnsalamander; Jefferson-Salamander
f Salamandre de Jefferson; Ambystome de Jefferson; Salamandre de Jefferson à points bleus

116 ***Ambystoma laterale***
Caudata - Ambystomatidae
e Blue-spotted Salamander
d Blauflecken-Querzahnmolch; Blauflecken-Querzahnsalamander
f Salamandre à points bleus; Ambystome à points bleus; Ambystome maculée bleue; Salamandre maculée bleue

117 ***Ambystoma lermaense***
Caudata - Ambystomatidae
e Lake Lerma Salamander

118 ***Ambystoma mabeei***
Caudata - Ambystomatidae
e Mabee's Salamander
d Mabees Querzahnmolch; Mabees Querzahnsalamander; Mabees Salamander

119 ***Ambystoma macrodactylum***
Caudata - Ambystomatidae
e Long-toed Salamander
d Langzehen-Querzahnmolch; Langzehen-Querzahnsalamander; Langzehen-Salamander
f Salamandre à longs doigts; Ambystome à long doigts

120 ***Ambystoma macrodactylum columbianum***
Caudata - Ambystomatidae
e Eastern Long-toed Salamander; Central Long-toed Salamander (NA)

121 ***Ambystoma macrodactylum croceum***
Caudata - Ambystomatidae
e Santa Cruz Long-toed Salamander

122 ***Ambystoma macrodactylum krausei***
Caudata - Ambystomatidae
e Northern Long-toed Salamander; Eastern Long-toed Salamander (NA)

123 ***Ambystoma macrodactylum macrodactylum***
Caudata - Ambystomatidae
e Western Long-toed Salamander

124 ***Ambystoma macrodactylum sigillatum***
Caudata - Ambystomatidae
e Southern Long-toed Salamander

125 ***Ambystoma maculatum***
Caudata - Ambystomatidae
e Spotted Salamander (NA); American Spotted Salamander; Yellow-spotted Salamander
d Fleckenquerzahnmolch; Fleckensalamander; Fleckenquerzahnsalamander; Gefleckter Querzahnmolch; Gefleckter Querzahnsalamander
f Salamandre maculée; Ambystome maculé; Ambystome tacheté à points blancs; Salamandre taschetée à points blancs

126 ***Ambystoma mavortium***
Caudata - Ambystomatidae
e Barred Tiger Salamander
d Gebänderter Tigersalamander

127 ***Ambystoma mexicanum***
Caudata - Ambystomatidae
e Axolotl
d Axolotl; Mexikanischer Querzahnmolch; Mexikanischer Querzahnsalamander
f Axolotl
i Axolotl

128 ***Ambystoma nothogenes***
Caudata - Ambystomatidae
e Kelley's Island Salamander

129 ***Ambystoma opacum***
Caudata - Ambystomatidae
e Marbled Salamander
d Gebänderter Salamander; Gebänderter Querzahnmolch; Gebänderter Querzahnsalamander; Marmor-Querzahnsalamander; Marmorquerzahnmolch
f Salamandre marbrée; Salamandre opaque; Ambystome marbré; Ambystome opaque
i Salamandra marmorata

130 ***Ambystoma ordinarium***
Caudata - Ambystomatidae
e Puerto Hondo Stream Salamander; Michoacan Stream Salamander

131 ***Ambystoma platineum***
Caudata - Ambystomatidae
e Silvery Salamander

132 ***Ambystoma rosaceum***
Caudata - Ambystomatidae
e Rosy Salamander; Tarahumara Salamander
d Chihuahua-Querzahnmolch; Chihuahua-Querzahnsalamander

133 ***Ambystoma rosaceum nigrum***
Caudata - Ambystomatidae
e Durango Tarahumara Salamander

134 ***Ambystoma rosaceum rosaceum***
Caudata - Ambystomatidae
e Common Tarahumara Salamander

135 ***Ambystoma talpoideum***
Caudata - Ambystomatidae
e Mole Salamander
d Maulwurf-Querzahnmolch; Maulwurf-Querzahnsalamander; Maulwurf-Salamander
f Salamandre taupe; Ambystome taupe

136 ***Ambystoma taylori***
Caudata - Ambystomatidae
e Taylor's Salamander

137 ***Ambystoma texanum***
Caudata - Ambystomatidae
e Small-mouthed Salamader; Smallmouth Salamander
d Schmalkopf-Querzahnmolch; Schmalkopf-Quersahnsalamander; Engmaul-Querzahnmolch; Kurzmaul-Querzahnmolch; Engmaul-Querzahnsalamander; Kurzmaul-Querzahnsalamander; Texas-Querzahnmolch
f Ambystome à nez court; Salamandre à nez court

138 ***Ambystoma tigrinum***
Caudata - Ambystomatidae
e Tiger Salamander
d Tiger-Querzahnmolch; Tiger-Querzahnsalamander; Tigersalamander; Östlicher Tigerquerzahnmolch
f Ambystome tigré

139 ***Ambystoma tigrinum californiense***
Caudata - Ambystomatidae
e California Tiger Salamander

140 ***Ambystoma tigrinum diaboli***
Caudata - Ambystomatidae
e Grey Tiger Salamander

141 ***Ambystoma tigrinum mavortium***
Caudata - Ambystomatidae
e Barred Tiger Salamander
d Tiger-Querzahnmolch

142 ***Ambystoma tigrinum melanostictum***
Caudata - Ambystomatidae
e Blotched Tiger Salamander

143 ***Ambystoma tigrinum nebulosum***
Caudata - Ambystomatidae
e Arizona Tiger Salamander

144 ***Ambystoma tigrinum stebbinsi***
Caudata - Ambystomatidae
e Sonoran Tiger Salamander; Sonora Tiger Salamander

145 ***Ambystoma tigrinum tigrinum***
Caudata - Ambystomatidae
e Eastern Tiger Salamander; Eastern Tiger (NA)

146 ***Ambystoma tremblayi***
Caudata - Ambystomatidae
e Tremblay's Salamander
d Tremblays Salamander
f Salamandre de Tremblay

147 ***Ambystoma velascoi***
Caudata - Ambystomatidae
e Plateau Tiger Salamander

148 **Ambystomatidae**
Caudata
e Mole Salamanders; Ambystomids
d Querzahnmolche; Querzahnsalamander
f Ambystomatidés

149 ***Ametia***
Anura - Ranidae
e Aquatic River Frogs

150 ***Ametia vertebralis***
Anura - Ranidae
e Aquatic River Frog (CSA); Ice Frog; Water Rana (CSA)

151 ***Amolops***
Anura - Ranidae
e Sucker Frogs; Cascade Frogs
d Ostasiatischer Kaskadenfrösche; Kaskadenfrösche

152 ***Amolops chapaensis***
Anura - Ranidae
e Vietnam Sucker Frog

153 ***Amolops chunganensis***
Anura - Ranidae
e Chungan Sucker Frog
d Chinesischer Heuschreckenfrosch; Chungan-Kaskadenfrosch

154 ***Amolops formosus***
Anura - Ranidae
e Assam Sucker Frog

155 ***Amolops gerbillus***
Anura - Ranidae
e Yembung Sucker Frog

156 ***Amolops granulosus***
Anura - Ranidae
e Sichuan Sucker Frog

157 ***Amolops hainanensis***
Anura - Ranidae
e Hainan Sucker Frog

158 ***Amolops himalayanus***
Anura - Ranidae
e Himalaya Sucker Frog

159 ***Amolops hongkongensis***
Anura - Ranidae
e Hong Kong Sucker Frog; Hong Kong Cascade Frog

160 ***Amolops kangtingensis***
Anura - Ranidae
e Kanting Sucker Frog

161 ***Amolops kaulbacki***
Anura - Ranidae
e Burmese Sucker Frog

162 ***Amolops larutensis***
Anura - Ranidae
e Larut Sucker Frog; Larut Cascade Frog (SEA)
d Larut-Kaskadenfrosch

163 ***Amolops liangshanensis***
Anura - Ranidae
e Liangshan Sucker Frog

164 ***Amolops lifanensis***
Anura - Ranidae
e Lifan Sucker Frog

165 ***Amolops loloensis***
Anura - Ranidae
e Lolokou Sucker Frog

166 ***Amolops longimanus***
Anura - Ranidae
e Kambaiti Sucker Frog

167 ***Amolops mantzorum***
Anura - Ranidae
e Mouping Sucker Frog

168 ***Amolops marmoratus***
Anura - Ranidae
e Marbled Sucker Frog; Himalayan Torrent Frog (ISC)

169 ***Amolops monticola***
Anura - Ranidae
e Mountain Sucker Frog

170 ***Amolops nepalicus***
Anura - Ranidae
e Nepal Sucker Frog

171 ***Amolops ricketti***
Anura - Ranidae
e Chinese Sucker Frog

172 ***Amolops taiwanensis***
Anura - Ranidae
e Taiwan Sucker Frog

173 ***Amolops tormotus***
Anura - Ranidae
e Anhui Sucker Frog

174 ***Amolops torrentis***
Anura - Ranidae
e Torrent Sucker Frog

175 ***Amolops viridimaculatus***
Anura - Ranidae
e Dahaoping Sucker Frog

176 ***Amolops wyiensis***
Anura - Ranidae
e Sanchiang Sucker Frog

177 ***Amphiuma***
Caudata - Amphiumidae
e Amphiumas; Congo Eels; Eel Newts
d Aalmolche
f Amphiumes; Amphiumes du Congo; Amphiumes des fossés

178 ***Amphiuma means***
Caudata - Amphiumidae
e Amphiuma; Two-toed Amphiuma (NA); Two-toed Congo Eel; Congo Eel (NA); Congo Snake
d Zweizehen-Aalmolch
f Amphiume à deux doigts; Amphiume pénétrante
i Salamandra dalle due dite

179 ***Amphiuma pholeter***
Caudata - Amphiumidae
e One-toed Amphiuma; One-toed Congo Eel
d Einzehen-Aalmolch
f Amphiume à un doigt

180 ***Amphiuma tridactylum***
Caudata - Amphiumidae
e Three-toed Salamander; Three-toed Congo Eel; Three-toed Amphiuma (NA)
d Dreizehen-Aalmolch
f Amphiume à trois doigts; Amphiume tridactyle

181 **Amphiumidae**
Caudata
e Congo Eels; Amphiumas
d Aalmolche
f Amphiumes

182 ***Andinophryne***
Anura - Bufonidae
e Andes Toads
d Anden-Kröten
f Crapauds du Chili

183 ***Andinophryne atelopoides***
Anura - Bufonidae
e Western Andes Toad

184 ***Andinophryne colomai***
Anura - Bufonidae
e Carchi Andes Toad

185 ***Andinophryne ollalai***
Anura - Bufonidae
e Tandayapa Andes Toad

186 ***Andrias***
Caudata - Cryptobranchidae
e Giant Salamanders
d Asiatische Riesensalamander
f Salamandres géantes

187 ***Andrias davidianus***
Caudata - Cryptobranchidae
e Chinese Giant Salamander
d Chinesischer Riesensalamander
f Salamandre géante de Chine

188 ***Andrias japonicus***
Caudata - Cryptobranchidae
e Japanese Giant Salamander
d Japanischer Riesensalamander
f Salamandre géante du Japon
i Salamandra gigante

189 ***Aneides***
Caudata - Plethodontidae

e Climbing Salamanders
d Baumsalamander
f Salamandres des arbres

190 ***Aneides aenus***
Caudata - Plethodontidae
e Green Salamander
d Erzsalamander; Grüner Baumsalamander
f Salamandre verte

191 ***Aneides ferreus***
Caudata - Plethodontidae
e Clouded Salamander (NA)
d Baumsalamander; Gewölkter Baumsalamander
f Salamandre pommelée

192 ***Aneides flavipunctatus***
Caudata - Plethodontidae
e Black Salamander
d Schwarzer Baumsalamander; Weißpunktsalamander

193 ***Aneides flavipunctatus flavipunctus***
Caudata - Plethodontidae
e Speckled Black Salamander

194 ***Aneides flavipunctatus niger***
Caudata - Plethodontidae
e Santa Cruz Black Salamander

195 ***Aneides hardyi***
Caudata - Plethodontidae
e Sacramento Mountain Salamander
d Neumexikanischer Baumsalamander

196 ***Aneides lugubris***
Caudata - Plethodontidae
e Arboreal Salamander; Alligator Salamander
d Alligatorsalamander; Kalifornischer Baumsalamander; Düsterer Alligatoren-Salamander; Baumsalamander; Alligatorensalamander
f Salamandre des arbres

197 ***Aneides vagrans***
Caudata - Plethodontidae
e Wandering Salamander

198 ***Anhydrophryne***
Anura - Ranidae
e Rattray's Frogs; Hogsback Frogs
f Grenouilles de Rattray

199 ***Anhydrophryne rattrayi***
Anura - Ranidae
e Rattray's Frog; Rattray's Forest Frog; Hogsback Frog (CSA)
f Grenouille de Rattray

200 ***Anodonthyla***
Anura - Microhylidae
e Climbing Frogs

201 ***Anodonthyla boulengerii***
Anura - Microhylidae
e Boulenger's Climbing Frog

202 ***Anodonthyla montana***
Anura - Microhylidae
e Mountain Climbing Frog

203 ***Anodonthyla nigrigularis***
Anura - Microhylidae
e Black-throted Climbing Frog

204 ***Anodonthyla rouxae***
Anura - Microhylidae
e Madagascar Climbing Frog

205 ***Anotheca***
Anura - Hylidae
e Spiny-headed Tree Frogs; Spine-headed Tree Frogs; Spinyhead Tree Frogs
d Kronenlaubfrösche

206 ***Anotheca spinosa***
Anura - Hylidae
e Spiny-headed Tree Frog; Mexican Hylid; Spine-headed Tree Frog; Spinyhead Tree Frog
d Kronenlaubfrosch

207 ***Ansonia***
Anura - Bufonidae
e Stream Toads; Torrent Toads
d Bachkröten

208 ***Ansonia albomaculata***
Anura - Bufonidae
e Whitebelly Stream Toad

209 ***Ansonia fuliginea***
Anura - Bufonidae
e North Borneo Stream Toad

210 ***Ansonia guibei***
Anura - Bufonidae
e Mesilau Stream Toad

211 ***Ansonia hanitchi***
Anura - Bufonidae
e Kamaian Stream Toad

212 ***Ansonia latidisca***
Anura - Bufonidae
e Sambas Stream Toad

213 ***Ansonia leptopus***
Anura - Bufonidae
e Matang Stream Toad

214 ***Ansonia longidigita***
Anura - Bufonidae
e Long-fingered Stream Toad; Cricket-voiced Toad
d Zirpkröte
f Crapaud à voix de grillon

215 ***Ansonia malayana***
Anura - Bufonidae
e Malaya Stream Toad; Malayan Stream Toad (SEA)
d Malaiische Bachkröte

216 ***Ansonia mcgregori***
Anura - Bufonidae
e McGregor's Stream Toad

217 ***Ansonia minuta***
Anura - Bufonidae
e Tiny Stream Toad

218 ***Ansonia muelleri***
Anura - Bufonidae
e Müller's Stream Toad

219 ***Ansonia ornata***
Anura - Bufonidae
e Ornate Stream Toad; Malabar Torrent Toad (ISC)

220 ***Ansonia penangensis***
Anura - Bufonidae
e Penang Stream Toad
d Penang-Bachkröte

221 ***Ansonia platysoma***
Anura - Bufonidae
e Luidan Stream Toad

222 ***Ansonia rubrigina***
Anura - Bufonidae
e Kerala Stream Toad

223 ***Ansonia siamensis***
Anura - Bufonidae
e Thai Stream Toad; Siamese Stream Toad (SEA)
d Siam-Bachkröte

224 ***Ansonia spinulifer***
Anura - Bufonidae
e Kina Balu Stream Toad

225 ***Ansonia tiomanica***
Anura - Bufonidae
e Pulo Tioman Stream Toad; Tioman Stream Toad (SEA)
d Tioman-Bachkröte

226 ***Ansonia torrentis***
Anura - Bufonidae
e Gunung Mulu Stream Toad

227 ***Aparashpenodon***
Anura - Hylidae
e Casque-headed Tree Frog; Casque-headed Frog
d Panzerkopf-Laubfrosch

228 ***Aparashpenodon bokermanni***
Anura - Hylidae
e Bokermann's Casque-headed Frog

229 ***Aparashpenodon brunoi***
Anura - Hylidae
e Bruno's Casque-headed Frog; Bruno Bony-headed Frog
d Keilkopffrosch

230 ***Aparashpenodon venezolanos***
Anura - Hylidae
e Venezuela Casque-headed Frog

231 ***Aphantophryne***
Anura - Microhylidae
e Guinea Frogs

232 ***Aphantophryne minuta***
Anura - Microhylidae
e Myola Guinea Frog

233 ***Aphantophryne pansa***
Anura - Microhylidae
e Scratchley Guinea Frog

234 ***Aphantophryne sabini***
Anura - Microhylidae
e Guest House Guinea Frog

235 ***Aplastodiscus***
Anura - Hylidae
e Cane-brake Tree Frogs

236 ***Aplastodiscus perviridus***
Anura - Hylidae
e Cane-brake Tree Frog

237 ***Arcovomer***
Anura - Microhylidae
e Passarelli's Frogs

238 ***Arcovomer passarellii***
Anura - Microhylidae
e Passarelli's Frog

239 ***Arenophryne***
Anura - Myobatrachidae
e Australian Dumpy Frogs; Sandhill Frogs; Round Frogs
d Australische Sandfrösche

240 ***Arenophryne rotunda***
Anura - Myobatrachidae
e Australian Dumpy Frog; Dumpy Frog; Sandhill Frog; Round Frog
d Australischer Sandfrosch

241 ***Argenteohyla***
Anura - Hylidae
e Argentine Toads

242 ***Argenteohyla siermersi***
Anura - Hylidae
e Argentine Toad; Red-spotted Argentine Toad

243 ***Arlequinus***
Anura - Hyperoliidae
e Mubebgue Frogs

244 ***Arlequinus krebsi***
Anura - Hyperoliidae
e Mubebgue Frog

245 ***Aromobates***
Anura - Dendrobatidae
e Skunk Frogs

246 ***Aromobates nocturnus***
Anura - Dendrobatidae
e Skunk Frog

247 ***Arthroleptella***
Anura - Ranidae
e Bandit Frogs; Chirping Frogs (CSA); Squeakers (CSA)

248 ***Arthroleptella bicolor***
Anura - Ranidae
e Bainskloof Chirping Frog

249 ***Arthroleptella drewsii***
Anura - Ranidae
e Drews's Chirping Frog

250 ***Arthroleptella hewitti***
Anura - Ranidae
e Yellow Bandit Frog; Natal Chirping Frog (CSA); Chirping Frog (CSA)

251 ***Arthroleptella lightfooti***
Anura - Ranidae
e Brown Bandit Frog; Cape Chirping Frog (CSA); Tiny Chirping Frog

252 ***Arthroleptella ngongoniensis***
Anura - Ranidae
e Natal Bandit Frog; Mistbelt Chirping Frog (CSA)

253 ***Arthroleptella villiersi***
Anura - Ranidae
e Villiers's Chirping Frog

254 ***Arthroleptides***
Anura - Ranidae
e Rocky River Frogs

255 ***Arthroleptides dutoiti***
Anura - Ranidae
e Kenya Rocky River Frog; Elgon Montane Torrent Frog

256 ***Arthroleptides martiensseni***
Anura - Ranidae
e Tanzania Rocky River Frog; Usambara Montane Torrent Frog

257 ***Arthroleptis***
Anura - Arthroleptidae
e Screeching Frogs; Bush Squeakers (CSA); Squeakers (CSA); Humus Frogs (CSA); Squeaking Frogs (CSA)
d Langfingerfrösche; Quiekfrösche

258 ***Arthroleptis adelphus***
Anura - Arthroleptidae
e Foulassi Screeching Frog; Montane Squeaker

259 ***Arthroleptis adolfifriederici***
Anura - Arthroleptidae
e Rugegewald Screeching Frog; France's Squeaker

260 ***Arthroleptis affinis***
Anura - Arthroleptidae
e Screeching Frog

261 ***Arthroleptis bivittatus***
Anura - Arthroleptidae
e Tumbo Island Screeching Frog

262 ***Arthroleptis brevipalmatus***
Anura - Arthroleptidae
e Ahl's Screeching Frog

263 ***Arthroleptis brevipes***
Anura - Arthroleptidae
e Togo Screeching Frog

264 ***Arthroleptis carquejai***
Anura - Arthroleptidae
e Cambondo Screeching Frog

265 ***Arthroleptis crusculum***
Anura - Arthroleptidae
e Guinea Screeching Frog

266 ***Arthroleptis discodacatyla***
Anura - Arthroleptidae
e Laurent's Screeching Frog

267 ***Arthroleptis francei***
Anura - Arthroleptidae
e Ruo River Screeching Frog

268 ***Arthroleptis hematogaster***
Anura - Arthroleptidae
e Itombwe Screeching Frog

269 ***Arthroleptis lameerei***
Anura - Arthroleptidae
e Katanga Screeching Frog

270 ***Arthroleptis loveridgei***
Anura - Arthroleptidae
e Loveridge's Screeching Frog

271 ***Arthroleptis milletihorsini***
Anura - Arthroleptidae
e Mali Screeching Frog

272 ***Arthroleptis mossoensis***
Anura - Arthroleptidae
e Mosso Screeching Frog

273 ***Arthroleptis nimbaensis***
Anura - Arthroleptidae
e Mount Nimba Screeching Frog

274 ***Arthroleptis phrynoides***
Anura - Arthroleptidae
e Lomami Screeching Frog

275 ***Arthroleptis poecilonotus***
Anura - Arthroleptidae
e West African Screeching Frog

276 ***Arthroleptis pyrrhoscelis***
Anura - Arthroleptidae
e Kivu Screeching Frog

277 ***Arthroleptis reichei***
Anura - Arthroleptidae
e Poroto Mountains Screeching Frog; Large-toed Squeaker

278 ***Arthroleptis schubotzi***
Anura - Arthroleptidae
e Burundi Screeching Frog

279 ***Arthroleptis spinalis***
Anura - Arthroleptidae
e Tanganyika Screeching Frog

280 ***Arthroleptis stenodactylus***
Anura - Arthroleptidae
e Kithengo Screeching Frog; Common Squeaker; Shovel-footed Squeaker (CSA); Savanna Squeaking Frog (CSA); Dune Squeaker (CSA)

281 ***Arthroleptis sylvaticus***
Anura - Arthroleptidae
e Forest Screeching Frog

282 ***Arthroleptis taeniatus***
Anura - Arthroleptidae
e Striped Screeching Frog

283 ***Arthroleptis tanneri***
Anura - Arthroleptidae
e Tanzania Screeching Frog

284 ***Arthroleptis troglodytes***
Anura - Arthroleptidae
e South Rhodesia Screeching Frog; Troglodyte Frog; Cave Squeaker (CSA); Troglodyte Humus Frog (CSA)
d Höhlenfrosch
f Grenouille troglodyte

285 ***Arthroleptis tuberosus***
Anura - Arthroleptidae
e Rainforest Screeching Frog

286 ***Arthroleptis variabilis***
Anura - Arthroleptidae
e Buea Screeching Frog

287 ***Arthroleptis vercammeni***
Anura - Arthroleptidae
e Mwana Screeching Frog

288 ***Arthroleptis wahlbergi***
Anura - Arthroleptidae
e Wahlberg's Screeching Frog; Bush Squeaker (CSA); Wahlberg's Humus Frog (CSA); Wahlberg's Frog

289 ***Arthroleptis xenochirus***
Anura - Arthroleptidae
e Marimba Screeching Frog

290 ***Arthroleptis xenodactyloides***
Anura - Arthroleptidae
e Chirinda Screeching Frog; Chirinda Cricket Frog; Hewitt's Humus Frog (CSA)

291 ***Arthroleptis xenodactylus***
Anura - Arthroleptidae
e Amani Screeching Frog; Dwarf Squeaker; Northern Squeaker

292 ***Arthroleptis zimmeri***
Anura - Arthroleptidae
e Zimmer's Screeching Frog

293 **Arthroletptidae**
Anura
e Screeching Frogs
d Langfingerfrösche

294 **Ascaphidae**
Anura
e Tailed Frogs
d Schwanzfrösche

295 ***Ascaphus***
Anura - Ascaphidae
e Tailed Frogs
d Schwanzfrösche
f Grenouilles à queue

296 ***Ascaphus truei***
Anura - Ascaphidae
e Tailed Frog
d Schwanzfrosch
f Grenouille à queue

297 ***Ascaphus truei californicus***
Anura - Ascaphidae
e California Tailed Frog

298 ***Ascaphus truei montanus***
Anura - Ascaphidae
e Rocky Mountain Tailed Frog

299 ***Ascaphus truei truei***
Anura - Ascaphidae
e Olympic Tailed Frog

300 ***Assa***
Anura - Myobatrachidae
e Pouched Frogs
d Darlington-Beutelfrösche

301 ***Assa darlingtoni***
Anura - Myobatrachidae
e Pouched Frog; Hip-pocket Frog; Marsupial Frog
d Darlington-Beutelfrosch; Darlingtons Beutelfrosch

302 ***Asterophrys***
Anura - Microhylidae
e New Guinea Bush Frogs
d Papua-Engmundfrosch

303 ***Asterophrys turpicola***
Anura - Microhylidae
e New Guinea Bush Frog

304 ***Astylosternus***
Anura - Arthroleptidae
e Night Frogs

305 ***Astylosternus batesi***
Anura - Arthroleptidae
e Benito River Night Frog

306 ***Astylosternus corrugatus***
Anura - Arthroleptidae
e Southern Night Frog

307 ***Astylosternus diadermatus***
Anura - Arthroleptidae
e Victoria Night Frog

308 ***Astylosternus fallax***
Anura - Arthroleptidae
e Fopouanga Night Frog

309 ***Astylosternus laurenti***
Anura - Arthroleptidae
e Laurent's Night Frog

310 ***Astylosternus montanus***
Anura - Arthroleptidae
e Mountain Night Frog

311 ***Astylosternus nganhanus***
Anura - Arthroleptidae
e Nganha Night Frog

312 ***Astylosternus occidentalis***
Anura - Arthroleptidae
e Western Night Frog

313 ***Astylosternus perreti***
Anura - Arthroleptidae
e Perret's Night Frog

314 ***Astylosternus ranoides***
Anura - Arthroleptidae
e Central Night Frog

315 ***Astylosternus rheophilus***
Anura - Arthroleptidae
e Cameroon Range Night Frog

316 ***Astylosternus schiotzi***
Anura - Arthroleptidae
e Apouh Night Frog

317 ***Atelognathus***
Anura - Leptodactylidae
e Patagonia Frogs

318 ***Atelognathus grandisonae***
Anura - Leptodactylidae
e Puerto Eden Frog

319 ***Atelognathus nitoi***
Anura - Leptodactylidae
e Rio Negro Frog

320 ***Atelognathus patagonicus***
Anura - Leptodactylidae
e Patagonia Frog

321 ***Atelognathus praebasalticus***
Anura - Leptodactylidae
e Zapala Frog

322 ***Atelognathus reverberii***
Anura - Leptodactylidae
e Laguna Raimunda Frog

323 ***Atelognathus salai***
Anura - Leptodactylidae
e Portezuelo Frog

324 ***Atelognathus solitarius***
Anura - Leptodactylidae
e Las Bayas Creek Frog

325 ***Atelophrynicus***
Anura - Bufonidae
e Stream Toads

326 ***Atelophrynicus chrysophorus***
Anura - Bufonidae
e Rio Viejo Toad

327 ***Atelopus***
Anura - Bufonidae
e Stubfoot Toads; Harlequin Frogs; Atelopids
d Stummelfußfrösche; Stummelfüße; Harlekinfrösche
f Atélopes

328 ***Atelopus arthuri***
Anura - Bufonidae
e Arthur's Stubfoot Toad

329 ***Atelopus balios***
Anura - Bufonidae
e Rio Pescado Stubfoot Toad

330 ***Atelopus bomolochos***
Anura - Bufonidae
e Azuay Stubfoot Toad

331 ***Atelopus boulengeri***
Anura - Bufonidae
e Boulenger's Stubfoot Toad; Boulenger's Arrow-poison Frog
f Grenouille de Boulenger; Atélope de Boulenger

332 ***Atelopus carauta***
Anura - Bufonidae
e Rio Carauta Stubfoot Toad

333 ***Atelopus carbonerensis***
Anura - Bufonidae
e La Carbonera Stubfoot Toad

334 ***Atelopus carrikeri***
Anura - Bufonidae
e Guajira Stubfoot Toad

335 ***Atelopus certus***
Anura - Bufonidae
e Darien Stubfoot Toad

336 ***Atelopus chiriquiensis***
Anura - Bufonidae
e Lewis's Stubfoot Toad

337 ***Atelopus coynei***
Anura - Bufonidae
e Rio Faisanes Stubfoot Toad

338 ***Atelopus cruciger***
Anura - Bufonidae
e Veragua Stubfoot Toad

339 ***Atelopus ebenoides***
Anura - Bufonidae
e Huila Stubfoot Toad

340 ***Atelopus elegans***
Anura - Bufonidae
e Elegant Stubfoot Toad

341 ***Atelopus erythropus***
Anura - Bufonidae
e Carabaya Stubfoot Toad

342 ***Atelopus eusebianus***
Anura - Bufonidae
e Malvasa Stubfoot Toad

343 ***Atelopus farci***
Anura - Bufonidae
e Forest Stubfoot Toad

344 ***Atelopus flavescens***
Anura - Bufonidae
e Cayenne Stubfoot Toad
d Gelber Stummelfuß
f Atélope jaunâtre

345 ***Atelopus franciscus***
Anura - Bufonidae
e Central Coast Stubfoot Toad
f Atélope de Guyane française

346 ***Atelopus galactogaster***
Anura - Bufonidae
e Antado Stubfoot Toad

347 ***Atelopus glyphus***
Anura - Bufonidae
e Pirri Range Stubfoot Toad

348 ***Atelopus halihelos***
Anura - Bufonidae
e Morana-Santiago Stubfoot Toad

349 ***Atelopus ignescens***
Anura - Bufonidae
e Quito Stubfoot Toad

350 ***Atelopus longibrachius***
Anura - Bufonidae
e El Tambo Stubfoot Toad

351 ***Atelopus longirostris***
Anura - Bufonidae
e Longnose Stubfoot Toad

352 ***Atelopus lynchi***
Anura - Bufonidae
e Lynch's Stubfoot Toad

353 ***Atelopus mindoensis***
Anura - Bufonidae
e Mindo Stubfoot Toad

354 ***Atelopus minutulus***
Anura - Bufonidae
e Colombian Stubfoot Toad

355 ***Atelopus mucubajiensis***
Anura - Bufonidae
e Mucubaji Stubfoot Toad

356 ***Atelopus muisca***
Anura - Bufonidae
e La Arboleda Stubfoot Toad

357 ***Atelopus nepiozomus***
Anura - Bufonidae
e Gualecenita Stubfoot Toad

358 ***Atelopus nicefori***
Anura - Bufonidae
e Nicéforo's Stubfoot Toad

359 ***Atelopus oxyrhynchus***
Anura - Bufonidae
e Rednose Stubfoot Toad

360 ***Atelopus pachydermus***
Anura - Bufonidae
e Schmidt's Stubfoot Toad

361 ***Atelopus palmatus***
Anura - Bufonidae
e Andersson's Stubfoot Toad

362 ***Atelopus pedimarmoratus***
Anura - Bufonidae
e San Isidro Stubfoot Toad

363 ***Atelopus peruensis***
Anura - Bufonidae
e Peru Stubfoot Toad

364 ***Atelopus pictiventris***
Anura - Bufonidae
e Painted Stubfoot Toad

365 ***Atelopus pinangoi***
Anura - Bufonidae
e Pinango Stubfoot Toad

366 ***Atelopus planispina***
Anura - Bufonidae
e Napo Stubfoot Toad; Flat-spined Atelopus

367 ***Atelopus sanjosei***
Anura - Bufonidae
e Anori Stubfoot Toad

368 ***Atelopus seminifer***
Anura - Bufonidae
e Upper Amazon Stubfoot Toad

369 ***Atelopus senex***
Anura - Bufonidae
e Pass Stubfoot Toad

370 ***Atelopus sorianoi***
Anura - Bufonidae
e Cloud-forest Stubfoot Toad

371 ***Atelopus spumarius***
Anura - Bufonidae
e Pebas Stubfoot Toad

372 ***Atelopus spumarius barbotini***
Anura - Bufonidae
f Atélope du Père Barbotin

373 ***Atelopus spumarius hooghmoedi***
Anura - Bufonidae
f Atélope d'Hooghmoed

374 ***Atelopus spurelli***
Anura - Bufonidae
e Condoto Stubfoot Toad

375 ***Atelopus subornatus***
Anura - Bufonidae
e Bogota Stubfoot Toad

376 ***Atelopus tamaensis***
Anura - Bufonidae
e Venezuelan Stubfoot Toad

377 ***Atelopus tricolor***
Anura - Bufonidae
e Three-coloured Stubfoot Toad

378 ***Atelopus varius***
Anura - Bufonidae
e Veragoa Stubfoot Toad; Painted Frog; Panamanian Golden Frog
d Pracht-Stummelfuß

379 ***Atelopus walkeri***
Anura - Bufonidae
e Walker's Stubfoot Toad

380 ***Atelopus willimani***
Anura - Bufonidae
e Bolivian Stubfoot Toad

381 ***Atelopus zeteki***
Anura - Bufonidae
e Cerro Campana Stubfoot Toad; Harlequin Toad; Panama Golden Stub-footed Toad
d Panama-Stummelfuß
f Grenouille de Zetek

382 ***Atopophrynus***
Anura - Leptodactylidae
e Sonson Frogs

383 ***Atopophrynus syntomopus***
Anura - Leptodactylidae
e Sonson Frog

384 ***Atympanophrys***
Anura - Megophryidae
e Shaping Frogs

385 ***Atympanophrys shapingensis***
Anura - Megophryidae
e Shaping Frog

386 ***Aubria***
Anura - Ranidae
e Ball Frogs

387 ***Aubria masako***
Anura - Ranidae
e Masako Ball Frog

388 ***Aubria subsigillata***
Anura - Ranidae
e Brown Ball Frog; West African Brown Frog
f Grenouille brune d'Afrique

B

389 ***Balebriveceps***
Anura - Microhylidae
e Bale Mountains Tree Frogs

390 ***Balebriveceps hillmani***
Anura - Microhylidae
e Bale Mountains Tree Frog

391 ***Barbourula***
Anura - Bombinatoridae
e Jungle Toads
d Barbour-Frösche

392 ***Barbourula busuangensis***
Anura - Bombinatoridae
e Busuanga Jungle Toad; Busuanga Disk-tongued Frog
d Barbour-Frosch

393 ***Barbourula kalimantanensis***
Anura - Bombinatoridae
e Kalimantan Jungle Toad

394 ***Barycholos***
Anura - Leptodactylidae
e Chimbo Frogs

395 ***Barycholos pulcher***
Anura - Leptodactylidae
e Chimbo Frog

396 ***Barycholos savagei***
Anura - Leptodactylidae
e Savage's Goias Frog

397 ***Barygenys***
Anura - Microhylidae
e Papua Frogs

398 ***Barygenys atra***
Anura - Microhylidae
e Günther's Papua Frog

399 ***Barygenys cheesmanae***
Anura - Microhylidae
e Cheesman's Papua Frog

400 ***Barygenys exsul***
Anura - Microhylidae
e Zweifel Papua Frog

401 ***Barygenys flavigularis***
Anura - Microhylidae
e Yellow-throated Papua Frog

402 ***Barygenys maculata***
Anura - Microhylidae
e Spotted Papua Frog

403 ***Barygenys nana***
Anura - Microhylidae
e Highland Papua Frog

404 ***Barygenys parvula***
Anura - Microhylidae
e Mandang Papua Frog

405 ***Batrachophrynus***
Anura - Leptodactylidae
e Andes Smooth Frogs

406 ***Batrachophrynus macrostomus***
Anura - Leptodactylidae
e Andes Smooth Frog; Lake Junin Frog
d Junin-Frosch

407 ***Batrachoseps***
Caudata - Plethodontidae
e Slender Salamanders

d Wurmsalamander (pl); Schleichensalamander (pl)
f Batrachoseps

408 ***Batrachoseps aridus***
Caudata - Plethodontidae
e Desert Slender Salamander

409 ***Batrachoseps attenuatus***
Caudata - Plethodontidae
e California Slender Salamander
d Kalifornischer Wurmsalamander; Schleichensalamander; Gartensalamander; Schlangensalamander

410 ***Batrachoseps attenuatus attenuatus***
Caudata - Plethodontidae
e Northern Slender Salamander

411 ***Batrachoseps attenuatus leucopus***
Caudata - Plethodontidae
e Southern Slender Salamander

412 ***Batrachoseps campi***
Caudata - Plethodontidae
e Inyo Mountains Slender Salamander

413 ***Batrachoseps diabolicus***
Caudata - Plethodontidae
e Hell Hollow Slender Salamander

414 ***Batrachoseps gabrieli***
Caudata - Plethodontidae
e San Gabriel Slender Salamander

415 ***Batrachoseps gavilianensis***
Caudata - Plethodontidae
e Gabilan Mountains Slender Salamander

416 ***Batrachoseps gregarius***
Caudata - Plethodontidae
e Gregarious Slender Salamander

417 ***Batrachoseps incognitus***
Caudata - Plethodontidae
e San Simeon Slender Salamander

418 ***Batrachoseps kawia***
Caudata - Plethodontidae
e Sequoia Slender Salamander

419 ***Batrachoseps luciae***
Caudata - Plethodontidae
e Santa Lucia Slender Salamander

420 ***Batrachoseps major***
Caudata - Plethodontidae
e Garden Slender Salamander

421 ***Batrachoseps major major***
Caudata - Plethodontidae
e Garden Slender Salamander

422 ***Batrachoseps minor***
Caudata - Plethodontidae
e Lesser Slender Salamander

423 ***Batrachoseps nigriventris***
Caudata - Plethodontidae
e Blackbelly Slender Salamander; Black-bellied Slender Salamander (NA)

424 ***Batrachoseps pacificus***
Caudata - Plethodontidae
e Pacific Slender Salamander
d Pazifischer Schlanksalamander

425 ***Batrachoseps pacificus catalinae***
Caudata - Plethodontidae
e Santa Catalina Slender Salamander

426 ***Batrachoseps pacificus pacificus***
Caudata - Plethodontidae
e Channel Islands Slender Salamander (NA)

427 ***Batrachoseps regius***
Caudata - Plethodontidae
e Kings River Slender Salamander

428 ***Batrachoseps relictus***
Caudata - Plethodontidae
e Relictual Salamander; Relictual Slender Salamander (NA)

429 ***Batrachoseps simatus***
Caudata - Plethodontidae
e Kern Canyon Slender Salamander

430 ***Batrachoseps stebbinsi***
Caudata - Plethodontidae
e Tehachapi Slender Salamander

431 ***Batrachoseps wrighti***
Caudata - Plethodontidae
e Oregon Slender Salamander
d Oregon-Wurmsalamander
f Batrachoseps de Wright

432 ***Batrachuperus***
Caudata - Hynobiidae
e Mountain Salamanders
d Asiatische Gebirgsmolche; Gebirgsmolche

433 ***Batrachuperus gorganensis***
Caudata - Hynobiidae
e Gorgan Mountain Salamander

434 ***Batrachuperus karlschmidti***
Caudata - Hynobiidae
e Chiala Mountain Salamander
d Schmidt Gebirgsmolch

435 ***Batrachuperus mustersi***
Caudata - Hynobiidae
e Paghman Mountain Salamander
d Afghanischer Gebirgsmolch; Afghanistanischer Gebirgsmolch; Afghanischer Winkelzahnmolch; Afghanistanischer Winkelzahnmolch; Afghanistan Salamander

436 ***Batrachuperus persicus***
Caudata - Hynobiidae
e Persia Mountain Salamander

437 ***Batrachuperus pinchonii***
Caudata - Hynobiidae
e Western Chinese Mountain Salamander
d Sohlengebirgsmolch
f Poisson des sapins

438 ***Batrachuperus tibetanus***
Caudata - Hynobiidae
e Tibetan Mountain Salamander

439 ***Batrachuperus yenyuanensis***
Caudata - Hynobiidae
e Yenyuanhsien Mountain Salamander
d Chinesischer Gebirgsseemolch

440 ***Batrachyla***
Anura - Leptodactylidae
e South American Wood Frogs

441 ***Batrachyla antartandica***
Anura - Leptodactylidae
e Marbled Wood Frog

442 ***Batrachyla leptopus***
Anura - Leptodactylidae
e Grey Wood Frog

443 ***Batrachyla taeniata***
Anura - Leptodactylidae
e Banded Wood Frog

444 ***Batrachylodes***
Anura - Ranidae
e Sticky-toed Frogs

445 ***Batrachylodes elegans***
Anura - Ranidae
e Elegant Sticky-toed Frog

446 ***Batrachylodes gigas***
Anura - Ranidae
e Bougainville Sticky-toed Frog

447 ***Batrachylodes mediodiscus***
Anura - Ranidae
e Buka Sticky-toed Frog

448 ***Batrachylodes minutus***
Anura - Ranidae
e Tine Sticky-toed Frog

449 ***Batrachylodes montanus***
Anura - Ranidae
e Mountain Sticky-toed Frog

450 ***Batrachylodes trossulus***
Anura - Ranidae
e Torakina Sticky-toed Frog

451 ***Batrachylodes vertebralis***
Anura - Ranidae
e Fauro Sticky-toed Frog

452 ***Batrachylodes wolfi***
Anura - Ranidae
e Wolf's Sticky-toed Frog

453 ***Bolitoglossa***
Caudata - Plethodontidae
e Tropical Lungless Salamanders; Mushroomtongue Salamanders
d Echte Pilzzungensalamander (pl); Klettersalamander (pl)
f Boltiglosses

454 ***Bolitoglossa adspersa***
Caudata - Plethodontidae
e Peters's Mushroomtongue Salamander
d Kordilleren-Salamander

455 ***Bolitoglossa altamazonicca***
Caudata - Plethodontidae
e Nauta Mushroomtongue Salamander
d Amazonas Klettersalamander

456 ***Bolitoglossa alvaradoi***
Caudata - Plethodontidae
e Moravia de Chirripo Salamander

457 ***Bolitoglossa arborescandens***
Caudata - Plethodontidae
e Costa Rica Mushroomtongue Salamander
d Baumkletterer

458 ***Bolitoglossa biseriata***
Caudata - Plethodontidae
e Two-lined Mushroomtongue Salamander

459 ***Bolitoglossa borburata***
Caudata - Plethodontidae
e Carabobo Mushroomtongue Salamander

460 ***Bolitoglossa capitana***
Caudata - Plethodontidae
e Orphan Salamander

461 ***Bolitoglossa carri***
Caudata - Plethodontidae
e Cloud-forest Salamander

462 ***Bolitoglossa celaque***
Caudata - Plethodontidae
e Celaque Mushroomtongue Salamander

463 ***Bolitoglossa cerroensis***
Caudata - Plethodontidae
e Millville Mushroomtongue Salamander

464 ***Bolitoglossa chica***
Caudata - Plethodontidae
e Hotel Zaracay Salamander

465 ***Bolitoglossa colonnea***
Caudata - Plethodontidae
e La Loma Salamander

466 ***Bolitoglossa compacta***
Caudata - Plethodontidae
e Cerro Pando Salamander
d Braunroter Schleuderzungen-Salamander

467 ***Bolitoglossa conanti***
Caudata - Plethodontidae
e Conant's Mushroomtongue Salamander

468 ***Bolitoglossa cuchumatana***
Caudata - Plethodontidae
e Oak Forest Salamander

469 ***Bolitoglossa cuna***
Caudata - Plethodontidae
e Camp Sasardi Salamander

470 ***Bolitoglossa digitigrande***
Caudata - Plethodontidae
e Rio Santa Rosa Salamander

471 ***Bolitoglossa dofleini***
Caudata - Plethodontidae
e Alta Vera Paz Salamander; Palm Salamander
d Großer Palmensalamander

472 ***Bolitoglossa dunni***
Caudata - Plethodontidae
e Dunn's Mushroomtongue Salamander; Dunn's Salamander

473 ***Bolitoglossa engelhardti***
Caudata - Plethodontidae
e Engelhardt's Mushroomtongue Salamander
d Engelhardts Gebirgssalamander

474 ***Bolitoglossa epimela***
Caudata - Plethodontidae
e Tapanti Mushroomtongue Salamander

475 ***Bolitoglossa equatoriana***
Caudata - Plethodontidae
e Ecuador Mushroomtongue Salamander

476 ***Bolitoglossa flavimembris***
Caudata - Plethodontidae
e Yellow-legged Mushroomtongue Salamander

477 ***Bolitoglossa flaviventris***
Caudata - Plethodontidae
e Yellowbelly Mushroomtongue Salamander

478 ***Bolitoglossa franklini***
Caudata - Plethodontidae
e Franklin's Mushroomtongue Salamander

479 ***Bolitoglossa franklini franklini***
Caudata - Plethodontidae
e Franklin's Mushroomtongue Salamander

480 ***Bolitoglossa franklini nigrovlavescens***
Caudata - Plethodontidae
e Blackbelly Mushroomtongue Salamander

481 ***Bolitoglossa gracilis***
Caudata - Plethodontidae
e Rio Quiri Salamander

482 ***Bolitoglossa hartwegi***
Caudata - Plethodontidae

e Hartweg's Mushroomtongue Salamander

483 ***Bolitoglossa helmrichi***
Caudata - Plethodontidae
e Coban Mushroomtongue Salamander

484 ***Bolitoglossa hermosa***
Caudata - Plethodontidae
e Guerreran Mushroomtongue Salamander

485 ***Bolitoglossa hypacra***
Caudata - Plethodontidae
e Paramo Frontino Salamander

486 ***Bolitoglossa ignicolor***
Caudata - Plethodontidae
e Camron Mushroomtongue Salamander
d Marmor Klettersalamander

487 ***Bolitoglossa jacksoni***
Caudata - Plethodontidae
e Jackson's Mushroomtongue Salamander

488 ***Bolitoglossa lincolni***
Caudata - Plethodontidae
e Lincoln's Mushroomtongue Salamander

489 ***Bolitoglossa macrinii***
Caudata - Plethodontidae
e Oaxacan Mushroomtongue Salamander

490 ***Bolitoglossa marmorea***
Caudata - Plethodontidae
e Crater Salamander

491 ***Bolitoglossa medemi***
Caudata - Plethodontidae
e Finca Chibigui Salamander

492 ***Bolitoglossa meliana***
Caudata - Plethodontidae
e Guatemala Mushroomtongue Salamander

493 ***Bolitoglossa mexicana***
Caudata - Plethodontidae
e Mexican Mushroomtongue Salamander; Galliwasp
d Mexikanischer Pilzzungensalamander; Palmensalamander; Mexikanischer Klettersalamander
f Boltiglosse mexicain

494 ***Bolitoglossa morio***
Caudata - Plethodontidae
e Cope's Mushroomtongue Salamander
d Dunkler Klettersalamander

495 ***Bolitoglossa mulleri***
Caudata - Plethodontidae
e Müller's Mushroomtongue Salamander

496 ***Bolitoglossa nicefori***
Caudata - Plethodontidae
e San Gil Mushroomtongue Salamander

497 ***Bolitoglossa nigrescens***
Caudata - Plethodontidae
e Cordillera Central Salamander

498 ***Bolitoglossa occidentalis***
Caudata - Plethodontidae
e Southern Banana Salamander
d Mexikanischer Tieflandsalamander; Westlicher Klettersalamander

499 ***Bolitoglossa odonnelli***
Caudata - Plethodontidae
e O'Donnel's Salamander

500 ***Bolitoglossa orestes***
Caudata - Plethodontidae

e Culata Mushroomtongue Salamander; Blue Ridge Dusky Salamander

501 ***Bolitoglossa palmata***
Caudata - Plethodontidae
e Amazon Mushroomtongue Salamander

502 ***Bolitoglossa pandi***
Caudata - Plethodontidae
e Pandi Mushroomtongue Salamander

503 ***Bolitoglossa peruviana***
Caudata - Plethodontidae
e Peru Mushroomtongue Salamander
d Amazonischer Klettersalamander

504 ***Bolitoglossa phalarosoma***
Caudata - Plethodontidae
e Medellin Mushroomtongue Salamander
d Nebelsalamander

505 ***Bolitoglossa platydactyla***
Caudata - Plethodontidae
e Broadfoot Mushroomtongue Salamander
d Roter Klettersalamander

506 ***Bolitoglossa ramosi***
Caudata - Plethodontidae
e Ramos's Mushroomtongue Salamander

507 ***Bolitoglossa resplendens***
Caudata - Plethodontidae
e Splendid Mushroomtongue Salamander

508 ***Bolitoglossa riletti***
Caudata - Plethodontidae
e Rilett's Mushroomtongue Salamander

509 ***Bolitoglossa robusta***
Caudata - Plethodontidae
e Robust Mushroomtongue Salamander
d Rötlicher Klettersalamander

510 ***Bolitoglossa rostrata***
Caudata - Plethodontidae
e Longnose Mushroomtongue Salamander

511 ***Bolitoglossa rufescens***
Caudata - Plethodontidae
e Northern Banana Salamander

512 ***Bolitoglossa salvini***
Caudata - Plethodontidae
e Salvin's Mushroomtongue Salamander

513 ***Bolitoglossa savagei***
Caudata - Plethodontidae
e Savage's Mushroomtongue Salamander

514 ***Bolitoglossa schizodactyla***
Caudata - Plethodontidae
e Cocle Mushroomtongue Salamander

515 ***Bolitoglossa schmidti***
Caudata - Plethodontidae
e Schmidt's Mushroomtongue Salamander

516 ***Bolitoglossa silverstonei***
Caudata - Plethodontidae
e Silverstone's Mushroomtongue Salamander

517 ***Bolitoglossa sima***
Caudata - Plethodontidae
e Northwestern Mushroomtongue Salamander

518 ***Bolitoglossa sooyorum***
Caudata - Plethodontidae
e Cordillera Talamanca Salamander

519 ***Bolitoglossa striatula***
Caudata - Plethodontidae
e Cukra Mushroomtongue Salamander
d Gestreifter Klettersalamander

520 ***Bolitoglossa stuarti***
Caudata - Plethodontidae
e Stuart's Mushroomtongue Salamander

521 ***Bolitoglossa subpalmata***
Caudata - Plethodontidae
e La Palma Salamander; Palm Salamander
d Kostarika-Pilzzungensalamander

522 ***Bolitoglossa taylori***
Caudata - Plethodontidae
e Cerro Cituro Mushroomtongue Salamander

523 ***Bolitoglossa vallecula***
Caudata - Plethodontidae
e Yarumal Mushroomtongue Salamander

524 ***Bolitoglossa veracrucis***
Caudata - Plethodontidae
e Veracruz Mushroomtongue Salamander

525 ***Bolitoglossa walkeri***
Caudata - Plethodontidae
e Walker's Mushroomtongue Salamander

526 ***Bolitoglossa yucatana***
Caudata - Plethodontidae
e Yucatan Mushroomtongue Salamander

527 ***Bombina***
Anura - Bombinatoridae
e Firebelly Toads; Firebellies; Fire-bellied Toads
d Feuerkröten; Unken
f Crapauds sonneurs; Sonneurs

528 ***Bombina bombina***
Anura - Bombinatoridae
e Firebelly Toad; Fire-bellied Toad
d Rotbauchunke; Tieflandunke; Feuerkröte
f Sonneur à feu; Sonneur igné; Sonneur à ventre de feu; Sonneur à ventre rouge

529 ***Bombina fortinuptialis***
Anura - Bombinatoridae
e Guangxi Firebelly Toad

530 ***Bombina maxima***
Anura - Bombinatoridae
e Yunnan Firebelly Toad; Yellow-bellied Toad; Chinese Toad
d Chinesische Riesenunke; Riesenunke
f Sonneur géant

531 ***Bombina microdeladigitore***
Anura - Bombinatoridae
e Hubei Firebelly Toad

532 ***Bombina orientalis***
Anura - Bombinatoridae
e Oriental Firebelly Toad; Oriental Fire-bellied Toad; Korean Fire-bellied Toad; Eastern Fire-bellied Toad; Chinese Bell Toad
d Chinesische Rotbauchunke; Orientalische Rotbauchunke
f Sonneur asiatique; Sonneur asiatique à ventre rouge; Crapaud sonneur asiatique

533 ***Bombina pachypus***
Anura - Bombinatoridae
i Ululone meridionale; Ululone dal ventre giallo

534 ***Bombina variegata***
Anura - Bombinatoridae
e Yellowbelly Toad; Yellow-bellied Toad

d Gelbbauchunke; Bergunke
f Sonneur à ventre jaune; Sonneur à pieds épais; Crapaud sonneur; Crapaud sonneur à ventre jaune
i Ululone a ventre giallo; Ululone variegato

535 **Bombinatoridae**
Anura
e Firebelly Toads

536 ***Boophis***
Anura - Rhacophoridae
e Bright-eyed Frogs

537 ***Boophis albilabris***
Anura - Rhacophoridae
e White-lipped Bright-eyed Frog

538 ***Boophis albipunctatus***
Anura - Rhacophoridae
e Spotted Bright-eyed Frog

539 ***Boophis boehmei***
Anura - Rhacophoridae
e Böhme's Bright-eyed Frog

540 ***Boophis brachychir***
Anura - Rhacophoridae
e Nosy Be Bright-eyed Frog

541 ***Boophis difficilis***
Anura - Rhacophoridae
e Foizana Bright-eyed Frog

542 ***Boophis erythrodactylus***
Anura - Rhacophoridae
e Forest Bright-eyed Frog

543 ***Boophis goudoti***
Anura - Rhacophoridae
e Goudot's Bright-eyed Frog

544 ***Boophis granulosus***
Anura - Rhacophoridae
e Warty Bright-eyed Frog

545 ***Boophis hillenii***
Anura - Rhacophoridae
e Perinet's Bright-eyed Frog

546 ***Boophis idea***
Anura - Rhacophoridae
e Ida's Bright-eyed Frog

547 ***Boophis jaegeri***
Anura - Rhacophoridae
e Jaeger's Bright-eyed Frog

548 ***Boophis laurentii***
Anura - Rhacophoridae
e Andringitra Bright-eyed Frog

549 ***Boophis luteus***
Anura - Rhacophoridae
e Ankafana Bright-eyed Frog

550 ***Boophis madagascariensis***
Anura - Rhacophoridae
e Madagascar Bright-eyed Frog

551 ***Boophis majori***
Anura - Rhacophoridae
e Ambohimitombo Bright-eyed Frog

552 ***Boophis mandraka***
Anura - Rhacophoridae
e Mandraka Bright-eyed Frog

553 ***Boophis microtis***
Anura - Rhacophoridae
e Guibe's Bright-eyed Frog

554 ***Boophis microtympanum***
Anura - Rhacophoridae
e Imerina Bright-eyed Frog

555 ***Boophis miniatus***
Anura - Rhacophoridae

e Tiny Bright-eyed Frog

556 ***Boophis ophistodon***
Anura - Rhacophoridae
e Eastern Bright-eyed Frog

557 ***Boophis pauliani***
Anura - Rhacophoridae
e Moramanga Bright-eyed Frog

558 ***Boophis rappioles***
Anura - Rhacophoridae
e Central Bright-eyed Frog

559 ***Boophis reticulatus***
Anura - Rhacophoridae
e Reticulate Bright-eyed Frog

560 ***Boophis rhodoscelis***
Anura - Rhacophoridae
e Betsileo Bright-eyed Frog

561 ***Boophis tephraeomystax***
Anura - Rhacophoridae
e Dumeril's Bright-eyed Frog

562 ***Boophis viridis***
Anura - Rhacophoridae
e Green Bright-eyed Frog

563 ***Boophis williamsi***
Anura - Rhacophoridae
e Williams's Bright-eyed Frog

564 ***Boulengerula***
Gymnophiona - Caeciliaidae
e Boulenger's Caecilians; Usambara Bluish-grey Caecilians

565 ***Boulengerula boulengerula***
Gymnophiona - Caeciliaidae
e Boulenger's Caecilian; Usambara Bluish-grey Caecilian

566 **Brachycephalidae**
Anura
e Saddleback Toads; Brachycephalid Toads
d Sattelkröten
f Brachycéphalidés

567 ***Brachycephalus***
Anura - Brachycephalidae
e Saddleback Toads
d Sattelkröten
f Brachycéphales

568 ***Brachycephalus didactyla***
Anura - Brachycephalidae
e Serra Cantareira Saddleback Toad

569 ***Brachycephalus ephippium***
Anura - Brachycephalidae
e Spix's Saddleback Toad; Golden Frog; Gold Frog (NA)
d Sattelkröte; Gelbe Sattelkröte; Sattelfröschchen
f Brachycéphale porteselle; Grenouille d'or

570 ***Brachytarsophrys***
Anura - Megophryidae
e Karin Hills Frogs

571 ***Brachytarsophrys carinensis***
Anura - Megophryidae
e Karin Hills Frog

572 ***Bradytriton***
Caudata - Plethodontidae
e Finca Chiblac Salamanders

573 ***Bradytriton silus***
Caudata - Plethodontidae
e Finca Chiblac Salamander

574 ***Brasilotyphlus***
Gymnophiona - Caeciliaidae
e Brazilian Caecilians

575 ***Brasilotyphlus braziliensis***
Gymnophiona - Caeciliaidae
e Brazilian Caecilian

576 ***Breviceps***
Anura - Microhylidae
e Short-headed Frogs; Rain Frogs (CSA); Blaasops (CSA); Shortheads (CSA)
d Kurzkopffrösche; Regenfrösche

577 ***Breviceps acutirostris***
Anura - Microhylidae
e Cape Short-headed Frog; Strawberry Rain Frog (CSA)

578 ***Breviceps adspersus***
Anura - Microhylidae
e Transvaal Short-headed Frog; Common Rain Frog (CSA); Bushveld Rain Frog (CSA); South African Shorthead; Blaasop (CSA); Rain Frog; Common Short-headed Frog (CSA)
d Gesprenkelter Kurzkopffrosch; Regenfrosch

579 ***Breviceps adspersus adspersus***
Anura - Microhylidae
e Bushveld Rain Frog (CSA)

580 ***Breviceps fuscus***
Anura - Microhylidae
e Brown Short-headed Frog; Plain Rain Frog (CSA); Jan Blom (CSA)

581 ***Breviceps gibbosus***
Anura - Microhylidae
e South African Short-headed Frog; South African Rain Frog; Cape Rain Frog (CSA); Cape Shorthead (CSA); Blaasop (CSA); Jan Blom (CSA); Linnaeus's Short-headed Frog (CSA)
f Crapaud bossu

582 ***Breviceps macrops***
Anura - Microhylidae
e Boulenger's Short-headed Frog; Desert Rain Frog (CSA); Melkpadda (CSA); Web-footed Rain Frog (CSA)

583 ***Breviceps maculatus***
Anura - Microhylidae
e Spotted Short-headed Frog; Spotted Rain Frog (CSA)

584 ***Breviceps montanus***
Anura - Microhylidae
e Mountain Short-headed Frog; Mountain Blaasop (CSA); Cape Mountain Rain Frog (CSA)

585 ***Breviceps mossambicus***
Anura - Microhylidae
e Mozambique Short-headed Frog; Mozambique Rain Frog (CSA); Flat-faced Frog; Mocambique Rain Frog (CSA)
d Gemeiner Regenfrosch

586 ***Breviceps mossambicus mossambicus***
Anura - Microhylidae
e Common Rain Frog (CSA); Common Short-headed Frog (CSA)

587 ***Breviceps namaquensis***
Anura - Microhylidae
e Namaqualand Short-headed Frog; Namaqua Rain Frog (CSA)

588 ***Breviceps poweri***
Anura - Microhylidae
e Power's Short-headed Frog; Power's Rain Frog

589 ***Breviceps rosei***
Anura - Microhylidae
e Rose's Short-headed Frog; Rose's Blaasop (CSA); Sand Rain Frog (CSA)

590 ***Breviceps sylvestris***
Anura - Microhylidae
e Forest Short-headed Frog; Transvaal Forest Rain Frog (CSA)

591 ***Breviceps sylvestris sylvestris***
Anura - Microhylidae
e Woodbush Rain Frog (CSA); Woodbush Short-headed Frog (CSA)

592 ***Breviceps sylvestris taeniatus***
Anura - Microhylidae
e Soutspansberg Rain Frog (CSA)

593 ***Breviceps verrucosus***
Anura - Microhylidae
e Natal Short-headed Frog; Plaintive Rain Frog (CSA); Verruculose Rain Frog (CSA); Verruculose Short-headed Frog (CSA); Rough-skinned Rain Frog (CSA)

594 ***Buergeria***
Anura - Rhacophoridae
e Bürger's Frogs
d Singfrösche
f Rainettes de Bürger

595 ***Buergeria buergeri***
Anura - Rhacophoridae
e Bürger's Frog
d Singfrosch
f Rainette de Bürger

596 ***Buergeria japonica***
Anura - Rhacophoridae
e Japanese Bürger's Frog

597 ***Buergeria pollicaris***
Anura - Rhacophoridae
e Taiwan Bürger's Frog

598 ***Buergeria robusta***
Anura - Rhacophoridae
e Robust Bürger's Frog

599 ***Bufo***
Anura - Bufonidae
e True Toads; Toads (CSA); Ground Toads; Typical Toads
d Landkröten; Kröten; Echte Kröten
f Crapauds; Vrais crapauds

600 ***Bufo abatus***
Anura - Bufonidae
e Darjeeling Toad

601 ***Bufo achalensis***
Anura - Bufonidae
e Cordoba Toad

602 ***Bufo ailaoanus***
Anura - Bufonidae
e Ejia Toad

603 ***Bufo alvarius***
Anura - Bufonidae
e Colorado River Toad; Colorado River Frog; Sonoran Desert Toad (NA)
d Colorado-Kröte
f Crapaud du Colorado

604 ***Bufo amatolicus***
Anura - Bufonidae
e Amatola Toad

605 ***Bufo amboroensis***
Anura - Bufonidae
e Cochabamba Toad

606 ***Bufo americanus***
Anura - Bufonidae
e American Toad; Southern Toad (NA); Hudson Bay Toad
d Amerikanische Kröte
f Crapaud américain; Crapaud d'Amérique

607 ***Bufo americanus americanus***
Anura - Bufonidae

e Eastern American Toad

608 ***Bufo americanus charlesmithi***
Anura - Bufonidae
e Dwarf American Toad

609 ***Bufo anderssoni***
Anura - Bufonidae
e Andersson's Toad

610 ***Bufo andrewsi***
Anura - Bufonidae
e Andrews's Toad

611 ***Bufo angusticeps***
Anura - Bufonidae
e Sand Toad (CSA); Common Cape Toad (CSA)

612 ***Bufo arabicus***
Anura - Bufonidae
e Arabian Toad

613 ***Bufo arborescandens***
Anura - Bufonidae
e Mendoza Pass Toad

614 ***Bufo arenarium***
Anura - Bufonidae
e Common Toad (NA); Toad (NA); Argentine Toad
d Sandkröte

615 ***Bufo arequipensis***
Anura - Bufonidae
e Rio Chili Toad

616 ***Bufo arunco***
Anura - Bufonidae
e Concepcion Toad

617 ***Bufo asmarae***
Anura - Bufonidae
e Asmara Toad

618 ***Bufo asper***
Anura - Bufonidae
e Java Toad; River Toad
d Rauhkröte; Flußkröte
f Crapaud âpre

619 ***Bufo atacamensis***
Anura - Bufonidae
e Vallenar Toad

620 ***Bufo atukoralei***
Anura - Bufonidae
e Yala Toad

621 ***Bufo bankorensis***
Anura - Bufonidae
e Central Formosa Toad

622 ***Bufo baxteri***
Anura - Bufonidae
e Wyoming Toad

623 ***Bufo beddomii***
Anura - Bufonidae
e Beddome's Toad

624 ***Bufo beebei***
Anura - Bufonidae
e Rivero Toad

625 ***Bufo beiranus***
Anura - Bufonidae
e Beira Toad

626 ***Bufo berghei***
Anura - Bufonidae
e Makamba Toad

627 ***Bufo biporcatus***
Anura - Bufonidae
e Indonesian Toad
f Crapaud à deux arètes

628 ***Bufo blanfordii***
Anura - Bufonidae
e Blanford's Toad

629 ***Bufo blombergi***
Anura - Bufonidae
e Colombian Giant Toad; Blomberg's Toad
d Kolumbische Riesenkröte; Kolumbianische Riesenkröte; Blombergs Kröte
f Crapaud de Blomberg

630 ***Bufo bocourti***
Anura - Bufonidae
e Bocourt's Toad

631 ***Bufo boreas***
Anura - Bufonidae
e Western Toad
d Nordkröte; Salzkröte; Schwarzkröte
f Crapaud du Nord; Crapaud de l'Ouest; Crapaud boréal

632 ***Bufo boreas boreas***
Anura - Bufonidae
e Boreal Toad; Western Boreal Toad

633 ***Bufo boreas halophilus***
Anura - Bufonidae
e California Toad

634 ***Bufo boreas nelsoni***
Anura - Bufonidae
e Amargosa Toad

635 ***Bufo brauni***
Anura - Bufonidae
e Braun's Toad; Dead-leaf Toad

636 ***Bufo brevirostris***
Anura - Bufonidae
e Kemholey Toad

637 ***Bufo brongersmai***
Anura - Bufonidae
e Tiznit Toad; Brongersma's Toad
f Crapaud de Brongersma

638 ***Bufo buchneri***
Anura - Bufonidae
e Buchner's Toad

639 ***Bufo bufo***
Anura - Bufonidae
e Common European Toad (NA); Common Toad; Toad; European Common Toad; European Toad; Eurasian Common Toad
d Feldkröte; Erdkröte; Gemeine Kröte; Kröte; Lark
f Crapaud commun; Crapaud commun d'Europe; Crapaud; Crapaud brun; Crapaud cendré; Crapaud ordinaire
i Rospo comune; Rosp; Birinaza

640 ***Bufo bufo spinosus***
Anura - Bufonidae
e Common Toad
f Crapaud commun

641 ***Bufo burmanus***
Anura - Bufonidae
e Burma Toad

642 ***Bufo calamita***
Anura - Bufonidae
e Natterjack Toad; Natterjack; Running Toad; Goldenback
d Kreuzkröte; Rohrkröte; Hausunke
f Crapaud calamite; Crapaud des joncs; Crapaud des roseaux
i Rospo calamita; Rospo crociato

643 ***Bufo californicus***
Anura - Bufonidae
e Mexican Arroyo Toad; Arroyo Toad (NA)

644 ***Bufo camterunensis***
Anura - Bufonidae
e Oban Toad

645 ***Bufo canaliferus***
Anura - Bufonidae
e Dwarf Toad

646 ***Bufo canorus***
Anura - Bufonidae
e Yosemite Toad

647 ***Bufo castaneoticus***
Anura - Bufonidae
e Para Toad

648 ***Bufo cavifrons***
Anura - Bufonidae
e Mountain Toad

649 ***Bufo celebensis***
Anura - Bufonidae
e Celebes Toad

650 ***Bufo ceratophrys***
Anura - Bufonidae
e Horned Toad

651 ***Bufo chappuisi***
Anura - Bufonidae
e Turukana Toad

652 ***Bufo chlorogaster***
Anura - Bufonidae
e Greenbelly Toad
f Crapaud à ventre jaune

653 ***Bufo chudeau***
Anura - Bufonidae
e Bata Marsh Toad

654 ***Bufo claviger***
Anura - Bufonidae
e Benkulen Toad

655 ***Bufo coccifer***
Anura - Bufonidae
e Southern Roundgland Toad

656 ***Bufo cognatus***
Anura - Bufonidae
e Great Plains Toad; Plains Toad
d Prairiekröte; Texas-Kröte
f Crapaud de prairie

657 ***Bufo compactilis***
Anura - Bufonidae
e Plateau Toad

658 ***Bufo conifer***
Anura - Bufonidae
e Evergreen Toad

659 ***Bufo cophotis***
Anura - Bufonidae
e Paramo Toad

660 ***Bufo corynetes***
Anura - Bufonidae
e Abra Malaga Toad

661 ***Bufo cristatus***
Anura - Bufonidae
e Large-crested Toad

662 ***Bufo cristiglans***
Anura - Bufonidae
e Tingi Hills Toad

663 ***Bufo crucifer***
Anura - Bufonidae
e Striped Toad
d Schmuckkröte
f Crapaud à oreilles noires

664 ***Bufo cruciger***
Anura - Bufonidae
e Cape Toad

665 ***Bufo cycladen***
Anura - Bufonidae
e Northern Roundgland Toad

666 ***Bufo danielae***
Anura - Bufonidae
e Ivory Coast Toad

667 ***Bufo dapsilis***
Anura - Bufonidae
e Bom Jardim Toad

668 ***Bufo debilis***
Anura - Bufonidae
e Green Toad

669 ***Bufo debilis debilis***
Anura - Bufonidae
e Eastern Green Toad

670 ***Bufo debilis insidior***
Anura - Bufonidae
e Western Green Toad

671 ***Bufo dhufarensis***
Anura - Bufonidae
e Oman Toad

672 ***Bufo diptychus***
Anura - Bufonidae
e Cope's Toad

673 ***Bufo dodsoni***
Anura - Bufonidae
e Dodson's Toad
d Dodsons Kröte
f Crapaud de Dobson

674 ***Bufo dombensis***
Anura - Bufonidae
e Dombe Toad

675 ***Bufo dorbignyi***
Anura - Bufonidae
e D'Orbigny's Toad

676 ***Bufo exsul***
Anura - Bufonidae
e Black Toad; Inyo County Toad; Deep Springs Toad
d Schwarze Kröte

677 ***Bufo fastidiosus***
Anura - Bufonidae
e Pico Blanco Toad

678 ***Bufo fenoulheti***
Anura - Bufonidae
e Newington Toad; Northern Pygmy Toad (CSA); Dwarf Toad (CSA); Flat Toad (CSA)

679 ***Bufo fergusonii***
Anura - Bufonidae
e Ferguson's Toad
d Fergusons Kröte
f Crapaud de Ferguson

680 ***Bufo fernadezae***
Anura - Bufonidae
e Bella Vista Toad

681 ***Bufo fissipes***
Anura - Bufonidae
e Carabaya Toad

682 ***Bufo flaveolineatus***
Anura - Bufonidae
e Capillococha Toad

683 ***Bufo fowleri***
Anura - Bufonidae
e Fowler's Toad

684 ***Bufo fuliginatus***
Anura - Bufonidae
e Shaba Province Toad

685 ***Bufo funereus***
Anura - Bufonidae
e Angola Toad

686 ***Bufo gabbi***
Anura - Bufonidae
e Talamanca Toad

687 ***Bufo galeatus***
Anura - Bufonidae
e Gamboja Toad

688 ***Bufo gallardoi***
Anura - Bufonidae
e Gallardo's Toad

689 ***Bufo gargarizans***
Anura - Bufonidae
e Chusan Island Toad; Asiatic Toad; Asiatic Common Toad; Chinese Toad

690 ***Bufo gariepensis***
Anura - Bufonidae
e Karoo Toad (CSA); Mountain Toad; Gariep Toad (CSA); Inyangani Toad (CSA)

691 ***Bufo garmani***
Anura - Bufonidae
e Garman's Toad; Olive Toad (CSA); Garman's Square-marked Toad (CSA); Light-nosed Toad (CSA); Northern Mottled Toad (CSA); Eastern Olive Toad (CSA)

692 ***Bufo gemmifer***
Anura - Bufonidae
e Jeweled Toad

693 ***Bufo glaberrimus***
Anura - Bufonidae
e Cyndinamarca Toad

694 ***Bufo gnustae***
Anura - Bufonidae
e Rio Grande Toad

695 ***Bufo gracilipes***
Anura - Bufonidae
e French Congo Toad

696 ***Bufo grandisonae***
Anura - Bufonidae
e Mossamedes Toad

697 ***Bufo granulosus***
Anura - Bufonidae
e Common Lesser Toad; Lesser Toad; Granular Toad
d D'Orbignys Kröte
f Crapaud goitreux; Crapaud granuleux

698 ***Bufo guttatus***
Anura - Bufonidae
e Spotted Toad; Smooth-sided Toad
f Crapaud à taches jaunes; Crapaud de Leschenault

699 ***Bufo guttularis***
Anura - Bufonidae
e Lobatsi Toad; Guttural Toad (CSA); Common Toad (CSA); Savannah Toad; Common African Toad (CSA); Square-marked Toad (CSA)

700 ***Bufo hadramautinus***
Anura - Bufonidae
e Suia Hadramaut Toad

701 ***Bufo haematiticus***
Anura - Bufonidae
e Truando Toad

702 ***Bufo hemiophrys***
Anura - Bufonidae
e Dakota Toad; Canadian Toad
d Manitoba-Kröte

703 ***Bufo hemiophrys hemiophrys***
Anura - Bufonidae
e Canadian Toad

704 ***Bufo himalayanus***
Anura - Bufonidae
e Himalaya Toad; Himalayan Toad (ISC)

705 ***Bufo hoeschi***
Anura - Bufonidae
e Okahandja Toad

706 ***Bufo holdridgei***
Anura - Bufonidae
e Holdridge's Toad

707 ***Bufo hololius***
Anura - Bufonidae
e Malabar Toad

708 ***Bufo houstonensis***
Anura - Bufonidae
e Houston Toad
d Houston-Kröte
f Crapaud d'Houston

709 ***Bufo hypomelas***
Anura - Bufonidae
e Choco Toad

710 ***Bufo ibarrai***
Anura - Bufonidae
e Jalapa Toad

711 ***Bufo ictericus***
Anura - Bufonidae
e Yellow Cururu Toad

712 ***Bufo inca***
Anura - Bufonidae
e Inca Toad

713 ***Bufo intermedius***
Anura - Bufonidae
e Gunther's Tropical Toad

714 ***Bufo iserni***
Anura - Bufonidae
e Rio Perene Toad

715 ***Bufo japonicus***
Anura - Bufonidae
e Japanese Toad; Japanese Common Toad
d Japan-Kröte
f Crapaud commun du Japon; Crapaud du Japon

716 ***Bufo jordani***
Anura - Bufonidae
e Namibia Toad

717 ***Bufo justinianoi***
Anura - Bufonidae
e El Chape Toad

718 ***Bufo juxtasper***
Anura - Bufonidae
e Sungei Tawan Toad

719 ***Bufo kassassi***
Anura - Bufonidae
e Damietta Toad

720 ***Bufo kavangensis***
Anura - Bufonidae
e Khwai River Toad; Kavanga Pygmy Toad (CSA)

721 ***Bufo kavirensis***
Anura - Bufonidae
e Teheran Toad

722 ***Bufo kelaartii***
Anura - Bufonidae
e Kelaart's Toad

723 ***Bufo kellogi***
Anura - Bufonidae
e Little Mexican Toad

724 ***Bufo kerinyage***
Anura - Bufonidae
e Keith's Toad

725 ***Bufo kisoloensis***
Anura - Bufonidae
e Kisolo Toad

726 ***Bufo koynayensis***
Anura - Bufonidae
e Humbali Village Toad

727 ***Bufo langanoensis***
Anura - Bufonidae
e Lake Langano Toad

728 ***Bufo latastii***
Anura - Bufonidae
e Ladakh Toad

729 ***Bufo latifrons***
Anura - Bufonidae
e High Tropical Forest Toad

730 ***Bufo lemairii***
Anura - Bufonidae
e Pweto Toad; Lemaire's Toad (CSA); Yellow Swamp Toad

731 ***Bufo lemur***
Anura - Bufonidae
e Ridge-headed Toad (WI)

732 ***Bufo limensis***
Anura - Bufonidae
e Peru Coast Toad

733 ***Bufo lindneri***
Anura - Bufonidae
e Dar-es-Salaam Toad

734 ***Bufo lonnbergi***
Anura - Bufonidae
e Lonnberg's Toad; Dwarf Toad (CSA)

735 ***Bufo luetkenii***
Anura - Bufonidae
e Yellow Toad

736 ***Bufo lughensis***
Anura - Bufonidae
e Lugh Toad

737 ***Bufo macrotis***
Anura - Bufonidae
e Big-eared Toad
d Großohrkröte

738 ***Bufo maculatus***
Anura - Bufonidae
e Hallowell's Toad; Flat-backed Toad (CSA); Lesser Square-marked Toad CSA); Striped Toad; Egyptian Toad; Reuss's Toad; Desert Toad; Square-marked Toad
d Pantherkröte
f Crapaud panthérin; Crapaud jaspé; Crapaud à taches carrées

739 ***Bufo marinus***
Anura - Bufonidae
e Cane Toad; Giant Toad; Marine Toad
d Riesenkröte; Aga; Aga-Kröte
f Agua (Ants); Crapaud géant; Crapaud géant d'Amérique du Sud; Crapaud agua (Ants); Crapaud-boeuf; Crapaud buffle; Crapaud marin

740 ***Bufo marmoreus***
Anura - Bufonidae
e Marbled Toad; Wiegmann's Toad
d Wiegmans Kröte
f Crapaud marbré du Mexique; Crapaud de Wiegmann

741 ***Bufo mauritanicus***
Anura - Bufonidae
e Berber Toad; Moroccan Toad; Mauritanian Toad

d Maurische Kröte; Berberkröte
f Crapaud panthérin; Crapaud panthérin de Maurétanie; Crapaud de Maurétanie; Crapaud berbère

742 ***Bufo mazatlanensis***
Anura - Bufonidae
e Sinaloa Toad

743 ***Bufo melanochloris***
Anura - Bufonidae
e Dark Green Toad

744 ***Bufo melanogaster***
Anura - Bufonidae
e Blackbelly Toad

745 ***Bufo melanopleura***
Anura - Bufonidae
e Kakunde Toad

746 ***Bufo melanostictus***
Anura - Bufonidae
e Black-spined Toad; Black-spotted Toad; Asian Common Toad; Indian Common Toad; Common Indian Toad (ISC); South Asian Garden Toad; Common Toad (SEA)
d Schwarznarbenkröte
f Crapaud rude; Crapaud masqué

747 ***Bufo microscaphus***
Anura - Bufonidae
e Arizona Toad; Southwestern Toad; Arroya Toad

748 ***Bufo microscaphus californicus***
Anura - Bufonidae
e Southwestern Arroyo Toad; Arroya Toad (NA)

749 ***Bufo microscaphus mexicanus***
Anura - Bufonidae
e Mexican Spadefoot Toad

750 ***Bufo microscaphus microscaphus***
Anura - Bufonidae
e Arizona Toad

751 ***Bufo microtympanum***
Anura - Bufonidae
e Small-eared Toad; Southern Eared Toad

752 ***Bufo minshanicus***
Anura - Bufonidae
e Gansu Toad

753 ***Bufo nasicus***
Anura - Bufonidae
e Werner's Toad

754 ***Bufo nesiotes***
Anura - Bufonidae
e Laguna Toad

755 ***Bufo occidentalis***
Anura - Bufonidae
e Pine Toad

756 ***Bufo ocellatus***
Anura - Bufonidae
e Ocellated Toad

757 ***Bufo olivaceus***
Anura - Bufonidae
e Olive Toad; Baluchistan Coastal Toad; Makran Toad

758 ***Bufo pageoti***
Anura - Bufonidae
e Tonkin Toad

759 ***Bufo pantherinus***
Anura - Bufonidae
e Panther Toad; Western Leopard Toad (CSA)

760 ***Bufo paracnemis***
Anura - Bufonidae
e Cururu Toad; Rococo Toad

d Rokoko-Kröte; Parana-Kröte

761 ***Bufo pardalis***
Anura - Bufonidae
e Gleniffer Toad; Leopard Toad (CSA); Snoring Toad (CSA); August Toad (CSA); Eastern Leopard Toad (CSA)
d Pantherkröte; Südafrikanische Pantherkröte

762 ***Bufo parietalis***
Anura - Bufonidae
e Indian Toad

763 ***Bufo parkeri***
Anura - Bufonidae
e Parker's Toad

764 ***Bufo parvus***
Anura - Bufonidae
e Lesser Malacca Toad; Dwarf Toad (SEA)
d Kleine Kröte

765 ***Bufo pentoni***
Anura - Bufonidae
e Shaata Gardens Toad

766 ***Bufo periglenes***
Anura - Bufonidae
e Alajuela Toad; Orange Toad; Golden Toad
d Rote Kröte; Goldkröte

767 ***Bufo peripatetes***
Anura - Bufonidae
e Almirante Trail Toad

768 ***Bufo perplexus***
Anura - Bufonidae
e Confusing Toad

769 ***Bufo perreti***
Anura - Bufonidae
e Perret's Toad

770 ***Bufo poeppigii***
Anura - Bufonidae
e Grey Toad

771 ***Bufo poweri***
Anura - Bufonidae
e Kimberley Toad; Western Olive Toad (CSA)

772 ***Bufo punctatus***
Anura - Bufonidae
e Red-spotted Toad; Baird's Spotted Toad
d Rottupfenkröte; Rotfleckenkröte
f Crapaud tacheté de rouge

773 ***Bufo pygmaeus***
Anura - Bufonidae
e Rio Parahyba Toad

774 ***Bufo quadriporcatus***
Anura - Bufonidae
e Greater Malacca Toad; Red Swamp Toad (SEA)
d Rote Sumpfkröte

775 ***Bufo quechua***
Anura - Bufonidae
e Incachaca Toad

776 ***Bufo quercicus***
Anura - Bufonidae
e Oak Toad
d Eichenkröte
f Crapaud des chènes

777 ***Bufo raddei***
Anura - Bufonidae
e Tengger Desert Toad; Radde's Toad; Siberian Toad; Siberian Sand Toad
d Wüstenkröte; Asiatische Wüstenkröte; Mongolische Kröte

778 ***Bufo rangerei***
Anura - Bufonidae
e Kei Road Toad; Raucous Toad (CSA); Ranger's Toad (CSA)

779 ***Bufo reesi***
Anura - Bufonidae
e Merera Toad

780 ***Bufo regularis***
Anura - Bufonidae
e African Square-marked Toad; Egyptian Square-marked Toad; Square-marked Toad; Common Egyptian Toad; Leopard Toad; Reuss's Toad; African Toad (CSA); Panther Toad; Common Square-marked Toad; Common Toad (CSA)
d Pantherkröte
f Crapaud panthérin; Crapaud léopard

781 ***Bufo retiformis***
Anura - Bufonidae
e Sonoran Green Toad; Reticulated Toad; Sonora Green Toad
d Grüne Kröte; Amerikanische Kröte

782 ***Bufo robinsoni***
Anura - Bufonidae
e Paradise Toad

783 ***Bufo rubropunctatus***
Anura - Bufonidae
e Red-spotted Toad

784 ***Bufo rufus***
Anura - Bufonidae
e Rusty Toad

785 ***Bufo rumbolli***
Anura - Bufonidae
e Salta Toad

786 ***Bufo schneideri***
Anura - Bufonidae
e Schneider's Toad

787 ***Bufo scortecci***
Anura - Bufonidae
e Scortecci's Toad

788 ***Bufo sibiliai***
Anura - Bufonidae
e Ethiopia Toad

789 ***Bufo silentvalleyensis***
Anura - Bufonidae
e Silent Valley Toad

790 ***Bufo simus***
Anura - Bufonidae
e Flatnose Toad

791 ***Bufo speciosus***
Anura - Bufonidae
e Texas Toad

792 ***Bufo spinulosus***
Anura - Bufonidae
e Warty Toad; Spiny Toad
d Anden-Kröte

793 ***Bufo steindachneri***
Anura - Bufonidae
e Steindachner's Toad

794 ***Bufo stejnegeri***
Anura - Bufonidae
e Stejneger's Toad; Korean Toad; Korean Water Toad

795 ***Bufo sternosignatus***
Anura - Bufonidae
e Falcon Toad

796 ***Bufo stomaticus***
Anura - Bufonidae
e Assam Toad; Indus Toad; Indus Marbled Toad; Marbled Toad (ISC); North Indian Toad

f Crapaud de l'Inde

797 ***Bufo stuarti***
Anura - Bufonidae
e Stuart's Toad

798 ***Bufo sumatranus***
Anura - Bufonidae
e Sumatra Toad

799 ***Bufo superciliaris***
Anura - Bufonidae
e Zaire Toad; Cameroon Toad
d Zipfelkröte
f Crapaud aux paupières relevées

800 ***Bufo surdus***
Anura - Bufonidae
e Pakistan Toad; Luristan Toad; Lorestan Toad

801 ***Bufo tacanensis***
Anura - Bufonidae
e Volvan Tacana Toad

802 ***Bufo taitanus***
Anura - Bufonidae
e Taita Toad; Dwarf Toad; Black-chested Dwarf Toad

803 ***Bufo taitanus nyikae***
Anura - Bufonidae
e Nyika Dwarf Toad

804 ***Bufo taitanus taitanus***
Anura - Bufonidae
e Teita Dwarf Toad (CSA)

805 ***Bufo terrestris***
Anura - Bufonidae
e Southern Toad; Common American Toad
d Nordamerikanische Erdkröte; Karolina Kröte; Südkröte
f Crapaud criard

806 ***Bufo tibetanus***
Anura - Bufonidae
e Tibetan Toad

807 ***Bufo tienhoensis***
Anura - Bufonidae
e Tien-Ho Toad

808 ***Bufo tihamicus***
Anura - Bufonidae
e Balletto's Toad

809 ***Bufo togoensis***
Anura - Bufonidae
e Togo Toad

810 ***Bufo torrenticola***
Anura - Bufonidae
e Honshu Toad; Japanese Stream Toad

811 ***Bufo trifolium***
Anura - Bufonidae
e Huanuco Toad

812 ***Bufo tuberculatus***
Anura - Bufonidae
e Quinghai Lake Toad

813 ***Bufo tuberosus***
Anura - Bufonidae
e Fernando Po Toad
d Warzenkröte

814 ***Bufo turkanae***
Anura - Bufonidae
e Lake Turkana Toad

815 ***Bufo typhonius***
Anura - Bufonidae
e South American Common Toad; South American Toad; Mitred Toad
f Crapaud perlé; Crapaud à oreilles; Crapaud à petite crête

816 ***Bufo urunguensis***
Anura - Bufonidae
e Urungu Toad

817 ***Bufo ushoranus***
Anura - Bufonidae
e Ushora Toad

818 ***Bufo uzunguensis***
Anura - Bufonidae
e Uzungwe Toad

819 ***Bufo valhallae***
Anura - Bufonidae
e Pulo Weh Toad

820 ***Bufo valliceps***
Anura - Bufonidae
e Gulf Coast Toad; Nebulous Toad
d Golfkröte

821 ***Bufo valliceps macrocristatus***
Anura - Bufonidae
e Large Crested Toad

822 ***Bufo valliceps valliceps***
Anura - Bufonidae
e Gulf Coast Toad

823 ***Bufo variegatus***
Anura - Bufonidae
e Eden Harbour Toad; Patagonian Toad

824 ***Bufo vellardi***
Anura - Bufonidae
e Alto Maranon Toad

825 ***Bufo veraguensis***
Anura - Bufonidae
e Veragua Toad

826 ***Bufo verrucoissimo***
Anura - Bufonidae
e Caucasus Toad

827 ***Bufo vertebralis***
Anura - Bufonidae
e African Dwarf Toad; Pygmy Toad (CSA); Southern Pygmy Toad (CSA); Dwarf Toad; Flat Toad (CSA)
f Crapaud des steppes

828 ***Bufo villiersi***
Anura - Bufonidae
e Villier's Toad

829 ***Bufo viridis***
Anura - Bufonidae
e European Green Toad (NA); Green Toad; Variegated Toad
d Wechselkröte; Grüne Kröte
f Crapaud vert; Crapaud vert d'Europe
i Rospo smeraldino; Rospo verde; Zampèlgh

830 ***Bufo vittatus***
Anura - Bufonidae
e Banded Toad; Degen's Toad

831 ***Bufo wazae***
Anura - Bufonidae
e Waza Toad

832 ***Bufo woodhousii***
Anura - Bufonidae
e Woodhouse's Toad
d Woodhouses Kröte
f Crapaud de Woodhouse

833 ***Bufo woodhousii australis***
Anura - Bufonidae
e Southwestern Woodhouse's Toad

834 ***Bufo woodhousii velatus***
Anura - Bufonidae
e East Texas Toad

835 ***Bufo woodhousii woodhousii***
Anura - Bufonidae
e Woodhouse's Toad; Western Woodhouse's Toad; Rocky Mountain Toad

836 ***Bufo xeros***
Anura - Bufonidae
e Subsahara Toad

837 ***Bufoides***
Anura - Bufonidae
e Mawblang Toads

838 ***Bufoides meghalayanus***
Anura - Bufonidae
e Mawblang Toad

839 **Bufonidae**
Anura
e True Toads
d Kröten; Echte Kröten; Bufoniden
f Crapauds; Vrais crapauds; Bufonidés

C

840 ***Cacosternum***
Anura - Ranidae
e Metal Frogs; Cacos (CSA); Froglets (CSA); Dainty Frogs (CSA)

841 ***Cacosternum boettgeri***
Anura - Ranidae
e Boettger's Metal Frog; Comman Caco (CSA); Caco (CSA); Common Dainty Frog; Dainty Frog (CSA); Boettger's Frog; Boettger's Froglet (CSA); Boettger's Dainty Frog (CSA)

842 ***Cacosternum capense***
Anura - Ranidae
e Cape Metal Frog; Cape Caco (CSA); Cross-marked Frog (CSA); Cape Froglet (CSA)

843 ***Cacosternum leleupi***
Anura - Ranidae
e Katanga Metal Frog

844 ***Cacosternum namaquense***
Anura - Ranidae
e Namaqua Metal Frog; Namaqua Caco (CSA); Namaqua Froglet (CSA)

845 ***Cacosternum nanum***
Anura - Ranidae
e Mozambique Metal Frog; Bronze Caco (CSA); Dainty Frog; Dark-throated Froglet (CSA); Small Froglet

846 ***Cacosternum nanum nanum***
Anura - Ranidae
e Lesser Dainty Frog (CSA)

847 ***Cacosternum nanum parvum***
Anura - Ranidae
e Lesser Dainty Frog (CSA)

848 ***Cacosternum poyntoni***
Anura - Ranidae
e Nursery Metal Frog; Poynton's Caco (CSA)

849 ***Cacosternum striatus***
Anura - Ranidae
e Striped Metal Frog; Striped Caco (CSA)

850 ***Caecilia***
Gymnophiona - Caeciliaidae
e Common Caecilians
f Cécilies

851 ***Caecilia abitaguae***
Gymnophiona - Caeciliaidae
e Abitagua Caecilian

852 ***Caecilia albiventris***
Gymnophiona - Caeciliaidae
e Whitebelly Caecilian
f Cécilie à ventre blanc

853 ***Caecilia antioquiaensis***
Gymnophiona - Caeciliaidae
e Antioquia Caecilian

854 ***Caecilia armata***
Gymnophiona - Caeciliaidae
e Armoured Caecilian

855 ***Caecilia attenuata***
Gymnophiona - Caeciliaidae
e Santa Rosa Caecilian

856 ***Caecilia bokermanni***
Gymnophiona - Caeciliaidae

e Bokerman's Caecilian

857 ***Caecilia caribea***
Gymnophiona - Caeciliaidae
e Pensilvania Caecilian

858 ***Caecilia corpulenta***
Gymnophiona - Caeciliaidae
e Solid Caecilian

859 ***Caecilia crassisquama***
Gymnophiona - Caeciliaidae
e Normandia Caecilian

860 ***Caecilia degenerata***
Gymnophiona - Caeciliaidae
e Gartagoa Caecilian

861 ***Caecilia disossea***
Gymnophiona - Caeciliaidae
e Santiago Caecilian

862 ***Caecilia dunni***
Gymnophiona - Caeciliaidae
e Dunn's Caecilian

863 ***Caecilia flavopunctata***
Gymnophiona - Caeciliaidae
e Yellow-spotted Caecilian

864 ***Caecilia gracilis***
Gymnophiona - Caeciliaidae
e Surinam Caecilian
f Cécilie gracile; Cécilie lombricoïde

865 ***Caecilia guntheri***
Gymnophiona - Caeciliaidae
e Gunther's Caecilian

866 ***Caecilia inca***
Gymnophiona - Caeciliaidae
e Fundo Sinchona Caecilian

867 ***Caecilia leucocephala***
Gymnophiona - Caeciliaidae
e White-headed Caecilian

868 ***Caecilia marcusi***
Gymnophiona - Caeciliaidae
e Villa Tunari Caecilian

869 ***Caecilia mertensi***
Gymnophiona - Caeciliaidae
e Mertens's Caecilian

870 ***Caecilia nigricans***
Gymnophiona - Caeciliaidae
e Rio Lita Caecilian

871 ***Caecilia occidentalis***
Gymnophiona - Caeciliaidae
e Cauca Caecilian

872 ***Caecilia orientalis***
Gymnophiona - Caeciliaidae
e La Bonita Caecilian

873 ***Caecilia pachynema***
Gymnophiona - Caeciliaidae
e Intac Caecilian

874 ***Caecilia perdita***
Gymnophiona - Caeciliaidae
e Andagoya Caecilian

875 ***Caecilia pressula***
Gymnophiona - Caeciliaidae
e Marudi Mountains Caecilian

876 ***Caecilia subdermalis***
Gymnophiona - Caeciliaidae
e Caecilian Moscopan

877 ***Caecilia subnigricans***
Gymnophiona - Caeciliaidae
e Magdalena Valley Caecilian

878 ***Caecilia subterminalis***
Gymnophiona - Caeciliaidae
e Taylor's Ecuador Caecilian

879 ***Caecilia tentaculata***
Gymnophiona - Caeciliaidae
e Linnaeus Caecilian
f Cécilie tentaculée

880 ***Caecilia tenuissima***
Gymnophiona - Caeciliaidae
e Guayaquil Caecilian

881 ***Caecilia thompsoni***
Gymnophiona - Caeciliaidae
e Thompson's Caecilian

882 ***Caecilia volcani***
Gymnophiona - Caeciliaidae
e Cocle Caecilian

883 **Caeciliaidae**
Gymnophiona
e Caecilians; Common Caecilians; Non-tailed Caecilians
d Wurmwühlenverwandten
f Cécilidés

884 ***Callixalus***
Anura - Hyperoliidae
e Painted African Frogs

885 ***Callixalus pictus***
Anura - Hyperoliidae
e Painted African Frog

886 ***Calluella***
Anura - Microhylidae
e Squat Frogs

887 ***Calluella brooksi***
Anura - Microhylidae
e Brooks's Squat Frog

888 ***Calluella flava***
Anura - Microhylidae
e Borneo Squat Frog

889 ***Calluella guttulata***
Anura - Microhylidae
e Burmese Squat Frog; Spotted Spadefoot Frog (SEA)
d Gefleckter Grabfrosch

890 ***Calluella smithi***
Anura - Microhylidae
e Smith's Squat Frog

891 ***Calluella volzi***
Anura - Microhylidae
e Sumatra Squat Frog

892 ***Calluella yunnanensis***
Anura - Microhylidae
e Yunnan Squat Frog

893 ***Callulina***
Anura - Microhylidae
e Secret Frogs

894 ***Callulina kreffti***
Anura - Microhylidae
e Secret Frog; Kreft's Secret Frog

895 ***Callulops***
Anura - Microhylidae
e Callulops Frogs

896 ***Callulops boettgeri***
Anura - Microhylidae
e Boettger's Callulops Frog

897 ***Callulops doriae***
Anura - Microhylidae
e Doria's Callulops Frog

898 ***Callulops dubius***
Anura - Microhylidae

e Molucan Callulops Frog

899 ***Callulops eurydactylus***
Anura - Microhylidae
e Danowaria Callulops Frog

900 ***Callulops fuscus***
Anura - Microhylidae
e Brown Callulops Frog

901 ***Callulops glandulosus***
Anura - Microhylidae
e Warty Callulops Frog

902 ***Callulops humicola***
Anura - Microhylidae
e Kotuni Callulops Frog

903 ***Callulops kopsteini***
Anura - Microhylidae
e Kopstein's Callulops Frog

904 ***Callulops personatus***
Anura - Microhylidae
e Maprik Callulops Frog

905 ***Callulops robustus***
Anura - Microhylidae
e Boulenger's Callulops Frog

906 ***Callulops slateri***
Anura - Microhylidae
e Slater's Callulops Frog

907 ***Callulops stictogaster***
Anura - Microhylidae
e Irumbofoie Callulops Frog

908 ***Callulops wilhelmanus***
Anura - Microhylidae
e Wilhelm Callulops Frog

909 ***Calyptahyla***
Anura - Hylidae
e Antilles Frogs

910 ***Calyptahyla crucialis***
Anura - Hylidae
e Antilles Frog; Harlan's Antilles Frog

911 ***Capensibufo***
Anura - Bufonidae
e Cape Toads

912 ***Capensibufo rosei***
Anura - Bufonidae
e Muizenberg Cape Toad; Cape Mountain Toad (CSA); Striped Mountain Toad; Rose's Toad (CSA); Rose's Mountain Toad (CSA)
d Gestreifte Bergkröte
f Crapaud de Rose

913 ***Capensibufo tradouwi***
Anura - Bufonidae
e Swellendam Cape Toad; Cape Mountain Toad; Tradouw Toad; Tradouw Mountain Toad (CSA)

914 ***Cardioglossa***
Anura - Arthroleptidae
e Long-fingered Frog
d Herzzüngler (pl)
f Cardioglosses

915 ***Cardioglossa aureoli***
Anura - Arthroleptidae
e Freetown Long-fingered Frog

916 ***Cardioglossa cyaneospila***
Anura - Arthroleptidae
e Mukuzira Long-fingered Frog

917 ***Cardioglossa dorsalis***
Anura - Arthroleptidae
e Yoruba Long-fingered Frog

918 ***Cardioglossa elegans***
Anura - Arthroleptidae

e Elegant Long-fingered Frog

919 ***Cardioglossa escalerae***
Anura - Arthroleptidae
e Equatorial Guinea Long-fingered Frog

920 ***Cardioglossa gracilis***
Anura - Arthroleptidae
e Rio Benito Long-fingered Frog

921 ***Cardioglossa gratiosa***
Anura - Arthroleptidae
e Ongot Long-fingered Frog

922 ***Cardioglossa leucomystax***
Anura - Arthroleptidae
e Silver Long-fingered Frog

923 ***Cardioglossa liberiensis***
Anura - Arthroleptidae
e Liberia Long-fingered Frog

924 ***Cardioglossa melanogaster***
Anura - Arthroleptidae
e Amiet's Long-fingered Frog

925 ***Cardioglossa nigromaculata***
Anura - Arthroleptidae
e Black-spotted Long-fingered Frog

926 ***Cardioglossa oreas***
Anura - Arthroleptidae
e Mount Okou Long-fingered Frog

927 ***Cardioglossa peternageli***
Anura - Arthroleptidae
e Nagel's Long-fingered Frog

928 ***Cardioglossa pulchra***
Anura - Arthroleptidae
e Black Long-fingered Frog

929 ***Cardioglossa schiotzi***
Anura - Arthroleptidae
e Acha Tugi Long-fingered Frog

930 ***Cardioglossa trifasciata***
Anura - Arthroleptidae
e Nsoung Long-fingered Frog

931 ***Cardioglossa venusta***
Anura - Arthroleptidae
e Highland Long-fingered Frog

932 ***Caudacaecilia***
Gymnophiona - Ichthyophiidae
e Tailed Caecilians; Caecilians (SEA)
d Fischwühlen

933 ***Caudacaecilia asplenia***
Gymnophiona - Ichthyophiidae
e Boven Mahakkam Caecilian; Malayan Caecilian (SEA)
d Malaiische Fischwühle

934 ***Caudacaecilia larutensis***
Gymnophiona - Ichthyophiidae
e Larut Hills Caecilian; Larut Caecilian (SEA)
d Larut-Fischwühle

935 ***Caudacaecilia nigroflava***
Gymnophiona - Ichthyophiidae
e Kuala Lumpur Caecilian; Black-and-yellow Caecilian (SEA)
d Schwarz-gelbe Fischwühle

936 ***Caudacaecilia paucidentula***
Gymnophiona - Ichthyophiidae
e Kapahiang Caecilian

937 ***Caudacaecilia weberi***
Gymnophiona - Ichthyophiidae
e Malatgan River Caecilian

938 ***Caudiverbera***
Anura - Leptodactylidae

e Helmeted Water Toads

939 ***Caudiverbera caudiverbera***
Anura - Leptodactylidae
e Helmeted Water Toad; Gay's Frog; Chilean Toad
d Helmkopffrosch; Helmkopf; Chilenischer Helmkopf

940 ***Centrolene***
Anura - Centrolenidae
e Giant Glass Frogs
d Gecko-Glasfrösche

941 ***Centrolene acanthidicophalum***
Anura - Centrolenidae
e Santander Giant Glass Frog

942 ***Centrolene altudinale***
Anura - Centrolenidae
e Mountain Giant Glass Frog

943 ***Centrolene andinum***
Anura - Centrolenidae
e Andes Giant Glass Frog

944 ***Centrolene antioquiense***
Anura - Centrolenidae
e Antioquia Giant Glass Frog

945 ***Centrolene audax***
Anura - Centrolenidae
e Napo Giant Glass Frog

946 ***Centrolene azulae***
Anura - Centrolenidae
e Blue Giant Glass Frog

947 ***Centrolene ballux***
Anura - Centrolenidae
e Burrowes's Giant Glass Frog

948 ***Centrolene buckleyi***
Anura - Centrolenidae
e Buckley's Giant Glass Frog

949 ***Centrolene fernandoi***
Anura - Centrolenidae
e Fernando Giant Glass Frog

950 ***Centrolene geckoideum***
Anura - Centrolenidae
e Pacific Giant Glass Frog
d Gecko-Glasfrosch

951 ***Centrolene gemmatum***
Anura - Centrolenidae
e Pampas Giant Glass Frog

952 ***Centrolene gorzulae***
Anura - Centrolenidae
e Bolivar Giant Glass Frog

953 ***Centrolene grandisonae***
Anura - Centrolenidae
e Giant Glass Frog

954 ***Centrolene heloderma***
Anura - Centrolenidae
e Pichincha Giant Glass Frog

955 ***Centrolene hesperium***
Anura - Centrolenidae
e Basecamp Giant Glass Frog

956 ***Centrolene hybrida***
Anura - Centrolenidae
e Eastern Giant Glass Frog

957 ***Centrolene ilex***
Anura - Centrolenidae
e Limon Giant Glass Frog

958 ***Centrolene lemniscatum***
Anura - Centrolenidae
e Rioja Giant Glass Frog

959 ***Centrolene lentiginosum***
Anura - Centrolenidae
e Guacharaquita Giant Glass Frog

960 ***Centrolene lynchi***
Anura - Centrolenidae
e Lynch's Giant Glass Frog

961 ***Centrolene mariae***
Anura - Centrolenidae
e Maria's Giant Glass Frog

962 ***Centrolene medemi***
Anura - Centrolenidae
e Medem Giant Glass Frog

963 ***Centrolene muelleri***
Anura - Centrolenidae
e Müller's Giant Glass Frog

964 ***Centrolene notostictum***
Anura - Centrolenidae
e Cordillera Giant Glass Frog

965 ***Centrolene paezorum***
Anura - Centrolenidae
e Colombian Giant Glass Frog

966 ***Centrolene peristictum***
Anura - Centrolenidae
e Tandapi Giant Glass Frog

967 ***Centrolene petrophilum***
Anura - Centrolenidae
e Boyaca Giant Glass Frog

968 ***Centrolene pipilatum***
Anura - Centrolenidae
e Amazon Giant Glass Frog

969 ***Centrolene prosoblepon***
Anura - Centrolenidae
e Nicaragua Giant Glass Frog

970 ***Centrolene puyoensis***
Anura - Centrolenidae
e Puyo Giant Glass Frog

971 ***Centrolene sanchez***
Anura - Centrolenidae
e Sanchez's Giant Glass Frog

972 ***Centrolene scirtetes***
Anura - Centrolenidae
e Tandayapa Giant Glass Frog

973 ***Centrolene tayrona***
Anura - Centrolenidae
e Madalene Giant Glass Frog

974 **Centrolenidae**
Anura
e Glass Frogs
d Glasfrösche

975 ***Ceratobatrachus***
Anura - Ranidae
e Triangle Frogs

976 ***Ceratobatrachus guentheri***
Anura - Ranidae
e Triangle Frog; Günther's Triangle Forg; Günther's Toothed Frog; Solomon Islands Horned Frog
d Zipfelfrosch; Salomonischer Zipfelfrosch

977 ***Ceratophrys***
Anura - Leptodactylidae
e Common Horned Toads; Horned Toads; Horned Frogs
d Hornfrösche; Südamerikanischer Hornfrösche; Schildfrösche
f Crapauds cornues; Grenouilles cornues

978 ***Ceratophrys aurita***
Anura - Leptodactylidae
e Brazilian Horned Toad; Wied's Frog; Brazilian Horned Frog

d Bunthornfrosch
f Grenouille de Wied; Grenouille cornue de Brésil

979 ***Ceratophrys calcarata***
Anura - Leptodactylidae
e Colombian Horned Toad; Colombian Horned Frog
f Grenouille cornue de Colombie

980 ***Ceratophrys cornuta***
Anura - Leptodactylidae
e Surinam Horned Toad; Amazonian Horned Frog; Brazilian Horned Toad
d Hornfrosch; Gehörnter Hornfrosch; Großer Hornfrosch
f Grenouille à cornes; Grenouille cornue de l'Amazon; Cératophore cornu

981 ***Ceratophrys cranwelli***
Anura - Leptodactylidae
e Cranwell's Horned Toad

982 ***Ceratophrys joazeirenis***
Anura - Leptodactylidae
e Joazeiro Horned Toad

983 ***Ceratophrys ornata***
Anura - Leptodactylidae
e Ornate Horned Toad; Argentine Horned Frog; Ornate Horned Frog; Bell's Ceratophrys; Horned Frog; Barking Toad; Escuerzo
d Schmuckhornfrosch; Escuerzo
f Grenouille cornue de Bell; Escuerzo; Grenouille cornue ornée

984 ***Ceratophrys ornata x cornuta***
Anura - Leptodactylidae
e Bell's Frog
f Grenouille à grande bouche; Grenouille de Bell

985 ***Ceratophrys stolzmanni***
Anura - Leptodactylidae
e Stolzmann's Horned Toad; Escuerzo
d Escuerzo
f Escuerzo

986 ***Ceratophrys testudo***
Anura - Leptodactylidae
e Ecuador Horned Toad

987 ***Chacophrys***
Anura - Leptodactylidae
e Chaco Horned Toads

988 ***Chacophrys pierotti***
Anura - Leptodactylidae
e Chaco Horned Toad

989 ***Chaparana***
Anura - Ranidae
e Asian Frogs

990 ***Chaparana aenea***
Anura - Ranidae
e Doi Chang Asian Frog; Doi-Dang Frog (SEA)
d Doi-Dang-Frosch

991 ***Chaparana delacouri***
Anura - Ranidae
e Tonkin Asian Frog

992 ***Chaparana fansipani***
Anura - Ranidae
e Fan-si-pan Asian Frog

993 ***Chaparana quadranus***
Anura - Ranidae
e Kwang-yang Asian Frog

994 ***Chaparana sikimensis***
Anura - Ranidae
e Sikkim Asian Frog

995 ***Chaparana unculuanus***
Anura - Ranidae
e Yunnan Asian Frog

996 ***Chaperina***
Anura - Microhylidae
e Thorny Frogs

997 ***Chaperina fusca***
Anura - Microhylidae
e Thorny Frog; Brown Thorny Frog; Yellow-bellied Frog (SEA)
d Gelbbauchfrosch

998 ***Chiasmocleis***
Anura - Microhylidae
e Humming Frogs
f Chiasmocles

999 ***Chiasmocleis albopunctata***
Anura - Microhylidae
e White-spotted Humming Frog

1000 ***Chiasmocleis anatipes***
Anura - Microhylidae
e Santa Cecelia Humming Frog

1001 ***Chiasmocleis bassleri***
Anura - Microhylidae
e Bassler's Humming Frog

1002 ***Chiasmocleis bicegoi***
Anura - Microhylidae
e Sao Paulo Humming Frog

1003 ***Chiasmocleis centralis***
Anura - Microhylidae
e Central Humming Frog

1004 ***Chiasmocleis hudsoni***
Anura - Microhylidae
e Hudson's Humming Frog
f Chiasmocle de Hudson

1005 ***Chiasmocleis leucosticta***
Anura - Microhylidae
e Santa Catarina Humming Frog

1006 ***Chiasmocleis panamensis***
Anura - Microhylidae
e Panama Humming Frog

1007 ***Chiasmocleis schubarti***
Anura - Microhylidae
e Schubart's Humming Frog

1008 ***Chiasmocleis shudikarensis***
Anura - Microhylidae
e Guyana Humming Frog
f Chiasmocle de Shudikar

1009 ***Chiasmocleis urbanae***
Anura - Microhylidae
e Bokerman's Humming Frog

1010 ***Chiasmocleis ventrimaculata***
Anura - Microhylidae
e Dotted Humming Frog

1011 ***Chioglossa***
Caudata - Salamandridae
e Gold-striped Salamanders; Golden-striped Salamanders
d Goldstreifensalamander (pl); Scheidenzüngler (pl)
f Chioglosses

1012 ***Chioglossa lusitanica***
Caudata - Salamandridae
e Gold-striped Salamander; Golden-striped Salamander
d Goldstreifensalamander; Scheidenzüngler
f Chioglosse portugais; Salamandrine portugaise; Salamandre à bandes dorées; Salamandrine du Portugal; Chioglosse
i Chioglossa

1013 ***Chirixalus***
Anura - Rhacophoridae
e Asian Tree Frogs; Bush Frogs (SEA)
d Buschfrösche

1014 ***Chirixalus doriae***
Anura - Rhacophoridae
e Doria's Asian Tree Frog; Dori's Bush Frog (SEA)
d Doris Buschfrosch

1015 ***Chirixalus eiffingeri***
Anura - Rhacophoridae
e Eiffinger's Asian Tree Frog

1016 ***Chirixalus hasenae***
Anura - Rhacophoridae
e Hansen's Asian Tree Frog; Hasens's Bush Frog (SEA)
d Hasens Buschfrosch

1017 ***Chirixalus idiootocus***
Anura - Rhacophoridae
e Temple Asian Tree Frog

1018 ***Chirixalus laevis***
Anura - Rhacophoridae
e Smith's Asian Tree Frog

1019 ***Chirixalus nongkhorensis***
Anura - Rhacophoridae
e Nongkhor Asian Tree Frog; Nong Khor Bush Frog (SEA)
d Nong-Khor-Buschfrosch

1020 ***Chirixalus simus***
Anura - Rhacophoridae
e Assam Asian Tree Frog

1021 ***Chirixalus vittatus***
Anura - Rhacophoridae
e Striped Asian Tree Frog; Striped Bush Frog (SEA)
d Gestreifter Buschfrosch

1022 ***Chiromantis***
Anura - Rhacophoridae
e Foam-nest Tree Frogs; Foam-nest Frogs; Grey Tree Frogs (CSA)
d Graue Baumfrösche; Baumfrösche; Greiffrösche

1023 ***Chiromantis petersii***
Anura - Rhacophoridae
e Peters's Foam-nest Tree Frog; Central Foam-nest Tree Frog; Peters's Grey Tree Frog; Great Grey Tree Frog (CSA)

1024 ***Chiromantis rufescens***
Anura - Rhacophoridae
e African Foam-nest Tree Frog; Western Foam-nest Tree Frog; Western Grey Tree Frog
d Rauhäutiger Baumfrosch; Westlicher Grauer Baumfrosch
f Grenouille arboricole

1025 ***Chiromantis xerampelina***
Anura - Rhacophoridae
e African Grey Tree Frog; Foam-nest Frog (CSA); Grey Tree Frog (CSA); Southern Foam-nest Tree Frog (CSA); South-eastern Foam-nest Tree Frog; Great Grey Tree Frog; Great African Grey Tree Frog (CSA); Foam-nest Tree Frog (CSA); Large African Grey Tree Frog (CSA)
d Grauer Baumfrosch; Afrikanischer Greiffrosch

1026 ***Chiropterotriton***
Caudata - Plethodontidae
e Splayfoot Salamanders
d Schwielensalamander (pl)

1027 ***Chiropterotriton arboreus***
Caudata - Plethodontidae
e Arboreal Splayfoot Salamander

1028 ***Chiropterotriton chiropterus***
Caudata - Plethodontidae
e Common Splayfoot Salamander; Splayfoot Salamander
d Kleiner Schwielensalamander

1029 ***Chiropterotriton chondrostega***
Caudata - Plethodontidae
e Gristle-headed Splayfoot

Salamander; Gristlehead Splayfoot Salamander
d Taylors Schwielensalamander

1030 ***Chiropterotriton chondrostega chondrostega***
Caudata - Plethodontidae
e Gristlehead Splayfoot Salamander

1031 ***Chiropterotriton chondrostega cracens***
Caudata - Plethodontidae
e Graceful Splayfoot Salamander

1032 ***Chiropterotriton chondrostega terrestris***
Caudata - Plethodontidae
e Terrestrial Splayfoot Salamander

1033 ***Chiropterotriton dimidiatus***
Caudata - Plethodontidae
e Dwarf Splayfoot Salamander

1034 ***Chiropterotriton lavae***
Caudata - Plethodontidae
e Pygmy Splayfoot Salamander

1035 ***Chiropterotriton magnipes***
Caudata - Plethodontidae
e Bigfoot Splayfoot Salamander
d Großfuß-Schwielensalamander

1036 ***Chiropterotriton mosaueri***
Caudata - Plethodontidae
e Cave Splayfoot Salamander

1037 ***Chiropterotriton multidentatus***
Caudata - Plethodontidae
e Toothy Splayfoot Salamander
d Vielzahn-Schwielensalamander

1038 ***Chiropterotriton priscus***
Caudata - Plethodontidae
e Primeval Splayfoot Salamander

1039 ***Chlorolius***
Anura - Hyperoliidae
e Koehlera Green Frogs

1040 ***Chlorolius koehleri***
Anura - Hyperoliidae
e Koehlera Green Frog

1041 ***Choerophryne***
Anura - Microhylidae
e Torreicelli Mountain Frogs

1042 ***Choerophryne rostellifer***
Anura - Microhylidae
e Torreicelli Mountain Frog

1043 ***Chrysobatrachus***
Anura - Hyperoliidae
e Itombwe Golden Frogs

1044 ***Chrysobatrachus cupreonitens***
Anura - Hyperoliidae
e Itombwe Golden Frog

1045 ***Chthonerpeton***
Gymnophiona - Typhlonectidae
e Peters's Caecilians

1046 ***Chthonerpeton braestrupi***
Gymnophiona - Typhlonectidae
e Braestrup's Caecilian

1047 ***Chthonerpeton exile***
Gymnophiona - Typhlonectidae
e Bahia Caecilian

1048 ***Chthonerpeton indistinctum***
Gymnophiona - Typhlonectidae
e Argentine Caecilian

1049 ***Chthonerpeton onorei***
Gymnophiona - Typhlonectidae
e El Reventador Caecilian

1050 ***Chthonerpeton perrisodus***
Gymnophiona - Typhlonectidae
e Minas Gerals Caecilian

1051 ***Chthonerpeton viviparum***
Gymnophiona - Typhlonectidae
e Santa Caterina Caecilian

1052 ***Cochranella***
Anura - Centrolenidae
e Cochran Frogs
f Cochranelles

1053 ***Cochranella adiazeta***
Anura - Centrolenidae
e Western Cochran Frog

1054 ***Cochranella albomaculata***
Anura - Centrolenidae
e White-spotted Cochran Frog

1055 ***Cochranella ametarisia***
Anura - Centrolenidae
e Headwater Cochran Frog

1056 ***Cochranella anomala***
Anura - Centrolenidae
e Napo Cochran Frog

1057 ***Cochranella balionota***
Anura - Centrolenidae
e Mindo Cochran Frog

1058 ***Cochranella bejaranoi***
Anura - Centrolenidae
e Bolivian Cochran Frog

1059 ***Cochranella chancas***
Anura - Centrolenidae
e Peru Cochran Frog

1060 ***Cochranella cochranae***
Anura - Centrolenidae
e Cochran Frog

1061 ***Cochranella croccopodes***
Anura - Centrolenidae
e San Martin Cochran Frog

1062 ***Cochranella daidalea***
Anura - Centrolenidae
e Alban Cochran Frog

1063 ***Cochranella duidaena***
Anura - Centrolenidae
e Duida Cochran Frog

1064 ***Cochranella euhystrix***
Anura - Centrolenidae
e Ridge Cochran Frog

1065 ***Cochranella euknemos***
Anura - Centrolenidae
e San Jose Cochran Frog

1066 ***Cochranella flavopunctata***
Anura - Centrolenidae
e Yellow-spotted Cochran Frog

1067 ***Cochranella geijskesi***
Anura - Centrolenidae
e Wilhelmina Cochran Frog

1068 ***Cochranella granulosa***
Anura - Centrolenidae
e Grainy Cochran Frog

1069 ***Cochranella griffithsi***
Anura - Centrolenidae
e Ecuador Cochran Frog

1070 ***Cochranella helenae***
Anura - Centrolenidae
e Venezuela Cochran Frog

1071 ***Cochranella ignota***
Anura - Centrolenidae
e Lynch's Cochran Frog

1072 ***Cochranella megacheira***
Anura - Centrolenidae
e Santa Rosa Cochran Frog

1073 ***Cochranella megistra***
Anura - Centrolenidae
e Urrao Cochran Frog

1074 ***Cochranella midas***
Anura - Centrolenidae
e Santa Cecilia Cochran Frog

1075 ***Cochranella nephelophila***
Anura - Centrolenidae
e Florencia Cochran Frog

1076 ***Cochranella ocellata***
Anura - Centrolenidae
e Spotted Cochran Frog

1077 ***Cochranella ocellifera***
Anura - Centrolenidae
e Paramba Cochran Frog

1078 ***Cochranella orejuela***
Anura - Centrolenidae
e El Tambo Cochran Frog

1079 ***Cochranella oreonympha***
Anura - Centrolenidae
e Caqueta Cochran Frog

1080 ***Cochranella oyampiensis***
Anura - Centrolenidae
e Zidok Cochran Frog
f Cochranelle des Oyampis

1081 ***Cochranella phenax***
Anura - Centrolenidae
e Tutumbaro Cochran Frog

1082 ***Cochranella pluvialis***
Anura - Centrolenidae
e Pistipata Cochran Frog

1083 ***Cochranella prasina***
Anura - Centrolenidae
e Rio Calima Cochran Frog

1084 ***Cochranella ramirezi***
Anura - Centrolenidae
e Ramirez's Cochran Frog

1085 ***Cochranella resplendens***
Anura - Centrolenidae
e Resplendant Cochran Frog

1086 ***Cochranella ritae***
Anura - Centrolenidae
e Rita's Cochran Frog

1087 ***Cochranella riveroi***
Anura - Centrolenidae
e Rivero's Cochran Frog

1088 ***Cochranella ruizi***
Anura - Centrolenidae
e Ruiz's Cochran Frog

1089 ***Cochranella savagei***
Anura - Centrolenidae
e Savage's Cochran Frog

1090 ***Cochranella saxiscandens***
Anura - Centrolenidae
e Tarapoto Cochran Frog

1091 ***Cochranella siren***
Anura - Centrolenidae
e Rio Coca Cochran Frog

1092 ***Cochranella solitaria***
Anura - Centrolenidae
e Lonely Cochran Frog

1093 ***Cochranella spiculata***
Anura - Centrolenidae
e Cuzco Cochran Frog

1094 ***Cochranella spinosa***
Anura - Centrolenidae
e Spiny Cochran Frog

1095 ***Cochranella tangarana***
Anura - Centrolenidae
e Tangarana Cochran Frog

1096 ***Cochranella truebae***
Anura - Centrolenidae
e Trueb's Cochran Frog

1097 ***Colostethus***
Anura - Dendrobatidae
e Rocket Frogs
d Falsche Giftfrösche
f Colostethes

1098 ***Colostethus abditaurantius***
Anura - Dendrobatidae
e Bello Rocket Frog

1099 ***Colostethus agilis***
Anura - Dendrobatidae
e Cauca Rocket Frog

1100 ***Colostethus alacrus***
Anura - Dendrobatidae
e Finca Primavera Rocket Frog

1101 ***Colostethus alagoanus***
Anura - Dendrobatidae
e Alagoas Rocket Frog

1102 ***Colostethus alboguttatus***
Anura - Dendrobatidae
e Whitebelly Rocket Frog

1103 ***Colostethus anthracinus***
Anura - Dendrobatidae
e South American Rocket Frog

1104 ***Colostethus argyrogaster***
Anura - Dendrobatidae
e Imaza Rocket Frog

1105 ***Colostethus beebei***
Anura - Dendrobatidae
e Beebe's Rocket Frog
f Colostethe de Beebe

1106 ***Colostethus bocagei***
Anura - Dendrobatidae
e Bocage's Rocket Frog

1107 ***Colostethus brachystriatus***
Anura - Dendrobatidae
e Stripe-throated Rocket Frog

1108 ***Colostethus breviquartus***
Anura - Dendrobatidae
e Urrao Rocket Frog

1109 ***Colostethus bromelicola***
Anura - Dendrobatidae
e Coastal Rocket Frog

1110 ***Colostethus brunneus***
Anura - Dendrobatidae
e Chupada Rocket Frog

1111 ***Colostethus capixaba***
Anura - Dendrobatidae
e Linhars Rocket Frog

1112 ***Colostethus capuriensis***
Anura - Dendrobatidae
e Sierra Nevada Rocket Frog

1113 ***Colostethus carioca***
Anura - Dendrobatidae
e Guanabara Rocket Frog

1114 ***Colostethus cevalosi***
Anura - Dendrobatidae
e Paland Rocket Frog

1115 ***Colostethus chacopis***
Anura - Dendrobatidae
e Ravine Rocket Frog

1116 ***Colostethus chocoensis***
Anura - Dendrobatidae
e Choco Rocket Frog

1117 ***Colostethus citreicola***
Anura - Dendrobatidae
e Limon Rocket Frog

1118 ***Colostethus degranvillei***
Anura - Dendrobatidae
e Degranville's Rocket Frog
f Colostethe de Granville

1119 ***Colostethus dunni***
Anura - Dendrobatidae
e Dunn's Rocket Frog

1120 ***Colostethus duranti***
Anura - Dendrobatidae
e Durant's Rocket Frog

1121 ***Colostethus edwardsi***
Anura - Dendrobatidae
e Edwards's Rocket Frog

1122 ***Colostethus elechyhistus***
Anura - Dendrobatidae
e Loja Rocket Frog

1123 ***Colostethus exasperatus***
Anura - Dendrobatidae
e Yapitya Rocket Frog

1124 ***Colostethus faciopunctulatus***
Anura - Dendrobatidae
e Puerto Narino Rocket Frog

1125 ***Colostethus fallax***
Anura - Dendrobatidae
e Cotopaxi Rocket Frog

1126 ***Colostethus festae***
Anura - Dendrobatidae
e Valle Santiago Rocket Frog

1127 ***Colostethus flotator***
Anura - Dendrobatidae
e Rainforest Rocket Frog

1128 ***Colostethus fraterdanieli***
Anura - Dendrobatidae
e Santa Rita Rocket Frog

1129 ***Colostethus fugax***
Anura - Dendrobatidae
e Pastaza Rocket Frog

1130 ***Colostethus fuliginosus***
Anura - Dendrobatidae
e Quijos Rocket Frog

1131 ***Colostethus goianus***
Anura - Dendrobatidae
e Goias Rocket Frog

1132 ***Colostethus haydeeae***
Anura - Dendrobatidae
e El Vivero Rocket Frog

1133 ***Colostethus humilis***
Anura - Dendrobatidae
e Bocono Rocket Frog

1134 ***Colostethus idiomelas***
Anura - Dendrobatidae
e Rivero's Rocket Frog

1135 ***Colostethus imbricolus***
Anura - Dendrobatidae
e Silverstone's Rocket Frog

1136 ***Colostethus infraguttatus***
Anura - Dendrobatidae
e Chimbo Rocket Frog

1137 ***Colostethus inguinalis***
Anura - Dendrobatidae
e Common Rocket Frog; Rocket Frog
d Panama-Baumsteiger; Raketenfrosch

1138 ***Colostethus intermedius***
Anura - Dendrobatidae
e Andersson's Rocket Frog

1139 ***Colostethus jacubpetersi***
Anura - Dendrobatidae
e Quito Rocket Frog

1140 ***Colostethus kingburyi***
Anura - Dendrobatidae
e Kingsbury's Rocket Frog

1141 ***Colostethus lacrimosus***
Anura - Dendrobatidae
e Lowland Rocket Frog

1142 ***Colostethus latinasus***
Anura - Dendrobatidae
e Truando Rocket Frog

1143 ***Colostethus lehmanni***
Anura - Dendrobatidae
e Lehmann's Rocket Frog

1144 ***Colostethus leopardalis***
Anura - Dendrobatidae
e Leopard Rocket Frog

1145 ***Colostethus littoralis***
Anura - Dendrobatidae
e Lima Rocket Frog

1146 ***Colostethus maculosus***
Anura - Dendrobatidae
e Spotted Rocket Frog

1147 ***Colostethus mandelorum***
Anura - Dendrobatidae
e Mount Turumiquire Rocket Frog

1148 ***Colostethus marchesianus***
Anura - Dendrobatidae
e Dull Rocket Frog

1149 ***Colostethus marmoreoventris***
Anura - Dendrobatidae
e Rio Negro Rocket Frog

1150 ***Colostethus mayorgai***
Anura - Dendrobatidae
e Mayorga Rocket Frog

1151 ***Colostethus mcdiarmidi***
Anura - Dendrobatidae
e McDiarmid's Rocket Frog

1152 ***Colostethus meridensis***
Anura - Dendrobatidae
e Merida Rocket Frog

1153 ***Colostethus mertensi***
Anura - Dendrobatidae
e Mertens's Rocket Frog

1154 ***Colostethus mittermeieri***
Anura - Dendrobatidae
e Mittermeier's Rocket Frog

1155 ***Colostethus molinarii***
Anura - Dendrobatidae
e Las Playitas Rocket Frog

1156 ***Colostethus mystax***
Anura - Dendrobatidae
e Cloud-forest Rocket Frog

1157 ***Colostethus nexipus***
Anura - Dendrobatidae
e Los Tayos Rocket Frog

1158 ***Colostethus nubicola***
Anura - Dendrobatidae
e Boquete Rocket Frog

1159 ***Colostethus olfersioides***
Anura - Dendrobatidae
e Rio Rocket Frog

1160 ***Colostethus orostoma***
Anura - Dendrobatidae
e Tachira Rocket Frog

1161 ***Colostethus palmatus***
Anura - Dendrobatidae
e Palm Rocket Frog

1162 ***Colostethus parcus***
Anura - Dendrobatidae
e Gualaceo Rocket Frog

1163 ***Colostethus peculiaris***
Anura - Dendrobatidae
e Funny Rocket Frog

1164 ***Colostethus peruvianus***
Anura - Dendrobatidae
e Peru Rocket Frog

1165 ***Colostethus pinguis***
Anura - Dendrobatidae
e Malvasa Rocket Frog

1166 ***Colostethus poecilontus***
Anura - Dendrobatidae
e Amazonas Rocket Frog

1167 ***Colostethus pratti***
Anura - Dendrobatidae
e Pratt's Rocket Frog

1168 ***Colostethus pulchellus***
Anura - Dendrobatidae
e Espada's Rocket Frog

1169 ***Colostethus pumilis***
Anura - Dendrobatidae
e San Vincente Rocket Frog

1170 ***Colostethus ramosi***
Anura - Dendrobatidae
e Ramos's Rocket Frog

1171 ***Colostethus ranoides***
Anura - Dendrobatidae
e Llanos Rocket Frog

1172 ***Colostethus ruizi***
Anura - Dendrobatidae
e Ruiz's Rocket Frog

1173 ***Colostethus saltuensis***
Anura - Dendrobatidae
e Salty Rocket Frog

1174 ***Colostethus sanmartini***
Anura - Dendrobatidae
e San Martin Rocket Frog

1175 ***Colostethus sauli***
Anura - Dendrobatidae
e Saint Cecilia Rocket Frog
f Grenouille peinte

1176 ***Colostethus serranus***
Anura - Dendrobatidae
e Pefaur's Rocket Frog

1177 ***Colostethus shrevi***
Anura - Dendrobatidae
e Shreve's Rocket Frog

1178 ***Colostethus shuar***
Anura - Dendrobatidae
e Santiago Rocket Frog

1179 ***Colostethus stepheni***
Anura - Dendrobatidae
e Stephen's Rocket Frog

1180 ***Colostethus subpunctatus***
Anura - Dendrobatidae
e Bogota Rocket Frog

1181 ***Colostethus sylvaticus***
Anura - Dendrobatidae
e Forest Rocket Frog

1182 ***Colostethus taeniatus***
Anura - Dendrobatidae
e Rio Cosanga Rocket Frog

1183 ***Colostethus talamancae***
Anura - Dendrobatidae
e Talamanca Rocket Frog

1184 ***Colostethus tergogranularis***
Anura - Dendrobatidae
e Faldas Rocket Frog

1185 ***Colostethus thorntoni***
Anura - Dendrobatidae
e Thornton's Rocket Frog

1186 ***Colostethus torrenticola***
Anura - Dendrobatidae
e Waterfall Rocket Frog

1187 ***Colostethus trilineatus***
Anura - Dendrobatidae
e Three-striped Rocket Frog

1188 ***Colostethus vergeli***
Anura - Dendrobatidae
e Hellmich's Rocket Frog

1189 ***Colostethus vertebralis***
Anura - Dendrobatidae
e Boulenger's Rocket Frog

1190 ***Colostethus whymperi***
Anura - Dendrobatidae
e Tanti Rocket Frog

1191 ***Conraua***
Anura - Ranidae
e Slippery Frogs

1192 ***Conraua alleni***
Anura - Ranidae
e Allen's Slippery Frog

1193 ***Conraua beccari***
Anura - Ranidae
e Filfil Slippery Frog

1194 ***Conraua crassipes***
Anura - Ranidae
e Abo Slippery Frog

1195 ***Conraua derooi***
Anura - Ranidae
e Togo Slippery Frog

1196 ***Conraua goliath***
Anura - Ranidae
e African Giant Frog; Giant Slippery Frog; Goliath Frog
d Goliathfrosch
f Grenouille goliath; Grenouille géante; Grenouille géante du Cameroun; Grenouille du Cameroun

1197 ***Conraua robusta***
Anura - Ranidae
e Cameroon Slippery Frog

1198 ***Cophilaxus***
Anura - Microhylidae
e Rainforest Frogs

1199 ***Cophilaxus ateles***
Anura - Microhylidae
e Papua Rainforest Frog

1200 ***Cophilaxus biroi***
Anura - Microhylidae
e Sattelberg Rainforest Frog; Biro's Frog

1201 ***Cophilaxus bombiens***
Anura - Microhylidae
e Buzzing Rainforest Frog

1202 ***Cophilaxus cheesmanae***
Anura - Microhylidae
e Kokoda Rainforest Frog

1203 ***Cophilaxus concinnus***
Anura - Microhylidae
e Rattling Rainforest Frog

1204 ***Cophilaxus crepitans***
Anura - Microhylidae
e Rusty Rainforest Frog

1205 ***Cophilaxus cryptotympanum***
Anura - Microhylidae
e Zweifel's Rainforest Frog

1206 ***Cophilaxus darlingtoni***
Anura - Microhylidae
e Darlington's Rainforest Frog

1207 ***Cophilaxus daymani***
Anura - Microhylidae
e Dayman's Rainforest Frog

1208 ***Cophilaxus exiguus***
Anura - Microhylidae
e Cooktown Rainforest Frog

1209 ***Cophilaxus infacetus***
Anura - Microhylidae
e Chocolate Rainforest Frog

1210 ***Cophilaxus kaindiensis***
Anura - Microhylidae
e Kaindi Rainforest Frog

1211 ***Cophilaxus mcdonaldi***
Anura - Microhylidae
e McDonald's Rainforest Frog

1212 ***Cophilaxus montana***
Anura - Microhylidae
e Mountain Rainforest Frog

1213 ***Cophilaxus neglectus***
Anura - Microhylidae
e Brown Rainforest Frog

1214 ***Cophilaxus nubicola***
Anura - Microhylidae
e Michael Rainforest Frog

1215 ***Cophilaxus ornatus***
Anura - Microhylidae
e Ornate Rainforest Frog

1216 ***Cophilaxus parkeri***
Anura - Microhylidae
e Parker's Rainforest Frog

1217 ***Cophilaxus peninsularis***
Anura - Microhylidae
e Peninsula Rainforest Frog

1218 ***Cophilaxus pipilans***
Anura - Microhylidae
e Sempi Rainforest Frog

1219 ***Cophilaxus riparius***
Anura - Microhylidae
e Wilhelm Rainforest Frog

1220 ***Cophilaxus saxatilis***
Anura - Microhylidae
e Black Mountain Rainforest Frog

1221 ***Cophilaxus shellyi***
Anura - Microhylidae
e Shelly's Rainforest Frog

1222 ***Cophilaxus sphagnicola***
Anura - Microhylidae
e Morobe Rainforest Frog

1223 ***Cophilaxus tagulensis***
Anura - Microhylidae
e Tagula Rainforest Frog

1224 ***Cophilaxus tuberculus***
Anura - Microhylidae
e Warty Rainforest Frog

1225 ***Cophilaxus variegatus***
Anura - Microhylidae
e Common Rainforest Frog; Rainforest Frog

1226 ***Cophilaxus verecundus***
Anura - Microhylidae
e Myola Rainforest Frog

1227 ***Cophilaxus verrucosus***
Anura - Microhylidae
e Moroka Rainforest Frog

1228 ***Cophlilaxus hosmeri***
Anura - Microhylidae
e Clicking Rainforest Frog

1229 ***Cophyla***
Anura - Microhylidae
e Whistling Tree Frogs

1230 ***Cophyla phyllodactyla***
Anura - Microhylidae
e Whistling Tree Frog

1231 ***Copiula***
Anura - Microhylidae
e Mehely Frogs

1232 ***Copiula fistulans***
Anura - Microhylidae
e Lae Mehely Frog

1233 ***Copiula minor***
Anura - Microhylidae
e Milne Bay Mehely Frog

1234 ***Copiula oxyrhina***
Anura - Microhylidae
e Misima Island Mehely Frog

1235 ***Copiula pipiens***
Anura - Microhylidae
e Wirui Mehely Frog

1236 ***Copiula tyleri***
Anura - Microhylidae
e Sepik Mehely Frog

1237 ***Corythomantis***
Anura - Hylidae
e Greening's Frogs

1238 ***Corythomantis greeningi***
Anura - Hylidae
e Greening's Frog

1239 ***Crepidophryne***
Anura - Bufonidae
e Cerro Utyum Toads
d Kostarika-Kröten

1240 ***Crepidophryne epiotica***
Anura - Bufonidae
e Cerro Utyum Toad
d Kostarika-Kröte

1241 ***Crinia***
Anura - Myobatrachidae
e Australian Froglets
d Zirpfrösche

1242 ***Crinia bilingua***
Anura - Myobatrachidae
e Bleating Froglet; Bilingual Frog

1243 ***Crinia deserticola***
Anura - Myobatrachidae
e Sparrow Froglet

1244 ***Crinia georgiana***
Anura - Myobatrachidae
e Red-legged Froglet

1245 ***Crinia glauerti***
Anura - Myobatrachidae
e Rattle Froglet

1246 ***Crinia insignifera***
Anura - Myobatrachidae
e White-throated Froglet; Sign-bearing Frog
d Australischer Zirpfrosch

1247 ***Crinia parinsignifera***
Anura - Myobatrachidae
e Murray River Froglet; Eastern Brown Frog

1248 ***Crinia pseudinsignifera***
Anura - Myobatrachidae
e Darling Plateau Froglet

1249 ***Crinia remota***
Anura - Myobatrachidae
e Pasperbark Froglet

1250 ***Crinia riparia***
Anura - Myobatrachidae
e Flinders Range Froglet

1251 ***Crinia signifera***
Anura - Myobatrachidae
e Common Eastern Froglet (ANZ); Eastern Froglet; Signet Southern Frog; Western Brown Frog
d Australischer Zirpfrosch

1252 ***Crinia sloanei***
Anura - Myobatrachidae
e Sloane's Froglet

1253 ***Crinia subinsignifera***
Anura - Myobatrachidae
e Squelching Froglet

1254 ***Crinia tasmaniensis***
Anura - Myobatrachidae
e Tasmanian Froglet; Spotted Marsh Frog

1255 ***Crinia tinnula***
Anura - Myobatrachidae
e Wallum Froglet

1256 ***Crossodactylodes***
Anura - Leptodactylidae
e Bromeliad Frogs

1257 ***Crossodactylodes bokermanni***
Anura - Leptodactylidae
e Bokermann's Bromeliad Frog

1258 ***Crossodactylodes izecksohni***
Anura - Leptodactylidae
e Izecksohn's Bromeliad Frog

1259 ***Crossodactylodes pintoi***
Anura - Leptodactylidae
e Brown Bromeliad Frog

1260 ***Crossodactylus***
Anura - Leptodactylidae
e Spinythumb Frogs

1261 ***Crossodactylus aeneus***
Anura - Leptodactylidae
e Müller's Spinythumb Frog

1262 ***Crossodactylus bokermanni***
Anura - Leptodactylidae
e Bokermann's Spinythumb Frog

1263 ***Crossodactylus dantei***
Anura - Leptodactylidae
e Murici Spinythumb Frog

1264 ***Crossodactylus dispar***
Anura - Leptodactylidae
e Rio Spinythumb Frog

1265 ***Crossodactylus gaudichaudii***
Anura - Leptodactylidae
e Brazilian Spinythumb Frog; Gaudichaud's Frog
f Grenouille de Gaudichaud

1266 ***Crossodactylus grandis***
Anura - Leptodactylidae
e Bahia Spinythumb Frog

1267 ***Crossodactylus lutzorum***
Anura - Leptodactylidae
e Lutz's Spinythumb Frog

1268 ***Crossodactylus schmidti***
Anura - Leptodactylidae
e Schmidt's Spinythumb Frog

1269 ***Crossodactylus trachystomus***
Anura - Leptodactylidae
e Minas Gerais Spinythumb Frog

1270 ***Crotaphatrema***
Gymnophiona - Scolecomorphidae
e Cameroon Caecilians

1271 ***Crotaphatrema bornmuelleri***
Gymnophiona - Scolecomorphidae
e Bornmüller's Caecilian

1272 ***Crotaphatrema lamottei***
Gymnophiona - Scolecomorphidae
e Mont Oku Caecilian

1273 ***Cryptobatrachus***
Anura - Hylidae
e Backpack Frogs

1274 ***Cryptobatrachus boulengeri***
Anura - Hylidae
e Boulenger's Backpack Frog

1275 ***Cryptobatrachus furmanni***
Anura - Hylidae
e Fuhrmann's Backpack Frog

1276 ***Cryptobatrachus nicefori***
Anura - Hylidae
e Colombian Backpack Frog

1277 **Cryptobranchidae**
Caudata
e Giant Salamanders ; Giant Salamanders and Hellbenders
d Riesensalamander (pl)
f Cryptobranchidés; Dérotrèmes; Salamandres géantes

1278 ***Cryptobranchus***
Caudata - Cryptobranchidae
e Hellbenders
d Amerikanische Riesensalamander (pl); Schlammteufel (pl)
f Ménopomes

1279 ***Cryptobranchus alleganiensis***
Caudata - Cryptobranchidae
e Hellbender
d Hellbender; Schlammteufel
f Ménopome

1280 ***Cryptobranchus alleganiensis alleganiensis***
Caudata - Cryptobranchidae
e Eastern Hellbender

1281 ***Cryptobranchus alleganiensis bishopi***
Caudata - Cryptobranchidae
e Ozark Hellbender

1282 ***Cryptothylax***
Anura - Hyperoliidae
e Wax Frogs

1283 ***Cryptothylax greshoffi***
Anura - Hyperoliidae
e Greshoff's Wax Frog

1284 ***Cryptothylax minutus***
Anura - Hyperoliidae
e Tiny Wax Frog

1285 ***Ctenophryne***
Anura - Microhylidae
e Egg Frogs

1286 ***Ctenophryne geayi***
Anura - Microhylidae
e Brown Egg Frog
f Ctenophryne de Geay

1287 ***Ctenophryne minor***
Anura - Microhylidae
e Colombian Egg Frog

1288 ***Cycloramphus***
Anura - Leptodactylidae
e Button Frogs
f Cycloramphes

1289 ***Cycloramphus asper***
Anura - Leptodactylidae
e Werner's Button Frog

1290 ***Cycloramphus bandeirensis***
Anura - Leptodactylidae
e Bandeira Button Frog

1291 ***Cycloramphus bolitoglossus***
Anura - Leptodactylidae
e Blumenau Button Frog

1292 ***Cycloramphus boraceiensis***
Anura - Leptodactylidae
e Boraceia Button Frog

1293 ***Cycloramphus brasiliensis***
Anura - Leptodactylidae
e Brazilian Button Frog

1294 ***Cycloramphus carvalhoi***
Anura - Leptodactylidae
e Carvalho's Button Frog

1295 ***Cycloramphus catarinensis***
Anura - Leptodactylidae
e Santa Catarina Button Frog

1296 ***Cycloramphus cedrensis***
Anura - Leptodactylidae
e Cedros Button Frog

1297 ***Cycloramphus diringshofeni***
Anura - Leptodactylidae
e Sao Pento Button Frog

1298 ***Cycloramphus dubius***
Anura - Leptodactylidae
e Sao Paulo Button Frog

1299 ***Cycloramphus duseni***
Anura - Leptodactylidae
e Ypiranga Button Frog

1300 ***Cycloramphus eleutherodactylus***
Anura - Leptodactylidae
e Alto Button Frog

1301 ***Cycloramphus fuliginosus***
Anura - Leptodactylidae
e Tschudi's Button Frog
f Cycloramphe fuligineux

1302 ***Cycloramphus granulosus***
Anura - Leptodactylidae
e Granular Button Frog

1303 ***Cycloramphus izecksohni***
Anura - Leptodactylidae
e Izecksohn's Button Frog

1304 ***Cycloramphus jordanensis***
Anura - Leptodactylidae
e Jordan's Button Frog

1305 ***Cycloramphus juimirim***
Anura - Leptodactylidae
e Rio Verde Button Frog

1306 ***Cycloramphus lutzorum***
Anura - Leptodactylidae
e Lutz's Button Frog

1307 ***Cycloramphus migueli***
Anura - Leptodactylidae
e Miguel's Button Frog

1308 ***Cycloramphus mirandariberoi***
Anura - Leptodactylidae
e Miranda's Button Frog

1309 ***Cycloramphus ohausi***
Anura - Leptodactylidae
e Wandolleck's Button Frog

1310 ***Cycloramphus rhyakonastes***
Anura - Leptodactylidae
e Heyer's Button Frog

1311 ***Cycloramphus semipalmatus***
Anura - Leptodactylidae
e Campo Grande Button Frog

1312 ***Cycloramphus stejnegeri***
Anura - Leptodactylidae
e Stejneger's Button Frog

1313 ***Cycloramphus valae***
Anura - Leptodactylidae
e Gruta Button Frog

1314 ***Cyclorana***
Anura - Hylidae
e Water-holding Frogs
d Australische Handfrösche
f Leptodactyles-citernes

1315 ***Cyclorana australis***
Anura - Hylidae
e Australian Water-holding Frog

1316 ***Cyclorana brevipes***
Anura - Hylidae
e Savannah Water-holding Frog

1317 ***Cyclorana cryptotis***
Anura - Hylidae
e Earless Water-holding Frog

1318 ***Cyclorana cultripes***
Anura - Hylidae
e Olive Water-holding Frog

1319 ***Cyclorana longipes***
Anura - Hylidae
e Kimberley Water-holding Frog

1320 ***Cyclorana maculosa***
Anura - Hylidae
e Spotted Water-holding Frog

1321 ***Cyclorana maini***
Anura - Hylidae
e Main's Water-holding Frog

1322 ***Cyclorana manya***
Anura - Hylidae
e Bleating Water-holding Frog

1323 ***Cyclorana novaehollandiae***
Anura - Hylidae
e Eastern Water-holding Frog

1324 ***Cyclorana platycephala***
Anura - Hylidae
e Common Water-holding Frog; Water-holding Frog (ANZ); Flat-headed Frog
d Wasserreservoirfrosch; Plattköpfiger Handfrosch
f Grenouille à tête plate

1325 ***Cyclorana vagita***
Anura - Hylidae

e Crying Water-holding Frog

1326 ***Cyclorana verrucosa***
Anura - Hylidae
e Woodland Water-holding Frog

1327 ***Cynops***
Caudata - Salamandridae
e Firebelly Newts
d Ostasiatische Wassermolche; Feuerbauchmolche
f Cynops; Tritons

1328 ***Cynops chenggongensis***
Caudata - Salamandridae
e Chenggong Newt
d Chenggong-Feuerbauchmolch

1329 ***Cynops cyanurus***
Caudata - Salamandridae
e Cyan Newt
d Blauschwanz-Feuerbauchmolch; Feuerbauchmolch; Tüpfelmolch

1330 ***Cynops ensicauda***
Caudata - Salamandridae
e Swordtail Newt; Sword-tailed Newt; Yellow-bellied Newt; Okinawa Newt
d Schwertschwanzmolch
f Triton à queue d'épée

1331 ***Cynops orientalis***
Caudata - Salamandridae
e Chinese Dwarf Newt; Chinese Fire-bellied Newt
d Chinesischer Zwergmolch; Chinesischer Feuerbauchmolch
f Triton oriental; Triton nain de Chine

1332 ***Cynops orphicus***
Caudata - Salamandridae
e Dayang Newt
d Dayang-Molch

1333 ***Cynops pyrrhogaster***
Caudata - Salamandridae
e Japanese Firebelly Newt; Japanese Newt; Japanese Fire-bellied Newt
d Japanischer Feuerbauchmolch; Feuerbauchmolch
f Triton du Japon; Triton du Japon à ventre de feu; Cynops au ventre poupre
i Tritone giapponese

1334 ***Cynops shataukokensis***
Caudata - Salamandridae
e Sha Tau Kok Newt

1335 ***Cynops wolterstorffi***
Caudata - Salamandridae
e Yunnan Newt
d Wolterstorffs Molch
f Triton de Wolterstorff

D

1336 ***Dasypops***
Anura - Microhylidae
e Rio Mutum Frogs

1337 ***Dasypops schirchi***
Anura - Microhylidae
e Rio Mutum Frog

1338 ***Dendrobates***
Anura - Dendrobatidae
e Poison Frogs; Poison-arrow Frogs; Poison-dart Frogs
d Pfeilgiftfrösche; Baumsteiger; Baumsteigerfrösche
f Dendrobates

1339 ***Dendrobates arboreus***
Anura - Dendrobatidae
e Polkadot Poison Frog

1340 ***Dendrobates auratus***
Anura - Dendrobatidae
e Green-and-black Poison Frog; Green-and-golden Poison Frog; Green Poison-arrow Frog; Golden Poison-arrow Frog; Green-and-black Poison-arrow Frog; Green-and-black Poison-dart Frog; Black-and-green Poison-dart Frog; Green-and-black Dart-Poison Frog
d Goldbaumsteiger; Smaragd-Baumsteigerfrosch
f Dendrobate doré

1341 ***Dendrobates azureus***
Anura - Dendrobatidae
e Blue Poison Frog; Blue Poison-arrow Frog; Blue Poison-dart Frog
d Blauer Pfeilgiftfrosch; Azurblauer Pfeilgiftfrosch; Blauer Baumsteiger

1342 ***Dendrobates biolat***
Anura - Dendrobatidae
e Biolat Poison Frog; Biolat Poison-arrow Frog

1343 ***Dendrobates captivus***
Anura - Dendrobatidae
e Rio Santiago Poison Frog; Rio Santiago Poison-arrow Frog

1344 ***Dendrobates castaneoticus***
Anura - Dendrobatidae
e Brazilnut Poison Frog; Brazilnut Poison-arrow Frog

1345 ***Dendrobates fantasticus***
Anura - Dendrobatidae
e Red-headed Poison Frog; Red-headed Poison-arrow Frog

1346 ***Dendrobates galactonotus***
Anura - Dendrobatidae
e Splashback Poison Frog; Splashback Poison-arrow Frog; Splash-backed Poison-arrow Frog

1347 ***Dendrobates granulifera***
Anura - Dendrobatidae
e Granular Poison Frog; Granular Poison-arrow Frog; Granulated Poison-dart Frog
d Granulierter Pfeilgiftfrosch; Granulierter Baumsteiger

1348 ***Dendrobates histrionicus***
Anura - Dendrobatidae
e Harlequin Poison Frog; Harlequin Poison-arrow Frog; Common Poison-arrow Frog; Kokoe-pa
d Punktierter Pfeilgiftfrosch; Punktierter Baumsteiger

1349 ***Dendrobates imitator***
Anura - Dendrobatidae

e Mimic Poison Frog; Mimic Poison-arrow Frog

1350 ***Dendrobates lamasi***
Anura - Dendrobatidae
e Pasco Poison Frog; Pasco Poison-arrow Frog

1351 ***Dendrobates lehmanni***
Anura - Dendrobatidae
e Lehmann's Poison Frog; Lehmann's Poison-arrow Frog

1352 ***Dendrobates leucomelas***
Anura - Dendrobatidae
e Yellow-banded Poison Frog; Yellow-banded Poison-arrow Frog
d Gelbgebänderter Pfeilgiftfrosch; Schwarzgelber Baumsteiger; Gelbstreifiger Baumsteiger

1353 ***Dendrobates mysteriosus***
Anura - Dendrobatidae
e Maranon Poison Frog; Maranon Poison-arrow Frog

1354 ***Dendrobates occultator***
Anura - Dendrobatidae
e La Brea Poison Frog; La Brea Poison-arrow Frog

1355 ***Dendrobates pumilio***
Anura - Dendrobatidae
e Strawberry Poison Frog; Strawberry Poison-arrow Frog; Flaming Poison-arrow Frog; Strawberry Poison-dart Frog
d Zwerg-Panama-Baumsteiger; Erdbeerfröschchen; Kleiner Erdbeerfrosch
f Grenouille des fraises; Dendrobate nain de Panama

1356 ***Dendrobates quinquevittatus***
Anura - Dendrobatidae
e Rio Madeira Poison Frog; Rio Madeira Poison-arrow Frog; Amazon Poison-arrow Frog
d Amazonas-Baumsteiger

1357 ***Dendrobates reticulatus***
Anura - Dendrobatidae
e Redback Poison Frog; Redback Poison-arrow Frog

1358 ***Dendrobates rufulus***
Anura - Dendrobatidae
e Gorzula's Poison Frog; Gorzula's Poison-arrow Frog

1359 ***Dendrobates sirensis***
Anura - Dendrobatidae
e Sira Poison Frog; Sira Poison-arrow Frog

1360 ***Dendrobates speciosus***
Anura - Dendrobatidae
e Splelndid Poison Frog; Splelndid Poison-arrow Frog
d Glänzender Pfeilgiftfrosch; Glänzender Baumsteiger

1361 ***Dendrobates tinctorius***
Anura - Dendrobatidae
e Dyeing Poison Frog; Dyeing Poison-arrow Frog; Dyeing Poison-dart Frog
d Färberfrosch; Dreifarbiger Baumsteiger
f Rainette à tapirer; Dendrobate à tapirer; Grenouille vénéneuse; Dendrobate vénéneuse; Grenouille teinté du Bresil

1362 ***Dendrobates truncatus***
Anura - Dendrobatidae
e Yellow-striped Poison Frog; Yellow-striped Poison-arrow Frog

1363 ***Dendrobates vanzolinii***
Anura - Dendrobatidae
e Brazilian Poison Frog; Brazilian Poison-arrow Frog

1364 ***Dendrobates ventrimaculatus***
Anura - Dendrobatidae

e Amazonian Poison Frog; Amazonian Poison-arrow Frog
f Dendrobate à ventre tacheté

1365 Dendrobatidae
Anura
e Arrow-poison Frogs; Poison Frogs; Poison-arrow Frogs; Poison-dart Frogs
d Baumsteigerfrösche; Färberfrösche; Pfeilgiftfrösche; Dendrobatiden; Baumsteiger
f Dendrobatidés; Grenouilles peintes; Grenouilles à flèches

1366 *Dendrophryniscus*
Anura - Bufonidae
e Tree Toads
d Baumkröten

1367 *Dendrophryniscus brevipollicatus*
Anura - Bufonidae
e Coastal Tree Toad
d Kurzdaumige Baumkröte

1368 *Dendrophryniscus leucomystax*
Anura - Bufonidae
e Lowland Tree Toad

1369 *Dendrophryniscus minutus*
Anura - Bufonidae
e Tiny Tree Toad

1370 *Dendrotriton*
Caudata - Plethodontidae
e Bromeliad Salamanders
d Schwielensalamander (pl)

1371 *Dendrotriton bromeliacius*
Caudata - Plethodontidae
e Volcan Tajumulca Bromeliad Salamander

1372 *Dendrotriton cuchumatanus*
Caudata - Plethodontidae
e Forest Bromeliad Salamander

1373 *Dendrotriton megarhinus*
Caudata - Plethodontidae
e Longnose Bromeliad Salamander
d Großnasen-Schwielensalamander

1374 *Dendrotriton rabbi*
Caudata - Plethodontidae
e Guatemala Bromeliad Salamander

1375 *Dendrotriton xolocalcae*
Caudata - Plethodontidae
e Xolacalca Bromeliad Salamander
d Ovando-Schwielensalamander

1376 *Dermatonotus*
Anura - Microhylidae
e Termite Frogs

1377 *Dermatonotus muelleri*
Anura - Microhylidae
e Termite Frog; Müller's Termite Frog

1378 *Dermophis*
Gymnophiona - Caeciliaidae
e Mexican Caecilians
d Blindwühlen

1379 *Dermophis mexicanus*
Gymnophiona - Caeciliaidae
e Mexican Caecilian
d Mexikanische Blindwühle; Mexikanische Hautwühle

1380 *Dermophis oaxacae*
Gymnophiona - Caeciliaidae
e Oaxacan Caecilian
d Lafrentz Blindwühle; Lafrentz Hautwühle

1381 *Dermophis parviceps*
Gymnophiona - Caeciliaidae
e La Loma Caecilian

1382 ***Desmognathus***
Caudata - Plethodontidae
e Dusky Salamanders
d Bachsalamander (pl)

1383 ***Desmognathus aeneus***
Caudata - Plethodontidae
e Seepage Salamander; Cherokee Salamander (NA); Alabama Salamander
d Cherokee-Bachsalamander

1384 ***Desmognathus apalachicolae***
Caudata - Plethodontidae
e Apalachicola Dusky Salamander

1385 ***Desmognathus auriculatus***
Caudata - Plethodontidae
e Southern Dusky Salamander

1386 ***Desmognathus brimleyorum***
Caudata - Plethodontidae
e Ouachita Dusky Salamander; Central Dusky Salamander
d Ouachita-Bachsalamander; Brimleys Bachsalamander

1387 ***Desmognathus carolinensis***
Caudata - Plethodontidae
e Carolina Dusky Salamander

1388 ***Desmognathus conanti***
Caudata - Plethodontidae
e Spotted Dusky Salamander

1389 ***Desmognathus fuscus***
Caudata - Plethodontidae
e Dusky Salamander (NA)
d Brauner Bachsalamander
f Salamandre brune; Salamandre sombre

1390 ***Desmognathus fuscus carri***
Caudata - Plethodontidae
e Peninsula Dusky Salamander

1391 ***Desmognathus fuscus fuscus***
Caudata - Plethodontidae
e Northern Dusky Salamander

1392 ***Desmognathus imitator***
Caudata - Plethodontidae
e Imitator Salamander

1393 ***Desmognathus marmoratus***
Caudata - Plethodontidae
e Shovel-nosed Salamander
d Marmorierter Bachsalamander

1394 ***Desmognathus monticola***
Caudata - Plethodontidae
e Seal Salamander
d Robben-Salamander; Robben-Bachsalamander
f Salamandre phoque

1395 ***Desmognathus monticola***
Caudata - Plethodontidae
e Appalachian Seal Salamander

1396 ***Desmognathus monticola jeffersoni***
Caudata - Plethodontidae
e Virginia Seal Salamander

1397 ***Desmognathus ochrophaeus***
Caudata - Plethodontidae
e Mountain Dusky Salamander; Mountain Salamander
d Alleghany-Salamander; Alleghany-Bachsalamander; Bachsalamander
f Salamandre sombre des montagnes (Qué)

1398 ***Desmognathus ochrophaeus carolinensis***
Caudata - Plethodontidae
e Blue Ridge Mountain Salamander

1399 ***Desmognathus ochrophaeus ochrophaeus***
Caudata - Plethodontidae
e Allegheny Mountain Dusky

Salamander; Allegheny Salamander (NA); Allegheny Mountain Salamander (NA)

1400 ***Desmognathus ocoee***
Caudata - Plethodontidae
e Ocoee Salamander

1401 ***Desmognathus quadramaculatus***
Caudata - Plethodontidac
e Black-bellied Salamander; Blackbelly Salamander
d Schwarzbäuchiger Bachsalamander
f Salamandre à ventre noir

1402 ***Desmognathus santeetlah***
Caudata - Plethodontidae
e Santeetlah Dusky Salamander

1403 ***Desmognathus welteri***
Caudata - Plethodontidae
e Black Mountain Dusky Salamander; Black Mountain Salamander

1404 ***Desmognathus wrighti***
Caudata - Plethodontidae
e Pygmy Salamander
d Zwergbachsalamander
f Salamandre pygmée

1405 ***Dicamptodon***
Caudata - Dicamptodontidae
e American Giant Salamanders; Giant Salamanders (NA)
d Riesenquerzahnmolche; Riesenquerzahnsalamander (pl)
f Salamandres géantes

1406 ***Dicamptodon aterrimus***
Caudata - Dicamptodontidae
e Idaho Giant Salamander

1407 ***Dicamptodon copei***
Caudata - Dicamptodontidae
e Cope's Giant Salamander
d Copes Pazifik-Salamander
f Salamandre géante de Cope

1408 ***Dicamptodon ensatus***
Caudata - Dicamptodontidae
e California Giant Salamander; Pacific Giant Salamander
d Pazifischer Querzahnmolch; Pazifischer Riesenquerzahnmolch; Riesenquerzahnmolch
f Salamandre géante du Pacifique; Grande salamandre

1409 ***Dicamptodon tenebrosus***
Caudata - Dicamptodontidae
e Pacific Giant Salamander; Coastal Giant Salamander (NA); Pacific Salamander
d Pazifischer Riesenquerzahnmolch

1410 **Dicamptodontidae**
Caudata
e American Giant Salamanders
d Riesenquerzahnmolche; Riesensalamander (pl)

1411 ***Dicroglossus***
Anura - Ranidae
e Western Bullfrogs

1412 ***Dicroglossus occipitalis***
Anura - Ranidae
e Western Bullfrog
d Afrikanischer Ochsenfrosch; Afrikanischer Tigerfrosch
f Grenouille tigré africaine

1413 ***Didynamipus***
Anura - Bufonidae
e Four-digit Toads

1414 ***Didynamipus sjostedti***
Anura - Bufonidac
e Four-digit Toad

1415 ***Dimorphognathus***
Anura - Ranidae
e African Swamp Frogs

1416 ***Dimorphognathus africanus***
Anura - Ranidae
e African Swamp Frog

1417 ***Dischidodactylus***
Anura - Leptodactylidae
e Mount Duida Frogs

1418 ***Dischidodactylus colonelloi***
Anura - Leptodactylidae
e Ayarzaguena's Frog

1419 ***Dischidodactylus duidensis***
Anura - Leptodactylidae
e Mount Duida Frog

1420 ***Discodeles***
Anura - Ranidae
e Webbed Frogs

1421 ***Discodeles bufoniformis***
Anura - Ranidae
e Treasury Island Webbed Frog

1422 ***Discodeles guppyi***
Anura - Ranidae
e Shortland Island Webbed Frog

1423 ***Discodeles malakuna***
Anura - Ranidae
e Malakuna Webbed Frog

1424 ***Discodeles opisthodon***
Anura - Ranidae
e Faro Webbed Frog; Rear-toothed Frog
d Salomonen-Frosch

1425 ***Discodeles vogti***
Anura - Ranidae
e Admiralty Islands Webbed Frog

1426 **Discoglossidae**
Anura
e Painted Frogs; Disk-tongued Frogs
d Scheibenzüngler (pl); Discoglossiden
f Discoglossidés; Discoglosses

1427 ***Discoglossus***
Anura - Discoglossidae
e Painted Frogs; Discoglossus Frogs
d Eigentlicher Scheibenzüngler (pl); Echter Scheibenzüngler (pl); Scheibenzüngler (pl)
f Discoglosses
i Discoglossi

1428 ***Discoglossus galganoi***
Anura - Discoglossidae
e Portugal Painted Frog; West Iberian Painted Frog; Iberian Painted Frog
d Iberischer Scheibenzüngler
f Discoglosse de Galgano
i Discoglosso di Galgano; Discoglosso iberico

1429 ***Discoglossus jeanneae***
Anura - Discoglossidae
e Spanish Painted Frog
f Discoglosse sudibérique

1430 ***Discoglossus montalenti***
Anura - Discoglossidae
e Corsica Painted Frog; Corsican Painted Frog
d Korsischer Scheibenzüngler
f Discoglosse de Montalent; Discoglosse corse
i Discoglosso di Montalenti; Discoglosso corso

1431 ***Discoglossus nigriventer***
Anura - Discoglossidae
e Israel Painted Frog; Israel Discoglossus Frog
d Schwarzbäuchiger Scheibenzüngler

f Discoglosse à ventre noir

1432 ***Discoglossus pictus***
Anura - Discoglossidae
e Painted Frog
d Gemalter Scheibenzüngler; Gefleckter Scheibenzüngler; Bunte Scheibenzüngler
f Discoglosse peint; Discogloss à ventre blanc; Grenouille peinte
i Discoglosso dipinto

1433 ***Discoglossus pictus pictus***
Anura - Discoglossidae
e Sicilian Painted Frog
d Sizilianischer Scheibenzüngler
f Discoglosse sicilien
i Discoglosso siciliano; Rana sarda; Rana dipinta

1434 ***Discoglossus sardus***
Anura - Discoglossidae
e Sardinia Painted Frog; Tyrrhenian Frog; Tyrrhenian Painted Frog
d Sardicher Scheibenzüngler
f Discoglosse sarde
i Discoglosso sardo; Rana sarda; Discoglosso tirrenico

1435 ***Discophus***
Anura - Microhylidae
e Tomato Frogs
d Tomatenfrösche; Taubfrösche
f Discophus

1436 ***Discophus antongilii***
Anura - Microhylidae
e Tomato Frog
d Tomatenfrosch

1437 ***Discophus guineti***
Anura - Microhylidae
e Sambava Tomato Frog

1438 ***Discophus insularis***
Anura - Microhylidae
e Antsouhy Tomato Frog

1439 ***Duellmanohyla***
Anura - Hylidae
e Brook Frogs; Mountain Brook Frogs

1440 ***Duellmanohyla chamulae***
Anura - Hylidae
e Chamula Mountain Brook Frog

1441 ***Duellmanohyla ignicolor***
Anura - Hylidae
e Sierra Juarez Brook Frog

1442 ***Duellmanohyla lythrodes***
Anura - Hylidae
e Savage's Brook Frog

1443 ***Duellmanohyla salvavida***
Anura - Hylidae
e Honduras Brook Frog

1444 ***Duellmanohyla schmidtorum***
Anura - Hylidae
e Schmidt's Mountain Brook Frog

1445 ***Duellmanohyla soralia***
Anura - Hylidae
e Copan Brook Frog

1446 ***Duellmanohyla uranochroa***
Anura - Hylidae
e Costa Rica Brook Frog

E

1447 ***Echinotriton***
Caudata - Salamandridae
e Mountain Newts

1448 ***Echinotriton andersoni***
Caudata - Salamandridae
e Anderson's Newt

1449 ***Echinotriton asperrimus***
Caudata - Salamandridae
e Mount Yao Newt

1450 ***Echinotriton chinhaiensis***
Caudata - Salamandridae
e Chinhai Newt

1451 ***Edalorhina***
Anura - Leptodactylidae
e Snouted Frogs

1452 ***Edalorhina nasula***
Anura - Leptodactylidae
e Snouted Frog; Common Snouted Frog

1453 ***Edalorhina perezi***
Anura - Leptodactylidae
e Perez's Snouted Frog

1454 ***Elachyglossa***
Anura - Ranidae
e Gyldenstolpe's Frogs

1455 ***Elachyglossa gyldenstolpei***
Anura - Ranidae
e Gyldenstolpe's Frog

1456 ***Elastichocleis***
Anura - Microhylidae
e Oval Frogs
f Elastichocles

1457 ***Elastichocleis bicolor***
Anura - Microhylidae
e Two-coloured Oval Frog

1458 ***Elastichocleis ovalis***
Anura - Microhylidae
e Common Oval Frog; Oval Frog
f Elastichocle ovale

1459 ***Elastichocleis piauiensis***
Anura - Microhylidae
e Piaui Oval Frog

1460 ***Elastichocleis surinamensis***
Anura - Microhylidae
e Surinam Oval Frog

1461 ***Eleutherodactylus***
Anura - Leptodactylidae
e Robber Frogs; Rain Frogs
d Antillen-Pfeiffrösche
f Eleutherodactyles

1462 ***Eleutherodactylus abbotti***
Anura - Leptodactylidae
e Abbott's Robber Frog
d Kalkali-Fröschchen

1463 ***Eleutherodactylus acatallelus***
Anura - Leptodactylidae
e Cauca Robber Frog

1464 ***Eleutherodactylus acerus***
Anura - Leptodactylidae
e Papallacta Robber Frog

1465 ***Eleutherodactylus achatinus***
Anura - Leptodactylidae
e Cachabi Robber Frog

1466 ***Eleutherodactylus acmonis***
Anura - Leptodactylidae
e El Yunque Robber Frog

1467 ***Eleutherodactylus actites***
Anura - Leptodactylidae
e Lesser Pilalo Robber Frog

1468 ***Eleutherodactylus acuminatus***
Anura - Leptodactylidae
e Canelos Robber Frog

1469 ***Eleutherodactylus acutirostris***
Anura - Leptodactylidae
e Sharpsnout Robber Frog

1470 ***Eleutherodactylus affinis***
Anura - Leptodactylidae
e Cundinamarca Robber Frog

1471 ***Eleutherodactylus alalocophus***
Anura - Leptodactylidae
e Quindio Robber Frog

1472 ***Eleutherodactylus alberchi***
Anura - Leptodactylidae
e Rio Pilaton Dwarf Robber Frog

1473 ***Eleutherodactylus albipes***
Anura - Leptodactylidae
e Cuban Dwarf Robber Frog

1474 ***Eleutherodactylus alcoae***
Anura - Leptodactylidae
e Hispaniola Dwarf Robber Frog

1475 ***Eleutherodactylus alfredi***
Anura - Leptodactylidae
e Alfred's Rain Frog

1476 ***Eleutherodactylus alfredi alfredi***
Anura - Leptodactylidae
e Northern Alfred's Rain Frog

1477 ***Eleutherodactylus alfredi conspicuus***
Anura - Leptodactylidae
e Southern Alfred's Rain Frog

1478 ***Eleutherodactylus altae***
Anura - Leptodactylidae
e Mountain Robber Frog

1479 ***Eleutherodactylus alticola***
Anura - Leptodactylidae
e Lynn's Robber Frog

1480 ***Eleutherodactylus amadeus***
Anura - Leptodactylidae
e Haitian Robber Frog

1481 ***Eleutherodactylus amazonicus***
Anura - Leptodactylidae
e Amazon Robber Frog

1482 ***Eleutherodactylus anatipes***
Anura - Leptodactylidae
e Anatipes Robber Frog

1483 ***Eleutherodactylus anciano***
Anura - Leptodactylidae
e Corquin Robber Frog

1484 ***Eleutherodactylus andi***
Anura - Leptodactylidae
e Atlantic Robber Frog

1485 ***Eleutherodactylus andicola***
Anura - Leptodactylidae
e Boettger's Robber Frog

1486 ***Eleutherodactylus andrewsi***
Anura - Leptodactylidae
e Andrews's Robber Frog

1487 ***Eleutherodactylus angelicus***
Anura - Leptodactylidae
e Angel Robber Frog

1488 ***Eleutherodactylus angustidigitorum***
Anura - Leptodactylidae
e Patzcuaro Stream Frog; Patucuaro Peeping Frog

1489 ***Eleutherodactylus anolirex***
Anura - Leptodactylidae
e Santander Robber Frog

1490 ***Eleutherodactylus anomalus***
Anura - Leptodactylidae
e Choco Robber Frog

1491 ***Eleutherodactylus anonymus***
Anura - Leptodactylidae
e Secret Robber Frog

1492 ***Eleutherodactylus anotis***
Anura - Leptodactylidae
e Aragua Robber Frog

1493 ***Eleutherodactylus antillensis***
Anura - Leptodactylidae
e Antilles Robber Frog; Antillean Frog (WI)

1494 ***Eleutherodactylus apiculatus***
Anura - Leptodactylidae
e La Planada Robber Frog

1495 ***Eleutherodactylus apostates***
Anura - Leptodactylidae
e Apostates Robber Frog

1496 ***Eleutherodactylus appendiculatus***
Anura - Leptodactylidae
e Pacific Robber Frog

1497 ***Eleutherodactylus aptus***
Anura - Leptodactylidae
e Colonia Robber Frog

1498 ***Eleutherodactylus armstrongi***
Anura - Leptodactylidae
e Armstrong's Robber Frog

1499 ***Eleutherodactylus atkinsi***
Anura - Leptodactylidae
e Atkins's Robber Frog

1500 ***Eleutherodactylus atratus***
Anura - Leptodactylidae
e Santiago Robber Frog

1501 ***Eleutherodactylus audanti***
Anura - Leptodactylidae
e Peak Robber Frog

1502 ***Eleutherodactylus augusti***
Anura - Leptodactylidae
e Barking Frog; Robber Frog
d Mexikanischer Klippenfrosch

1503 ***Eleutherodactylus augusti augusti***
Anura - Leptodactylidae
e Common Barking Frog; Common Robber Frog

1504 ***Eleutherodactylus augusti cactorum***
Anura - Leptodactylidae
e Western Barking Frog; Western Robber Frog
d Westlicher Höhlenfrosch

1505 ***Eleutherodactylus augusti fuscofemora***
Anura - Leptodactylidae
e Zweifel's Barking Frog

1506 ***Eleutherodactylus augusti latrans***
Anura - Leptodactylidae
e Eastern Barking Frog; Eastern Robber Frog; Balcone's Barking Frog

1507 ***Eleutherodactylus auriculatoides***
Anura - Leptodactylidae
e Eared Robber Frog

1508 ***Eleutherodactylus auriculatus***
Anura - Leptodactylidae
e Eastern Cuba Robber Frog

1509 ***Eleutherodactylus aurilegulus***
Anura - Leptodactylidae
e Rio Viejo Robber Frog

1510 ***Eleutherodactylus azueroensis***
Anura - Leptodactylidae
e Azuero Robber Frog

1511 ***Eleutherodactylus babax***
Anura - Leptodactylidae
e Narino Robber Frog

1512 ***Eleutherodactylus bacchus***
Anura - Leptodactylidae
e Wine Robber Frog

1513 ***Eleutherodactylus bakeri***
Anura - Leptodactylidae
e Baker's Robber Frog

1514 ***Eleutherodactylus balionotus***
Anura - Leptodactylidae
e Crest Robber Frog

1515 ***Eleutherodactylus barlagnei***
Anura - Leptodactylidae
e Matouba Robber Frog

1516 ***Eleutherodactylus bartonsmithi***
Anura - Leptodactylidae
e Barton's Robber Frog

1517 ***Eleutherodactylus baryecuus***
Anura - Leptodactylidae
e Morona Robber Frog

1518 ***Eleutherodactylus batrachylus***
Anura - Leptodactylidae
e Tarmauliupan Arboreal Robber Frog

1519 ***Eleutherodactylus bearsei***
Anura - Leptodactylidae
e Bearse's Robber Frog

1520 ***Eleutherodactylus bellorna***
Anura - Leptodactylidae
e Murri Robber Frog

1521 ***Eleutherodactylus berkenbuschii***
Anura - Leptodactylidae
e Berkenbusch's Stream Frog

1522 ***Eleutherodactylus bernali***
Anura - Leptodactylidae
e Argelia Robber Frog

1523 ***Eleutherodactylus bicolor***
Anura - Leptodactylidae
e Two-coloured Robber Frog

1524 ***Eleutherodactylus bicumulus***
Anura - Leptodactylidae
e Caracas Robber Frog

1525 ***Eleutherodactylus bilineatus***
Anura - Leptodactylidae
e Two-lined Robber Frog

1526 ***Eleutherodactylus binotatus***
Anura - Leptodactylidae
e Clay Robber Frog

1527 ***Eleutherodactylus biporcatus***
Anura - Leptodactylidae
e Veragua Robber Frog

1528 ***Eleutherodactylus bockermanni***
Anura - Leptodactylidae
e Bockermann's Robber Frog

1529 ***Eleutherodactylus boconoensis***
Anura - Leptodactylidae
e Bocono Robber Frog

1530 ***Eleutherodactylus bocourti***
Anura - Leptodactylidae
e Bocourt's Robber Frog

1531 ***Eleutherodactylus bogotensis***
Anura - Leptodactylidae
e Bogota Robber Frog

1532 ***Eleutherodactylus bolbodactylus***
Anura - Leptodactylidae
e Rio Robber Frog

1533 ***Eleutherodactylus boulengeri***
Anura - Leptodactylidae
e Boulenger's Robber Frog

1534 ***Eleutherodactylus bransfordii***
Anura - Leptodactylidae
e Bransford's Robber Frog

1535 ***Eleutherodactylus bresslerae***
Anura - Leptodactylidae
e Rio Yumuri Robber Frog

1536 ***Eleutherodactylus brevifrons***
Anura - Leptodactylidae
e Cali Robber Frog

1537 ***Eleutherodactylus brevirostris***
Anura - Leptodactylidae
e Shortsnout Robber Frog

1538 ***Eleutherodactylus briceni***
Anura - Leptodactylidae
e Merida Robber Frog

1539 ***Eleutherodactylus brittoni***
Anura - Leptodactylidae
e Forester's Cabin Robber Frog; Grass Coqui

1540 ***Eleutherodactylus brocchi***
Anura - Leptodactylidae
e Guatemala Robber Frog

1541 ***Eleutherodactylus bromeliaceus***
Anura - Leptodactylidae
e Bromeliad Robber Frog

1542 ***Eleutherodactylus buckleyi***
Anura - Leptodactylidae
e Buckley's Robber Frog

1543 ***Eleutherodactylus bufoniformis***
Anura - Leptodactylidae
e Rusty Robber Frog

1544 ***Eleutherodactylus cabrerai***
Anura - Leptodactylidae
e Valle Robber Frog

1545 ***Eleutherodactylus cacao***
Anura - Leptodactylidae
e Cacau Robber Frog

1546 ***Eleutherodactylus cadena***
Anura - Leptodactylidae
e Nutibara Robber Frog

1547 ***Eleutherodactylus cajamarcensis***
Anura - Leptodactylidae
e Cajamarca Robber Frog

1548 ***Eleutherodactylus calcaratus***
Anura - Leptodactylidae
e San Antonio Robber Frog

1549 ***Eleutherodactylus calcarulatus***
Anura - Leptodactylidae
e Spurred Robber Frog

1550 ***Eleutherodactylus caprifer***
Anura - Leptodactylidae
e La Palma Robber Frog

1551 ***Eleutherodactylus caribe***
Anura - Leptodactylidae
e Caribbean Robber Frog

1552 ***Eleutherodactylus carmelitae***
Anura - Leptodactylidae
e Carmelita's Robber Frog

1553 ***Eleutherodactylus carvalhoi***
Anura - Leptodactylidae
e Carvalho's Robber Frog

1554 ***Eleutherodactylus caryophyllaceus***
Anura - Leptodactylidae
e La Lorna Robber Frog
d Blatt-Antillenfrosch

1555 ***Eleutherodactylus cavernicola***
Anura - Leptodactylidae
e Portland Cave Robber Frog

1556 ***Eleutherodactylus celator***
Anura - Leptodactylidae
e La Delicia Robber Frog

1557 ***Eleutherodactylus cerasinus***
Anura - Leptodactylidae
e Limon Robber Frog

1558 ***Eleutherodactylus cerastes***
Anura - Leptodactylidae
e Palma Real Robber Frog

1559 ***Eleutherodactylus ceuthospilus***
Anura - Leptodactylidae
e Wild's Robber Frog

1560 ***Eleutherodactylus chac***
Anura - Leptodactylidae
e Izabal's Robber Frog

1561 ***Eleutherodactylus chalceus***
Anura - Leptodactylidae
e Valley Robber Frog

1562 ***Eleutherodactylus cheiroplethus***
Anura - Leptodactylidae
e Rio Calles Robber Frog

1563 ***Eleutherodactylus chiastonotus***
Anura - Leptodactylidae
e Brownberg Robber Frog
f Eleutherodactyle porte-X

1564 ***Eleutherodactylus chloronotus***
Anura - Leptodactylidae
e Green Robber Frog

1565 ***Eleutherodactylus chlorophenax***
Anura - Leptodactylidae
e False Green Robber Frog

1566 ***Eleutherodactylus chlorosoma***
Anura - Leptodactylidae
e Zumbador Robber Frog

1567 ***Eleutherodactylus chrysozetes***
Anura - Leptodactylidae
e McCranie's Robber Frog

1568 ***Eleutherodactylus citriogaster***
Anura - Leptodactylidae
e Throated Robber Frog

1569 ***Eleutherodactylus cochranae***
Anura - Leptodactylidae
e Cochran's Robber Frog; Whistling Frog (WI)

1570 ***Eleutherodactylus colodactylus***
Anura - Leptodactylidae
e Piura Robber Frog

1571 ***Eleutherodactylus colostichos***
Anura - Leptodactylidae
e Las Canalejas Robber Frog

1572 ***Eleutherodactylus condor***
Anura - Leptodactylidae
e Condor Robber Frog

1573 ***Eleutherodactylus conspicillatus***
Anura - Leptodactylidae

e Chirping Robber Frog

1574 ***Eleutherodactylus cooki***
Anura - Leptodactylidae
e Cook's Robber Frog; Rock Frog (WI)

1575 ***Eleutherodactylus coqui***
Anura - Leptodactylidae
e Puerto Rican Coqui; Coqui Frog; Coqui; Common Coqui

1576 ***Eleutherodactylus cornutus***
Anura - Leptodactylidae
e Rio Suno Robber Frog
d Antillen-Pfeilfrosch

1577 ***Eleutherodactylus corona***
Anura - Leptodactylidae
e Caye Paul Robber Frog

1578 ***Eleutherodactylus cosnipatae***
Anura - Leptodactylidae
e Rio Cosnipata Robber Frog

1579 ***Eleutherodactylus counnouspeus***
Anura - Leptodactylidae
e Counou Robber Frog

1580 ***Eleutherodactylus crassidigitus***
Anura - Leptodactylidae
e Isla Bonita Robber Frog

1581 ***Eleutherodactylus cremnobates***
Anura - Leptodactylidae
e Rio Reventador Robber Frog

1582 ***Eleutherodactylus crenunguis***
Anura - Leptodactylidae
e Spring Robber Frog

1583 ***Eleutherodactylus cristinae***
Anura - Leptodactylidae
e Cristina's Robber Frog

1584 ***Eleutherodactylus croceoinguinus***
Anura - Leptodactylidae
e Santa Cecilia Robber Frog

1585 ***Eleutherodactylus crucifer***
Anura - Leptodactylidae
e Porvenir Robber Frog

1586 ***Eleutherodactylus cruentus***
Anura - Leptodactylidae
e Chiriqui Robber Frog
d Blutflecken-Antillenfrosch

1587 ***Eleutherodactylus cruentus***
Anura - Leptodactylidae
e La Paz Robber Frog

1588 ***Eleutherodactylus cruzi***
Anura - Leptodactylidae
e Cruz Robber Frog

1589 ***Eleutherodactylus cryophilius***
Anura - Leptodactylidae
e San Vincente Robber Frog

1590 ***Eleutherodactylus cryptomelas***
Anura - Leptodactylidae
e Cryptic Robber Frog

1591 ***Eleutherodactylus cuaquero***
Anura - Leptodactylidae
e Puntarenas Robber Frog

1592 ***Eleutherodactylus cubanus***
Anura - Leptodactylidae
e Cuban Robber Frog

1593 ***Eleutherodactylus cundalli***
Anura - Leptodactylidae
e Cundall's Robber Frog

1594 ***Eleutherodactylus cuneatus***
Anura - Leptodactylidae
e Juventu Robber Frog

1595 ***Eleutherodactylus curtipes***
Anura - Leptodactylidae
e Intac Robber Frog

1596 ***Eleutherodactylus cystignathoides***
Anura - Leptodactylidae
e Rio Grande Chirping Frog; Rio Grande Frog
d Rio-Grande-Frosch

1597 ***Eleutherodactylus cystignathoides campi***
Anura - Leptodactylidae
e Rio Grande Chirping Frog

1598 ***Eleutherodactylus cystignathoides cystignathoides***
Anura - Leptodactylidae
e Mexican Chirping Frog

1599 ***Eleutherodactylus danae***
Anura - Leptodactylidae
e Cuzco Robber Frog

1600 ***Eleutherodactylus darlingtoni***
Anura - Leptodactylidae
e Darlington's Robber Frog

1601 ***Eleutherodactylus daryi***
Anura - Leptodactylidae
e Ford's Robber Frog

1602 ***Eleutherodactylus decoratus***
Anura - Leptodactylidae
e Adorned Robber Frog
d Schmuck-Antillenfrosch

1603 ***Eleutherodactylus decoratus decoratus***
Anura - Leptodactylidae
e Adorned Robber Frog

1604 ***Eleutherodactylus decoratus purpurus***
Anura - Leptodactylidae
e Purple Adorned Robber Frog

1605 ***Eleutherodactylus delicatus***
Anura - Leptodactylidae
e Delicate Robber Frog

1606 ***Eleutherodactylus dennisi***
Anura - Leptodactylidae
e Dennis's Chirping Frog

1607 ***Eleutherodactylus devillei***
Anura - Leptodactylidae
e De Ville's Robber Frog

1608 ***Eleutherodactylus diadematus***
Anura - Leptodactylidae
e Diadem Robber Frog

1609 ***Eleutherodactylus diaphonus***
Anura - Leptodactylidae
e Rio Calima Robber Frog

1610 ***Eleutherodactylus diastema***
Anura - Leptodactylidae
e Caretta Robber Frog

1611 ***Eleutherodactylus dilatus***
Anura - Leptodactylidae
e Guerreran Peeping Frog

1612 ***Eleutherodactylus dimidiatus***
Anura - Leptodactylidae
e Miniature Robber Frog

1613 ***Eleutherodactylus discoidalis***
Anura - Leptodactylidae
e Tucuman Robber Frog

1614 ***Eleutherodactylus dixoni***
Anura - Leptodactylidae
e White-lipped Peeping Frog

1615 ***Eleutherodactylus dolomedes***
Anura - Leptodactylidae

e Hedges's Robber Frog

1616 ***Eleutherodactylus dolops***
Anura - Leptodactylidae
e Putumayo Robber Frog

1617 ***Eleutherodactylus dorsopictus***
Anura - Leptodactylidae
e Serna's Robber Frog

1618 ***Eleutherodactylus duellmani***
Anura - Leptodactylidae
e Duellman's Robber Frog

1619 ***Eleutherodactylus eileenae***
Anura - Leptodactylidae
e Eileen's Robber Frog; Cuban Free-fingered Frog

1620 ***Eleutherodactylus elassodiscus***
Anura - Leptodactylidae
e Cuyuja Robber Frog

1621 ***Eleutherodactylus elegans***
Anura - Leptodactylidae
e Elegant Robber Frog

1622 ***Eleutherodactylus emcelae***
Anura - Leptodactylidae
e Rio Claro Robber Frog

1623 ***Eleutherodactylus emiliae***
Anura - Leptodactylidae
e Emilia's Robber Frog

1624 ***Eleutherodactylus eneidae***
Anura - Leptodactylidae
e Villalba Robber Frog; Mottled Coqui (WI)

1625 ***Eleutherodactylus epipedus***
Anura - Leptodactylidae
e Santa Teresa Robber Frog

1626 ***Eleutherodactylus eremitus***
Anura - Leptodactylidae
e Chiriboga Robber Frog

1627 ***Eleutherodactylus eriphus***
Anura - Leptodactylidae
e Rio Jatuntinahua Robber Frog

1628 ***Eleutherodactylus ernesti***
Anura - Leptodactylidae
e Ernest's Robber Frog

1629 ***Eleutherodactylus erythromerus***
Anura - Leptodactylidae
e Soberbo Robber Frog

1630 ***Eleutherodactylus erythropleura***
Anura - Leptodactylidae
e Colombian Robber Frog

1631 ***Eleutherodactylus escoces***
Anura - Leptodactylidae
e Heredia Robber Frog

1632 ***Eleutherodactylus etheridgei***
Anura - Leptodactylidae
e Etheridge's Robber Frog

1633 ***Eleutherodactylus eunaster***
Anura - Leptodactylidae
e Les Cayes Robber Frog

1634 ***Eleutherodactylus eurydactylus***
Anura - Leptodactylidae
e Panguana Robber Frog

1635 ***Eleutherodactylus fenestratus***
Anura - Leptodactylidae
e Rio Mamore Robber Frog

1636 ***Eleutherodactylus fitzingeri***
Anura - Leptodactylidae
e Fitzinger's Robber Frog

1637 ***Eleutherodactylus flaqvescens***
Anura - Leptodactylidae
e Buenaventura Robber Frog
d Buschfrosch

1638 ***Eleutherodactylus fleischmanni***
Anura - Leptodactylidae
e Fleischmann's Robber Frog

1639 ***Eleutherodactylus fowleri***
Anura - Leptodactylidae
e Fowler's Robber Frog

1640 ***Eleutherodactylus frateri***
Anura - Leptodactylidae
e Meta Robber Frog

1641 ***Eleutherodactylus fraudator***
Anura - Leptodactylidae
e Cochamba Robber Frog

1642 ***Eleutherodactylus furyensis***
Anura - Leptodactylidae
e Furcy Robber Frog

1643 ***Eleutherodactylus fuscus***
Anura - Leptodactylidae
e Brown Robber Frog

1644 ***Eleutherodactylus gaigeae***
Anura - Leptodactylidae
e Port Randolph Robber Frog

1645 ***Eleutherodactylus galdi***
Anura - Leptodactylidae
e Espada's Robber Frog

1646 ***Eleutherodactylus ganonotus***
Anura - Leptodactylidae
e Yapitya Robber Frog

1647 ***Eleutherodactylus ginesi***
Anura - Leptodactylidae
e Rangel Robber Frog

1648 ***Eleutherodactylus gladiator***
Anura - Leptodactylidae
e Papallacta Valley Robber Frog

1649 ***Eleutherodactylus glandulifer***
Anura - Leptodactylidae
e Doris's Robber Frog

1650 ***Eleutherodactylus glanduliferoides***
Anura - Leptodactylidae
e La Viste Robber Frog

1651 ***Eleutherodactylus glandulosus***
Anura - Leptodactylidae
e Ecuador Robber Frog

1652 ***Eleutherodactylus glaphycompus***
Anura - Leptodactylidae
e Southwest Haiti Robber Frog

1653 ***Eleutherodactylus glaucoreius***
Anura - Leptodactylidae
e Portland Parish Robber Frog

1654 ***Eleutherodactylus glaucus***
Anura - Leptodactylidae
e Grey Robber Frog

1655 ***Eleutherodactylus gollmeri***
Anura - Leptodactylidae
e Evergreen Robber Frog

1656 ***Eleutherodactylus gossei***
Anura - Leptodactylidae
e Spaldings Robber Frog

1657 ***Eleutherodactylus grabhami***
Anura - Leptodactylidae
e Clarendon Parish Robber Frog

1658 ***Eleutherodactylus gracilis***
Anura - Leptodactylidae
e Pichinde Robber Frog

1659 ***Eleutherodactylus grahami***
Anura - Leptodactylidae
e Graham's Robber Frog

1660 ***Eleutherodactylus grandiceps***
Anura - Leptodactylidae
e Giant Robber Frog

1661 ***Eleutherodactylus grandis***
Anura - Leptodactylidae
e Great Peeping Frog

1662 ***Eleutherodactylus grandoculis***
Anura - Leptodactylidae
e Big-eyed Robber Frog

1663 ***Eleutherodactylus greggi***
Anura - Leptodactylidae
e Gregg's Robber Frog; Gregg's Stream Frog

1664 ***Eleutherodactylus greyi***
Anura - Leptodactylidae
e Grey's Robber Frog

1665 ***Eleutherodactylus gryllus***
Anura - Leptodactylidae
e Cricket Robber Frog; Cricket Coqui (WI)

1666 ***Eleutherodactylus gryphus***
Anura - Leptodactylidae
e Trelawny Parish Robber Frog

1667 ***Eleutherodactylus gualteri***
Anura - Leptodactylidae
e Organ Mountains Robber Frog

1668 ***Eleutherodactylus guanahacabibes***
Anura - Leptodactylidae
e Guanahacabibes Robber Frog

1669 ***Eleutherodactylus guantanamera***
Anura - Leptodactylidae
e Guantanamo Robber Frog

1670 ***Eleutherodactylus guentheri***
Anura - Leptodactylidae
e Steindachner's Robber Frog; Guenther's Frog
d Erdfröschchen
f Grenouille de Guenther

1671 ***Eleutherodactylus guerreroensis***
Anura - Leptodactylidae
e Guerreran Robber Frog

1672 ***Eleutherodactylus gularis***
Anura - Leptodactylidae
e Esmeralda's Robber Frog

1673 ***Eleutherodactylus gundlachi***
Anura - Leptodactylidae
e Gundlach's Robber Frog

1674 ***Eleutherodactylus guttilatus***
Anura - Leptodactylidae
e Spotted Chirping Frog; Mexican Cliff Frog

1675 ***Eleutherodactylus gutturalis***
Anura - Leptodactylidae
e Lower Matarony Robber Frog
f Eleutherodactyle à bande gutturale

1676 ***Eleutherodactylus haitianus***
Anura - Leptodactylidae
e Common Haitian Robber Frog

1677 ***Eleutherodactylus hamiotae***
Anura - Leptodactylidae
e Nono Robber Frog

1678 ***Eleutherodactylus hectus***
Anura - Leptodactylidae
e Dwarf Robber Frog

1679 ***Eleutherodactylus hedricki***
Anura - Leptodactylidae
e Hendrick's Robber Frog; Tree-hole Coqui (WI)

1680 ***Eleutherodactylus helonotus***
Anura - Leptodactylidae
e Rio Pitzara Robber Frog

1681 ***Eleutherodactylus heminola***
Anura - Leptodactylidae
e Shreve's Robber Frog

1682 ***Eleutherodactylus hernandezi***
Anura - Leptodactylidae
e Hernandez's Robber Frog

1683 ***Eleutherodactylus heterodactylus***
Anura - Leptodactylidae
e Caceres Robber Frog

1684 ***Eleutherodactylus hobartsmithi***
Anura - Leptodactylidae
e Smith's Pygmy Robber Frog

1685 ***Eleutherodactylus hoehnei***
Anura - Leptodactylidae
e Hoehn's Robber Frog

1686 ***Eleutherodactylus holti***
Anura - Leptodactylidae
e Holt's Robber Frog

1687 ***Eleutherodactylus hybotragus***
Anura - Leptodactylidae
e Agua Bonita Robber Frog

1688 ***Eleutherodactylus hylaeformis***
Anura - Leptodactylidae
e Pico Blanco Robber Frog

1689 ***Eleutherodactylus hypostenor***
Anura - Leptodactylidae
e Cabral Robber Frog

1690 ***Eleutherodactylus ignicolor***
Anura - Leptodactylidae
e Fire Robber Frog

1691 ***Eleutherodactylus imitatrix***
Anura - Leptodactylidae
e Imitator Robber Frog

1692 ***Eleutherodactylus incanus***
Anura - Leptodactylidae
e Rio Azuela Robber Frog

1693 ***Eleutherodactylus incomptus***
Anura - Leptodactylidae
e Santa Rosa Robber Frog

1694 ***Eleutherodactylus ingeri***
Anura - Leptodactylidae
e Inger's robber Frog

1695 ***Eleutherodactylus inguinalis***
Anura - Leptodactylidae
e New River Robber Frog
f Eleutherodactyle inguinal

1696 ***Eleutherodactylus inoptatus***
Anura - Leptodactylidae
e Diquini Robber Frog

1697 ***Eleutherodactylus insignitus***
Anura - Leptodactylidae
e Ground Robber Frog

1698 ***Eleutherodactylus intermedius***
Anura - Leptodactylidae
e Pico Turquino Robber Frog

1699 ***Eleutherodactylus interorbitalis***
Anura - Leptodactylidae
e Spectacled Chirping Frog

1700 ***Eleutherodactylus inusitatus***
Anura - Leptodactylidae
e Barking Robber Frog

1701 ***Eleutherodactylus ionthus***
Anura - Leptodactylidae
e La Maya Robber Frog

1702 ***Eleutherodactylus izecksohni***
Anura - Leptodactylidae
e Izecksohn's Robber Frog

1703 ***Eleutherodactylus jaimei***
Anura - Leptodactylidae
e Jaime's Robber Frog

1704 ***Eleutherodactylus jamaicensis***
Anura - Leptodactylidae
e Jamaica Robber Frog

1705 ***Eleutherodactylus jasperi***
Anura - Leptodactylidae
e Cayey Robber Frog; Golden Coqui (WI)

1706 ***Eleutherodactylus johannesdei***
Anura - Leptodactylidae
e Urrao Robber Frog

1707 ***Eleutherodactylus johnstonei***
Anura - Leptodactylidae
e Johnstone's Frog
f Eleutherodactyle de Johnston

1708 ***Eleutherodactylus jota***
Anura - Leptodactylidae
e Rio Changena Robber Frog

1709 ***Eleutherodactylus jugans***
Anura - Leptodactylidae
e Mable's Robber Frog

1710 ***Eleutherodactylus juipoca***
Anura - Leptodactylidae
e Souzas Robber Frog

1711 ***Eleutherodactylus junori***
Anura - Leptodactylidae
e Spaldings Robber Frog

1712 ***Eleutherodactylus karlschmidti***
Anura - Leptodactylidae
e Karl's Robber Frog; Web-footed Coqui (WI)

1713 ***Eleutherodactylus katopteroides***
Anura - Leptodactylidae
e Puyo Robber Frog

1714 ***Eleutherodactylus kirklandi***
Anura - Leptodactylidae
e Kirkland's Robber Frog

1715 ***Eleutherodactylus klinikowskii***
Anura - Leptodactylidae
e Klinikowski's Robber Frog

1716 ***Eleutherodactylus lacrimosus***
Anura - Leptodactylidae
e Jimenez's Robber Frog

1717 ***Eleutherodactylus lacteus***
Anura - Leptodactylidae
e Iguape Robber Frog

1718 ***Eleutherodactylus lamprotes***
Anura - Leptodactylidae
e Castillon Robber Frog

1719 ***Eleutherodactylus lancinii***
Anura - Leptodactylidae
e Lancini's Robber Frog

1720 ***Eleutherodactylus lanthanites***
Anura - Leptodactylidae
e Metallic Robber Frog

1721 ***Eleutherodactylus latens***
Anura - Leptodactylidae
e Boqueron Robber Frog

1722 ***Eleutherodactylus laticeps***
Anura - Leptodactylidae
e Broad-headed Rain Frog; Broad-headed Cave Frog

1723 ***Eleutherodactylus laticlavius***
Anura - Leptodactylidae
e Burrowes's Robber Frog

1724 ***Eleutherodactylus latidiscus***
Anura - Leptodactylidae
e Disc Robber Frog

1725 ***Eleutherodactylus leberi***
Anura - Leptodactylidae
e Leber's Robber Frog

1726 ***Eleutherodactylus lentiginosus***
Anura - Leptodactylidae
e Guacharaquita Robber Frog

1727 ***Eleutherodactylus lentus***
Anura - Leptodactylidae
e Virgin Islands Robber Frog; Mute Frog (WI)

1728 ***Eleutherodactylus leoncei***
Anura - Leptodactylidae
e Hispaniola Robber Frog

1729 ***Eleutherodactylus leoni***
Anura - Leptodactylidae
e Leon's Robber Frog

1730 ***Eleutherodactylus leprus***
Anura - Leptodactylidae
e Leprus Chirping Frog

1731 ***Eleutherodactylus leptolophus***
Anura - Leptodactylidae
e Volcano Robber Frog

1732 ***Eleutherodactylus leucopus***
Anura - Leptodactylidae
e Rio Chingual Robber Frog

1733 ***Eleutherodactylus limbatus***
Anura - Leptodactylidae
e Habana Robber Frog
d Kubanischer Zwergfrosch

1734 ***Eleutherodactylus lindae***
Anura - Leptodactylidae
e Santa Isabel Robber Frog

1735 ***Eleutherodactylus lineatus***
Anura - Leptodactylidae
e Montane Robber Frog

1736 ***Eleutherodactylus lividus***
Anura - Leptodactylidae
e Bruised Robber Frog

1737 ***Eleutherodactylus locustus***
Anura - Leptodactylidae
e Interior Robber Frog; Warty Coqui (WI)

1738 ***Eleutherodactylus longipes***
Anura - Leptodactylidae
e Longfoot Robber Frog; Long-footed Chirping Frog

1739 ***Eleutherodactylus longirostris***
Anura - Leptodactylidae
e Longsnout Robber Frog

1740 ***Eleutherodactylus loustes***
Anura - Leptodactylidae
e Maldonado Robber Frog

1741 ***Eleutherodactylus luceoi***
Anura - Leptodactylidae
e St. Nicholas Robber Frog

1742 ***Eleutherodactylus luteolateralis***
Anura - Leptodactylidae
e White-striped Robber Frog

1743 ***Eleutherodactylus luteolus***
Anura - Leptodactylidae
e Light Robber Frog

1744 ***Eleutherodactylus lutitus***
Anura - Leptodactylidae
e Rio Luisito Robber Frog

1745 ***Eleutherodactylus lymani***
Anura - Leptodactylidae
e Lyman's Robber Frog

1746 ***Eleutherodactylus lynchi***
Anura - Leptodactylidae
e Lynch's Robber Frog

1747 ***Eleutherodactylus lythrodes***
Anura - Leptodactylidae
e Lescure's Robber Frog

1748 ***Eleutherodactylus maculosus***
Anura - Leptodactylidae
e Spotted Robber Frog

1749 ***Eleutherodactylus malkini***
Anura - Leptodactylidae
e Rio Ampiyacu Robber Frog

1750 ***Eleutherodactylus mantipus***
Anura - Leptodactylidae
e Mantipus Robber Frog

1751 ***Eleutherodactylus mariposa***
Anura - Leptodactylidae
e Lily Robber Frog

1752 ***Eleutherodactylus marmoratus***
Anura - Leptodactylidae
e Marbled Robber Frog
f Eleutherodactyle marbré

1753 ***Eleutherodactylus marnockii***
Anura - Leptodactylidae
e Cliff Chirping Frog; Cliff Frog
d Marnocks Frosch

1754 ***Eleutherodactylus martiae***
Anura - Leptodactylidae
e Martia's Robber Frog

1755 ***Eleutherodactylus martinicensis***
Anura - Leptodactylidae
e Martinique Robber Frog
d Antillen-Frosch

1756 ***Eleutherodactylus matudae***
Anura - Leptodactylidae
e Matuda's Stream Frog

1757 ***Eleutherodactylus maurus***
Anura - Leptodactylidae
e Brown Peeping Frog

1758 ***Eleutherodactylus mausso***
Anura - Leptodactylidae
e Puerto Cabello Robber Frog

1759 ***Eleutherodactylus megalops***
Anura - Leptodactylidae
e San Lorenza Robber Frog

1760 ***Eleutherodactylus megalotympanum***
Anura - Leptodactylidae
e San Martin Robber Frog

1761 ***Eleutherodactylus melacara***
Anura - Leptodactylidae
e Estrada's Robber Frog

1762 ***Eleutherodactylus melanoproctus***
Anura - Leptodactylidae
e Black-marked Robber Frog

1763 ***Eleutherodactylus melanostictus***
Anura - Leptodactylidae
e Black-lined Robber Frog

1764 ***Eleutherodactylus mendax***
Anura - Leptodactylidae
e Pistipata Robber Frog

1765 ***Eleutherodactylus mercedesae***
Anura - Leptodactylidae
e Mercedes Robber Frog

1766 ***Eleutherodactylus merendonensis***
Anura - Leptodactylidae
e San Pedro Robber Frog

1767 ***Eleutherodactylus merostictus***
Anura - Leptodactylidae
e Bogotacito Robber Frog

1768 ***Eleutherodactylus mexicanus***
Anura - Leptodactylidae
e Mexican Robber Frog

1769 ***Eleutherodactylus milesi***
Anura - Leptodactylidae
e Miles's Robber Frog

1770 ***Eleutherodactylus minus***
Anura - Leptodactylidae
e Tilaran Robber Frog

1771 ***Eleutherodactylus minutus***
Anura - Leptodactylidae
e Tiny Robber Frog

1772 ***Eleutherodactylus miyatai***
Anura - Leptodactylidae
e Miyata's Robber Frog

1773 ***Eleutherodactylus modestus***
Anura - Leptodactylidae
e Blunt-toed Chirping Frog

1774 ***Eleutherodactylus modipeplus***
Anura - Leptodactylidae
e Urbina Robber Frog

1775 ***Eleutherodactylus molybrignus***
Anura - Leptodactylidae
e Uribe Robber Frog

1776 ***Eleutherodactylus mondolfii***
Anura - Leptodactylidae
e Mondolfi's Robber Frog

1777 ***Eleutherodactylus monensis***
Anura - Leptodactylidae
e Isla Mona Robber Frog; Mona Coqui (WI)

1778 ***Eleutherodactylus monnichorum***
Anura - Leptodactylidae
e Dunn's Robber Frog

1779 ***Eleutherodactylus montanus***
Anura - Leptodactylidae
e Dominican Mountain Robber Frog

1780 ***Eleutherodactylus moro***
Anura - Leptodactylidae
e La Hondura Robber Frog

1781 ***Eleutherodactylus muricatus***
Anura - Leptodactylidae
e Rio Faisanes Robber Frog

1782 ***Eleutherodactylus myersi***
Anura - Leptodactylidae
e Myers's Robber Frog

1783 ***Eleutherodactylus nasutus***
Anura - Leptodactylidae
e Pointy-nose Robber Frog
d Spitzschäuziger Antillenfrosch

1784 ***Eleutherodactylus nebulosus***
Anura - Leptodactylidae
e Huanuco Robber Frog

1785 ***Eleutherodactylus necerus***
Anura - Leptodactylidae

e Mindo Robber Frog

1786 ***Eleutherodactylus neodreptus***
Anura - Leptodactylidae
e Barahona Robber Frog

1787 ***Eleutherodactylus nicefori***
Anura - Leptodactylidae
e Nicéforo's Robber Frog

1788 ***Eleutherodactylus nigriventris***
Anura - Leptodactylidae
e Itatiaia Robber Frog

1789 ***Eleutherodactylus nigrogriseus***
Anura - Leptodactylidae
e Charcoal Robber Frog

1790 ***Eleutherodactylus nigrovittatus***
Anura - Leptodactylidae
e Black-banded Robber Frog

1791 ***Eleutherodactylus nitidus***
Anura - Leptodactylidae
e Shiny Peeping Frog

1792 ***Eleutherodactylus nitidus nitidus***
Anura - Leptodactylidae
e Shiny Peeping Frog

1793 ***Eleutherodactylus nitidus orarius***
Anura - Leptodactylidae
e Colima Shiny Peeping Frog

1794 ***Eleutherodactylus nitidus petersi***
Anura - Leptodactylidae
e Peters's Shiny Peeping Frog

1795 ***Eleutherodactylus nivicolimae***
Anura - Leptodactylidae
e Nevado de Colima Chirping Frog

1796 ***Eleutherodactylus noblei***
Anura - Leptodactylidae
e Noble's Robber Frog

1797 ***Eleutherodactylus nortoni***
Anura - Leptodactylidae
e Norton's Robber Frog

1798 ***Eleutherodactylus nubicola***
Anura - Leptodactylidae
e Portland Robber Frog

1799 ***Eleutherodactylus nyctophylax***
Anura - Leptodactylidae
e Tandapi Robber Frog

1800 ***Eleutherodactylus obmutescens***
Anura - Leptodactylidae
e Paramos Robber Frog

1801 ***Eleutherodactylus occidentalis***
Anura - Leptodactylidae
e Taylor's Barking Frog

1802 ***Eleutherodactylus ocellatus***
Anura - Leptodactylidae
e Ocellated Robber Frog

1803 ***Eleutherodactylus ockendeni***
Anura - Leptodactylidae
e Carabaya Robber Frog

1804 ***Eleutherodactylus ocreatus***
Anura - Leptodactylidae
e Carchi Robber Frog

1805 ***Eleutherodactylus octavioi***
Anura - Leptodactylidae
e Octavio's Robber Frog

1806 ***Eleutherodactylus oeus***
Anura - Leptodactylidae
e Espirito Santo Robber Frog

1807 ***Eleutherodactylus omiltemanus***
Anura - Leptodactylidae

e Günther's Robber Frog; Guerreran Robber Frog

1808 ***Eleutherodactylus oranobates***
Anura - Leptodactylidae
e Caldas Robber Frog

1809 ***Eleutherodactylus orcesi***
Anura - Leptodactylidae
e Bolivar Robber Frog

1810 ***Eleutherodactylus orcutti***
Anura - Leptodactylidae
e Arntully Robber Frog

1811 ***Eleutherodactylus orestes***
Anura - Leptodactylidae
e Urdaneta Robber Frog

1812 ***Eleutherodactylus ornatissimus***
Anura - Leptodactylidae
e Ornate Robber Frog

1813 ***Eleutherodactylus orphnolaimus***
Anura - Leptodactylidae
e Lage Agrio Robber Frog

1814 ***Eleutherodactylus oxyrhyncus***
Anura - Leptodactylidae
e Rednose Robber Frog

1815 ***Eleutherodactylus pallidus***
Anura - Leptodactylidae
e Pale Chirping Frog

1816 ***Eleutherodactylus palmeri***
Anura - Leptodactylidae
e Palmer's Robber Frog

1817 ***Eleutherodactylus pantoni***
Anura - Leptodactylidae
e Panton's Robber Frog

1818 ***Eleutherodactylus parabates***
Anura - Leptodactylidae
e Independencia Robber Frog

1819 ***Eleutherodactylus paramerus***
Anura - Leptodactylidae
e Paramo Robber Frog

1820 ***Eleutherodactylus parapelates***
Anura - Leptodactylidae
e Casillon Robber Frog

1821 ***Eleutherodactylus pardialis***
Anura - Leptodactylidae
e Leopard Robber Frog

1822 ***Eleutherodactylus parvillus***
Anura - Leptodactylidae
e Pichincha Robber Frog

1823 ***Eleutherodactylus parvus***
Anura - Leptodactylidae
e Leopard Robber Frog

1824 ***Eleutherodactylus pastazensis***
Anura - Leptodactylidae
e Pastaza Valley Robber Frog

1825 ***Eleutherodactylus patriciae***
Anura - Leptodactylidae
e Patricia's Robber Frog

1826 ***Eleutherodactylus paulodutrai***
Anura - Leptodactylidae
e Paulo's Robber Frog

1827 ***Eleutherodactylus paulsoni***
Anura - Leptodactylidae
e Paulson's Robber Frog

1828 ***Eleutherodactylus paululus***
Anura - Leptodactylidae
e Amazon Slope Robber Frog

1829 ***Eleutherodactylus pecki***
Anura - Leptodactylidae
e Peck's Robber Frog

1830 ***Eleutherodactylus pentasyringos***
Anura - Leptodactylidae
e Tube Robber Frog

1831 ***Eleutherodactylus peraticus***
Anura - Leptodactylidae
e Tenerife Robber Frog

1832 ***Eleutherodactylus percultus***
Anura - Leptodactylidae
e Zamora Robber Frog

1833 ***Eleutherodactylus peruvianus***
Anura - Leptodactylidae
e Peru Robber Frog

1834 ***Eleutherodactylus petersorum***
Anura - Leptodactylidae
e Peters's Robber Frog

1835 ***Eleutherodactylus petrobardus***
Anura - Leptodactylidae
e Huambos Robber Frog

1836 ***Eleutherodactylus pezopetrus***
Anura - Leptodactylidae
e Miranda Robber Frog

1837 ***Eleutherodactylus phoxocephalus***
Anura - Leptodactylidae
e Cotopaxi Robber Frog

1838 ***Eleutherodactylus phragmipleuron***
Anura - Leptodactylidae
e Sugar Robber Frog

1839 ***Eleutherodactylus pictissimus***
Anura - Leptodactylidae
e Painted Robber Frog

1840 ***Eleutherodactylus pinarensis***
Anura - Leptodactylidae
e Pinos Robber Frog

1841 ***Eleutherodactylus pinchoni***
Anura - Leptodactylidae
e Grand Cafe Robber Frog

1842 ***Eleutherodactylus pipilans***
Anura - Leptodactylidae
e Whistling Chirping Frog

1843 ***Eleutherodactylus pipilans nebulosus***
Anura - Leptodactylidae
e Nebulous Chirping Frog

1844 ***Eleutherodactylus pipilans pipilans***
Anura - Leptodactylidae
e Whistling Chirping Frog

1845 ***Eleutherodactylus pituinus***
Anura - Leptodactylidae
e La Horma Robber Frog

1846 ***Eleutherodactylus planirostris***
Anura - Leptodactylidae
e Greenhouse Frog
d Gewächshausfrosch
f Grenouille des serres

1847 ***Eleutherodactylus platydactylus***
Anura - Leptodactylidae
e Flat-fingered Robber Frog

1848 ***Eleutherodactylus pleurostriatus***
Anura - Leptodactylidae
e Side-striped Robber Frog

1849 ***Eleutherodactylus plicifer***
Anura - Leptodactylidae
e Iguarasse Robber Frog

1850 ***Eleutherodactylus podiciferus***
Anura - Leptodactylidae
e Cerro Utyum Robber Frog

1851 ***Eleutherodactylus polymniae***
Anura - Leptodactylidae
e Sierra Juarez Robber Frog

1852 ***Eleutherodactylus poolei***
Anura - Leptodactylidae
e Poole's Robber Frog

1853 ***Eleutherodactylus portoricensis***
Anura - Leptodactylidae
e Puerto Rican Robber Frog; Forest Coqui (VIP)

1854 ***Eleutherodactylus probolaeus***
Anura - Leptodactylidae
e Yuma Robber Frog

1855 ***Eleutherodactylus prolatus***
Anura - Leptodactylidae
e Hidden Robber Frog

1856 ***Eleutherodactylus prolixodiscus***
Anura - Leptodactylidae
e Bucaramanga Robber Frog

1857 ***Eleutherodactylus proserpens***
Anura - Leptodactylidae
e Sapote Robber Frog

1858 ***Eleutherodactylus pseudoacuminatus***
Anura - Leptodactylidae
e Sarayuca Robber Frog

1859 ***Eleutherodactylus pugnax***
Anura - Leptodactylidae
e Agua Robber Frog

1860 ***Eleutherodactylus pulidoi***
Anura - Leptodactylidae
e Matamula Robber Frog

1861 ***Eleutherodactylus pulvinatus***
Anura - Leptodactylidae
e Danto Robber Frog

1862 ***Eleutherodactylus punctariolus***
Anura - Leptodactylidae
e Bob's Robber Frog

1863 ***Eleutherodactylus pusillus***
Anura - Leptodactylidae
e Rio Mombucaba Robber Frog

1864 ***Eleutherodactylus pycnodermis***
Anura - Leptodactylidae
e Thickskin Robber Frog

1865 ***Eleutherodactylus pygmaeus***
Anura - Leptodactylidae
e Pygmy Robber Frog; Pygmy Free-fingered Frog

1866 ***Eleutherodactylus pyrrhomerus***
Anura - Leptodactylidae
e Lynch's Pilalo Robber Frog

1867 ***Eleutherodactylus quaquaveresus***
Anura - Leptodactylidae
e Rio Coca Robber Frog

1868 ***Eleutherodactylus quinquagesimus***
Anura - Leptodactylidae
e Zapadores Robber Frog

1869 ***Eleutherodactylus racemus***
Anura - Leptodactylidae
e Las Hermosas Robber Frog

1870 ***Eleutherodactylus ramagii***
Anura - Leptodactylidae
e Paraiba Robber Frog

1871 ***Eleutherodactylus randorum***
Anura - Leptodactylidae
e Boraceia Robber Frog

1872 ***Eleutherodactylus raniformis***
Anura - Leptodactylidae
e Robber Frog

1873 ***Eleutherodactylus rayo***
Anura - Leptodactylidae
e Sabana Robber Frog

1874 ***Eleutherodactylus repens***
Anura - Leptodactylidae
e Galeras Robber Frog

1875 ***Eleutherodactylus reticulatus***
Anura - Leptodactylidae
e Reticulate Robber Frog

1876 ***Eleutherodactylus rhabdolaemus***
Anura - Leptodactylidae
e Apurimac Robber Frog

1877 ***Eleutherodactylus rhodesi***
Anura - Leptodactylidae
e Rhodes Robber Frog

1878 ***Eleutherodactylus rhodopis***
Anura - Leptodactylidae
e Polymorphic Robber Frog

1879 ***Eleutherodactylus rhodoplichus***
Anura - Leptodactylidae
e Canchaque Robber Frog

1880 ***Eleutherodactylus richmondi***
Anura - Leptodactylidae
e Richmond's Robber Frog; Ground Coqui (WI)

1881 ***Eleutherodactylus ricordii***
Anura - Leptodactylidae
e Ricord's Robber Frog; Bahaman Tree Frog; Ricord's Frog
d Ricards Antillenfrosch; Kuba-Pfeiffrosch

1882 ***Eleutherodactylus ridens***
Anura - Leptodactylidae
e Rio San Juan Robber Frog

1883 ***Eleutherodactylus riveroi***
Anura - Leptodactylidae
e Rivero's Robber Frog

1884 ***Eleutherodactylus riveti***
Anura - Leptodactylidae
e Despax's Robber Frog

1885 ***Eleutherodactylus ronaldi***
Anura - Leptodactylidae
e Ronald's Robber Frog

1886 ***Eleutherodactylus rosadoi***
Anura - Leptodactylidae
e Rosado's Robber Frog

1887 ***Eleutherodactylus roseus***
Anura - Leptodactylidae
e Andagoya Robber Frog

1888 ***Eleutherodactylus rostralis***
Anura - Leptodactylidae
e Snouted Robber Frog

1889 ***Eleutherodactylus rozei***
Anura - Leptodactylidae
e Roze's Robber Frog

1890 ***Eleutherodactylus rubicundus***
Anura - Leptodactylidae
e Trench Robber Frog

1891 ***Eleutherodactylus rubrimaculatus***
Anura - Leptodactylidae
e Red-spotted Chirping Frog

1892 ***Eleutherodactylus rufescens***
Anura - Leptodactylidae
e Red Peeping Frog

1893 ***Eleutherodactylus rufifemoralis***
Anura - Leptodactylidae
e Red-legged Robber Frog

1894 ***Eleutherodactylus rugulosus***
Anura - Leptodactylidae
e Central American Rain Frog

1895 ***Eleutherodactylus ruidus***
Anura - Leptodactylidae
e Molleturo Robber Frog

1896 ***Eleutherodactylus ruizi***
Anura - Leptodactylidae
e Ruiz's Robber Frog

1897 ***Eleutherodactylus ruthae***
Anura - Leptodactylidae
e Ruth's Robber Frog

1898 ***Eleutherodactylus ruthveni***
Anura - Leptodactylidae
e Ruthven's Robber Frog

1899 ***Eleutherodactylus salaputium***
Anura - Leptodactylidae
e River Robber Frog

1900 ***Eleutherodactylus saltator***
Anura - Leptodactylidae
e Omilteme Robber Frog

1901 ***Eleutherodactylus sanctaemartae***
Anura - Leptodactylidae
e Santa Marta Robber Frog

1902 ***Eleutherodactylus sartori***
Anura - Leptodactylidae
e Chiapas Dwarf Robber Frog

1903 ***Eleutherodactylus savagei***
Anura - Leptodactylidae
e Pyburn's Robber Frog

1904 ***Eleutherodactylus saxatilis***
Anura - Leptodactylidae
e Marbled Peeping Frog

1905 ***Eleutherodactylus schmidti***
Anura - Leptodactylidae
e Schmidt's Robber Frog
d Schmidts Antillenfrosch

1906 ***Eleutherodactylus schultei***
Anura - Leptodactylidae
e Schulte's Robber Frog

1907 ***Eleutherodactylus schwartzi***
Anura - Leptodactylidae
e Schwartz Robber Frog; Virgin Islands Bo-Peep (WI)

1908 ***Eleutherodactylus sciagraphus***
Anura - Leptodactylidae
e Sud Robber Frog

1909 ***Eleutherodactylus scitulus***
Anura - Leptodactylidae
e Yuraccyacu Robber Frog

1910 ***Eleutherodactylus scoloblepharus***
Anura - Leptodactylidae
e Los Patos Robber Frog

1911 ***Eleutherodactylus scolodiscus***
Anura - Leptodactylidae
e Ricuarte Robber Frog

1912 ***Eleutherodactylus semipalmatus***
Anura - Leptodactylidae
e Foothill Robber Frog

1913 ***Eleutherodactylus sernai***
Anura - Leptodactylidae

e Pampas Robber Frog

1914 ***Eleutherodactylus sierramaestrae***
Anura - Leptodactylidae
e Sierra Maestra Robber Frog

1915 ***Eleutherodactylus silvicola***
Anura - Leptodactylidae
e Forest Robber Frog

1916 ***Eleutherodactylus simonbolivari***
Anura - Leptodactylidae
e Simon's Robber Frog

1917 ***Eleutherodactylus simoterus***
Anura - Leptodactylidae
e Albania Robber Frog

1918 ***Eleutherodactylus siopelus***
Anura - Leptodactylidae
e Reserve Robber Frog

1919 ***Eleutherodactylus sisyphodemus***
Anura - Leptodactylidae
e Quick Step Robber Frog

1920 ***Eleutherodactylus sobetes***
Anura - Leptodactylidae
e Sobetes Robber Frog

1921 ***Eleutherodactylus spanios***
Anura - Leptodactylidae
e Sao Paulo Robber Frog

1922 ***Eleutherodactylus spatulatus***
Anura - Leptodactylidae
e Spatulate Robber Frog

1923 ***Eleutherodactylus spilogaster***
Anura - Leptodactylidae
e Gambita Robber Frog

1924 ***Eleutherodactylus spinosus***
Anura - Leptodactylidae

e Spiny Robber Frog

1925 ***Eleutherodactylus stejnegerianus***
Anura - Leptodactylidae
e Stejneger's Robber Frog

1926 ***Eleutherodactylus stenodiscus***
Anura - Leptodactylidae
e Periquito Robber Frog

1927 ***Eleutherodactylus sternothylax***
Anura - Leptodactylidae
e Huancabamba Robber Frog

1928 ***Eleutherodactylus stuarti***
Anura - Leptodactylidae
e Stuart's Robber Frog

1929 ***Eleutherodactylus subsigillatus***
Anura - Leptodactylidae
e Salidero Robber Frog

1930 ***Eleutherodactylus suculus***
Anura - Leptodactylidae
e Channel Robber Frog

1931 ***Eleutherodactylus sulcatus***
Anura - Leptodactylidae
e Nauta Robber Frog

1932 ***Eleutherodactylus supernatis***
Anura - Leptodactylidae
e El Carmelo Robber Frog

1933 ***Eleutherodactylus surdus***
Anura - Leptodactylidae
e Cloud-forest Robber Frog

1934 ***Eleutherodactylus symingtoni***
Anura - Leptodactylidae
e Symington's Robber Frog

1935 ***Eleutherodactylus syristes***
Anura - Leptodactylidae

e Piping Peeping Frog

1936 ***Eleutherodactylus taeniatus***
Anura - Leptodactylidae
e Banded Robber Frog

1937 ***Eleutherodactylus talamancae***
Anura - Leptodactylidae
e Almirante Robber Frog

1938 ***Eleutherodactylus tamsitti***
Anura - Leptodactylidae
e San Adolfo Robber Frog

1939 ***Eleutherodactylus tarahumaraensis***
Anura - Leptodactylidae
e Tarahumara Barking Frog

1940 ***Eleutherodactylus taurus***
Anura - Leptodactylidae
e Golfito Robber Frog

1941 ***Eleutherodactylus taylori***
Anura - Leptodactylidae
e Taylor's Robber Frog

1942 ***Eleutherodactylus tayrona***
Anura - Leptodactylidae
e Cebolletta Robber Frog

1943 ***Eleutherodactylus tenebrionis***
Anura - Leptodactylidae
e Hotel Robber Frog

1944 ***Eleutherodactylus teretistes***
Anura - Leptodactylidae
e Whistling Frog

1945 ***Eleutherodactylus terraebolivaris***
Anura - Leptodactylidae
e Tobago Robber Frog

1946 ***Eleutherodactylus thectopternus***
Anura - Leptodactylidae

e Northern Cordilleras Robber Frog

1947 ***Eleutherodactylus thomasi***
Anura - Leptodactylidae
e Thomas's Robber Frog

1948 ***Eleutherodactylus thorectes***
Anura - Leptodactylidae
e Morne Macay Robber Frog

1949 ***Eleutherodactylus thymalopsoides***
Anura - Leptodactylidae
e Greater Pilalo Robber Frog

1950 ***Eleutherodactylus thymelensis***
Anura - Leptodactylidae
e Thymelen Robber Frog

1951 ***Eleutherodactylus toa***
Anura - Leptodactylidae
e Yarrey Robber Frog

1952 ***Eleutherodactylus toftae***
Anura - Leptodactylidae
e Pachitea Robber Frog

1953 ***Eleutherodactylus trachybloepharis***
Anura - Leptodactylidae
e El Topo Robber Frog

1954 ***Eleutherodactylus trepidotus***
Anura - Leptodactylidae
e Shy Robber Frog

1955 ***Eleutherodactylus tubernasus***
Anura - Leptodactylidae
e Pregonero Robber Frog

1956 ***Eleutherodactylus turquinensis***
Anura - Leptodactylidae
e Turquino Robber Frog

1957 ***Eleutherodactylus turumiquirensis***
Anura - Leptodactylidae

e Turumiquire Robber Frog

1958 ***Eleutherodactylus unicolor***
Anura - Leptodactylidae
e Dull Robber Frog; Burrow Coqui (WI)

1959 ***Eleutherodactylus unistrigatus***
Anura - Leptodactylidae
e Striped Robber Frog

1960 ***Eleutherodactylus uno***
Anura - Leptodactylidae
e Savage's Robber Frog

1961 ***Eleutherodactylus urichi***
Anura - Leptodactylidae
e Lesser Antilles Robber Frog

1962 ***Eleutherodactylus vanadise***
Anura - Leptodactylidae
e La Marca's Robber Frog

1963 ***Eleutherodactylus variabilis***
Anura - Leptodactylidae
e Variable Robber Frog

1964 ***Eleutherodactylus varians***
Anura - Leptodactylidae
e Cliff Robber Frog

1965 ***Eleutherodactylus varleyi***
Anura - Leptodactylidae
e Varley's Robber Frog

1966 ***Eleutherodactylus venancioi***
Anura - Leptodactylidae
e Venancio's Robber Frog

1967 ***Eleutherodactylus ventrilineatus***
Anura - Leptodactylidae
e La Hotte Robber Frog

1968 ***Eleutherodactylus ventrimarmoratus***
Anura - Leptodactylidae
e Tungurahua Robber Frog

1969 ***Eleutherodactylus verecundus***
Anura - Leptodactylidae
e Zacualtipan Robber Frog

1970 ***Eleutherodactylus verrucipes***
Anura - Leptodactylidae
e Big-eared Chirping Frog

1971 ***Eleutherodactylus verruculatus***
Anura - Leptodactylidae
e Least Chirping Frog

1972 ***Eleutherodactylus versicolor***
Anura - Leptodactylidae
e Loja Robber Frog

1973 ***Eleutherodactylus vertebralis***
Anura - Leptodactylidae
e Vertebral Robber Frog

1974 ***Eleutherodactylus vicarius***
Anura - Leptodactylidae
e Cononuco Robber Frog

1975 ***Eleutherodactylus vidua***
Anura - Leptodactylidae
e Mountain Crest Robber Frog

1976 ***Eleutherodactylus villarsi***
Anura - Leptodactylidae
e Rio Uaupes Robber Frog

1977 ***Eleutherodactylus vinhai***
Anura - Leptodactylidae
e Vinha Robber Frog

1978 ***Eleutherodactylus viridicans***
Anura - Leptodactylidae
e Cerro Munchique Robber Frog

1979 ***Eleutherodactylus vocalis***
Anura - Leptodactylidae
e Taylor's Stream Frog

1980 ***Eleutherodactylus vocator***
Anura - Leptodactylidae
e Agua Buena Robber Frog

1981 ***Eleutherodactylus walkeri***
Anura - Leptodactylidae
e Walker's Robber Frog

1982 ***Eleutherodactylus warreni***
Anura - Leptodactylidae
e Warren's Robber Frog

1983 ***Eleutherodactylus weinlandi***
Anura - Leptodactylidae
e Weinland's Robber Frog

1984 ***Eleutherodactylus wetmorei***
Anura - Leptodactylidae
e Wetmore's Robber Frog

1985 ***Eleutherodactylus wiensi***
Anura - Leptodactylidae
e Wiens's Robber Frog

1986 ***Eleutherodactylus wightmanae***
Anura - Leptodactylidae
e Wightman's Robber Frog; Wrinkled Coqui (WI)

1987 ***Eleutherodactylus w-nigrum***
Anura - Leptodactylidae
e Zurucuchu Robber Frog

1988 ***Eleutherodactylus xucanebi***
Anura - Leptodactylidae
e Xucaneb Robber Frog

1989 ***Eleutherodactylus yucatenensis***
Anura - Leptodactylidae
e Yucatan Rain Frog

1990 ***Eleutherodactylus zeuctotylus***
Anura - Leptodactylidae
e Sipalwini Robber Frog
f Eleutherodactyle zeucotyle

1991 ***Eleutherodactylus zeus***
Anura - Leptodactylidae
e Zeus Robber Frog

1992 ***Eleutherodactylus zimmermanae***
Anura - Leptodactylidae
e Zimmerman's Robber Frog

1993 ***Eleutherodactylus zugi***
Anura - Leptodactylidae
e Zug's Robber Frog

1994 ***Eleutherodactylus zygodacylus***
Anura - Leptodactylidae
e Danubio Robber Frog

1995 ***Ensatina***
Caudata - Plethodontidae
e Ensatinas; Eschscholtz's Salamanders
d Eschscholtz-Salamander (pl)

1996 ***Ensatina eschscholtzii***
Caudata - Plethodontidae
e Ensatina; Large-blotched Salamander; Eschscholtz's Salamander
d Eschscholtz-Salamander; Gelbaugensalamander
f Salamandre rouge; Salamandre rouge à larges taches; Salamandre rouge de Monterey; Salamandre rouge variable

1997 ***Ensatina eschscholtzii croceater***
Caudata - Plethodontidae
e Yellow-blotched Ensatina; Yellow-blotched Salamander

1998 ***Ensatina eschscholtzii eschscholtzii***
Caudata - Plethodontidae
e Monterey Ensatina; Monterey Salamander

1999 ***Ensatina eschscholtzii klauberi***
Caudata - Plethodontidae
e Large-blotched Salamander

2000 ***Ensatina eschscholtzii oregonensis***
Caudata - Plethodontidae
e Oregon Ensatina; Oregon Salamander
d Oregon-Salamander; Brauner Oregon-Salamander

2001 ***Ensatina eschscholtzii picta***
Caudata - Plethodontidae
e Painted Ensatina; Painted Salamander

2002 ***Ensatina eschscholtzii platensis***
Caudata - Plethodontidae
e Sierra Nevada Ensatina; Sierra Nevada Salamander

2003 ***Ensatina eschscholtzii xanthoptica***
Caudata - Plethodontidae
e Yellow-eyed Ensatina; Yellow-eyed Salamander

2004 ***Epicrionops***
Gymnophiona - Rhinatrematidae
e Beaked Caecilians

2005 ***Epicrionops bicolor***
Gymnophiona - Rhinatrematidae
e Two-coloured Caecilian

2006 ***Epicrionops columbianus***
Gymnophiona - Rhinatrematidae
e El Tambo Caecilian

2007 ***Epicrionops lativittatus***
Gymnophiona - Rhinatrematidae
e Eastern Caecilian

2008 ***Epicrionops marmoratus***
Gymnophiona - Rhinatrematidae
e Marbled Caecilian

2009 ***Epicrionops niger***
Gymnophiona - Rhinatrematidae
e Black Caecilian

2010 ***Epicrionops parkeri***
Gymnophiona - Rhinatrematidae
e Parker's Caecilian

2011 ***Epicrionops peruvianus***
Gymnophiona - Rhinatrematidae
e Marcapata Calley Caecilian

2012 ***Epicrionops petersi***
Gymnophiona - Rhinatrematidae
e Peters's Caecilian

2013 ***Epipedobates***
Anura - Dendrobatidae
e Phantasmal Poison Frogs; Phantasmal Poison-arrow Frogs
d Baumsteiger
f Epipedobates

2014 ***Epipedobates andinus***
Anura - Dendrobatidae
e La Planada Poison Frog; La Planada Poison-arrow Frog

2015 ***Epipedobates azureiventris***
Anura - Dendrobatidae
e Sky-blue Poison Frog; Sky-blue Poison-arrow Frog
d Blaubäuchiger Blattsteiger

2016 ***Epipedobates bassleri***
Anura - Dendrobatidae
e Pleasing Poison Frog; Pleasing Poison-arrow Frog

2017 ***Epipedobates billinguis***
Anura - Dendrobatidae
e Ecuador Poison Frog; Ecuador Poison-arrow Frog

2018 ***Epipedobates bolivianus***
Anura - Dendrobatidae
e Bolivian Poison Frog; Bolivian Poison-arrow Frog

2019 ***Epipedobates boulengeri***
Anura - Dendrobatidae
e Marbled Poison Frog; Marbled Poison-arrow Frog
d Marmorierter Baumsteiger

2020 ***Epipedobates cainarachi***
Anura - Dendrobatidae
e Cainarachi Poison Frog; Cainarachi Poison-arrow Frog

2021 ***Epipedobates erythromos***
Anura - Dendrobatidae
e Palenque Poison Frog; Palenque Poison-arrow Frog

2022 ***Epipedobates espinosai***
Anura - Dendrobatidae
e Espinosa Poison Frog; Espinosa Poison-arrow Frog; Turquoise-bellied Poison-arrow Frog

2023 ***Epipedobates femoralis***
Anura - Dendrobatidae
e Brilliant-thighed Poison Frog; Brilliant-thighed Poison-arrow Frog
f Epipedobate fémoral

2024 ***Epipedobates flavopictus***
Anura - Dendrobatidae
e Lutz's Poison Frog; Lutz's Poison-arrow Frog

2025 ***Epipedobates ingeri***
Anura - Dendrobatidae
e Nicéforo's Poison Frog; Nicéforo's Poison-arrow Frog; Brother Nicéforo's Poison-arrow Frog

2026 ***Epipedobates macero***
Anura - Dendrobatidae
e Manu Poison Frog; Manu Poison-arrow Frog

2027 ***Epipedobates maculatus***
Anura - Dendrobatidae
e Confusing Poison Frog; Confusing Poison-arrow Frog

2028 ***Epipedobates myersi***
Anura - Dendrobatidae
e Myers's Poison Frog; Myers's Poison-arrow Frog

2029 ***Epipedobates parvulus***
Anura - Dendrobatidae
e Ruby Poison Frog; Ruby Poison-arrow Frog
d Rubinroter Baumsteiger

2030 ***Epipedobates petersi***
Anura - Dendrobatidae
e Peru Poison Frog; Peru Poison-arrow Frog; Peruvian Poison-arrow Frog

2031 ***Epipedobates pictus***
Anura - Dendrobatidae
e Spot-legged Poison Frog; Spot-legged Poison-arrow Frog
d Gemalter Baumsteiger
f Dendrobate jaune; Dendrobate à taches jaunes

2032 ***Epipedobates pulchripectus***
Anura - Dendrobatidae
e Blue-breasted Poison Frog; Blue-breasted Poison-arrow Frog; Beautiful-breasted Poison-arrow Frog

2033 ***Epipedobates silverstonei***
Anura - Dendrobatidae
e Silverstone's Poison Frog; Silverstone's Poison-arrow Frog
d Roter Prachtgiftfrosch; Roter Baumsteiger

2034 ***Epipedobates smaragdinus***
Anura - Dendrobatidae
e Emerald Poison Frog; Emerald Poison-arrow Frog

2035 ***Epipedobates tricolor***
Anura - Dendrobatidae
e Phantasmal Poison Frog; Phantasmal Poison-arrow Frog
d Dreifarbiger Giftfrosch; Dreistreifen-Baumsteiger; Dreifarbiger Baumsteiger

2036 ***Epipedobates trivittatus***
Anura - Dendrobatidae
e Three-striped Poison Frog; Three-striped Poison-arrow Frog
d Dreistreifiger Baumsteiger; Grüner Riesengiftfrosch; Dreigestreifter Baumsteiger
f Dendrobate à trois raies

2037 ***Epipedobates zaparo***
Anura - Dendrobatidae
e Zaparo's Poison Frog; Zaparo's Poison-arrow Frog; Sanguine Poison-arrow Frog

2038 ***Ericabatrachus***
Anura - Ranidae
e Bale Mountains Frogs

2039 ***Ericabatrachus baleensis***
Anura - Ranidae
e Bale Mountains Frog

2040 ***Euparkerella***
Anura - Leptodactylidae
e Guanabara Frogs

2041 ***Euparkerella brasiliensis***
Anura - Leptodactylidae
e Brazilian Guanabara Frog

2042 ***Euparkerella cochranae***
Anura - Leptodactylidae
e Cochran's Guanabara Frog

2043 ***Euparkerella robusta***
Anura - Leptodactylidae
e Izecksohn's Guanabara Frog

2044 ***Euparkerella tridactyla***
Anura - Leptodactylidae
e Three-toed Guanabara Frog

2045 ***Euphlyctis***
Anura - Ranidae
e Five-fingered Frogs; Spine-footed Frogs (SEA)
d Spornfußfrösche

2046 ***Euphlyctis cornii***
Anura - Ranidae
e Eritrea Five-fingered Frog; Spotted Spine-footed Frog (SEA)
d Gefleckter Spornfußfrosch

2047 ***Euphlyctis cyanophlyctis***
Anura - Ranidae
e Cyan Five-fingered Frog; Skipper Frog; Skipping Frog; Skittering Frog (ISC)
d Indischer Wasserfrosch

2048 ***Euphlyctis ehrenbergi***
Anura - Ranidae
e Arabian Five-fingered Frog

2049 ***Euphlyctis hexadactylus***
Anura - Ranidae
e Indian Five-fingered Frog; Indian Pond Frog (ISC); Indian Green Frog
d Sechszehenfrosch

2050 ***Euproctus***
Caudata - Salamandridae
e European Mountain Salamanders
d Gebirgsmolche; Europäische Gebirgsmolche
f Euproctes; Tritons

i Euprotti

2051 ***Euproctus asper***
Caudata - Salamandridae
e Pyrennes Mountain Salamander; Pyrenean Mountain Newt; Pyrenean Brook Salamander
d Pyrenäen-Gebirgsmolch; Pyrenäen-Molch
f Euprocte pyrénéen; Triton pyrénéen; Triton des Pyrénées; Euprocte des Pyrénées
i Euprotto pirenaico; Tritone pirenaico

2052 ***Euproctus montanus***
Caudata - Salamandridae
e Corsican Mountain Salamander; Corsican Brook Salamander; Corsican Brook Newt; Corsican Mountain Newt
d Korsischer Gebirgsmolch
f Euprocte corse; Euprocte de Corse; Triton de Corse; Triton de montagne de Corse
i Euprotto corso; Tritone corso

2053 ***Euproctus platycephalus***
Caudata - Salamandridae
e Sardinia Mountain Salamander; Sardinian Mountain Newt; Sardinian Brook Salamander; Sardinian Brook Newt
d Hechtkopftriton; Hechtkopfgebirgsmolch; Sardischer Gebirgsmolch; Sardinischer Gebirgsmolch
f Euprocte sarde; Euprocte de Sardaigne; Triton de Sardaigne; Triton de montagne de Sardaigne
i Tritone sardo; Euprotto sardo

2054 ***Eupsophus***
Anura - Leptodactylidae
e Ground Frogs

2055 ***Eupsophus calcaratus***
Anura - Leptodactylidae
e Chiloe Island Ground Frog

2056 ***Eupsophus contulmoensis***
Anura - Leptodactylidae
e Contulmo Ground Frog

2057 ***Eupsophus emiliopugini***
Anura - Leptodactylidae
e Emilio's Ground Frog

2058 ***Eupsophus insularis***
Anura - Leptodactylidae
e Mocha Island Ground Frog

2059 ***Eupsophus migueli***
Anura - Leptodactylidae
e Miguel's Ground Frog

2060 ***Eupsophus nahuelbutensis***
Anura - Leptodactylidae
e Nahuelbuta Ground Frog

2061 ***Eupsophus roseus***
Anura - Leptodactylidae
e Rosy Ground Frog

2062 ***Eupsophus vertebralis***
Anura - Leptodactylidae
e Valdivia Ground Frog

2063 ***Eurycea***
Caudata - Plethodontidae
e American Brook Salamanders; Brook Salamanders (NA)
d Gelbsalamander (pl)
f Salamandres de ruisseau

2064 ***Eurycea aquatica***
Caudata - Plethodontidae
e Jefferson County Salamander; Dark-sided Salamander

2065 ***Eurycea bislineata***
Caudata - Plethodontidae

e Two-lined Salamander
d Zweistreifiger Gelbsalamander; Zweistreifensalamander; Zweistreifenmolch
f Salamandre à deux lignes

2066 ***Eurycea bislineata bisleneata***
Caudata - Plethodontidae
e Northern Two-lined Salamander
d Nördlicher Zweistreifiger Gelbsalamander
f Salamandre à deux lignes du Nord

2067 ***Eurycea bislineata rivicola***
Caudata - Plethodontidae
e Midwest Two-lined Salamander

2068 ***Eurycea bislineata wilderae***
Caudata - Plethodontidae
e Blue Ridge Two-lined Salamander

2069 ***Eurycea chisholmensis***
Caudata - Plethodontidae
e Chisholm Trail Salamander

2070 ***Eurycea cirrigera***
Caudata - Plethodontidae
e Southern Two-lined Salamander

2071 ***Eurycea guttolineata***
Caudata - Plethodontidae
e Three-lined Salamander

2072 ***Eurycea junaluska***
Caudata - Plethodontidae
e Junaluska Salamander
d Nord-Karolina-Gelbsalamander

2073 ***Eurycea longicauda***
Caudata - Plethodontidae
e Long-tailed Salamander; Longtail Salamander
d Gelbsalamander; Langschwänziger Gelbsalamander; Langschwanzsalamander; Langschwanzmolch
f Salamandre à longue queue

2074 ***Eurycea longicauda longicauda***
Caudata - Plethodontidae
e Longtail Salamander; Long-tailed Salamander
d Gelbsalamander; Langschwänziger Gelbsalamander; Langschwanzsalamander; Langschwanzmolch
f Salamandre à longue queue

2075 ***Eurycea longicauda melanopleura***
Caudata - Plethodontidae
e Dark-sided Salamander

2076 ***Eurycea lucifuga***
Caudata - Plethodontidae
e Cave Salamander
d Höhlengelbsalamander; Roter Höhlensalamander; Höhlensalamander
f Salamandre des cavernes

2077 ***Eurycea multiplicata***
Caudata - Plethodontidae
e Many-ribbed Salamander
d Rippengelbsalamander

2078 ***Eurycea multiplicata griseogaster***
Caudata - Plethodontidae
e Grey-bellied Salamader; Greybelly Salamander

2079 ***Eurycea multiplicata multiplicata***
Caudata - Plethodontidae
e Many-ribbed Salamander
d Rippengelbsalamander

2080 ***Eurycea nana***
Caudata - Plethodontidae
e San Marcos Salamander

2081 ***Eurycea naufragia***
Caudata - Plethodontidae
e Georgetown Salamander

2082 ***Eurycea neotene neotenes***
Caudata - Plethodontidae
e Bexar County Salamander

2083 ***Eurycea neotenes***
Caudata - Plethodontidae
e Texas Salamander
d Texas-Gelbsalamander; Texas-Höhlensalamander

2084 ***Eurycea neotenes latitans***
Caudata - Plethodontidae
e Cascade Cavern Salamander

2085 ***Eurycea pterophila***
Caudata - Plethodontidae
e Fern-bank Salamander

2086 ***Eurycea quadridigitata***
Caudata - Plethodontidae
e Dwarf Salamander
d Vierzehen-Gelbsalamander
f Salamandre naine

2087 ***Eurycea rathbuni***
Caudata - Plethodontidae
e Texas Blind Salamander
d Rathbuns Brunnenmolch
f Salamandre aveugle de Texas

2088 ***Eurycea sosorum***
Caudata - Plethodontidae
e Barton Springs Salamander

2089 ***Eurycea tonkawae***
Caudata - Plethodontidae
e Jollyville Plateau Salamander

2090 ***Eurycea tridentifera***
Caudata - Plethodontidae
e Comal Blind Salamander

2091 ***Eurycea troglodytes***
Caudata - Plethodontidae
e Valdina Farms Salamander

2092 ***Eurycea tynerensis***
Caudata - Plethodontidae
e Oklahoma Salamander
d Oklahoma-Gelbsalamander; Oklahoma-Salamander; Flachstirn-Gelbsalamander

2093 ***Eurycea wilderae***
Caudata - Plethodontidae
e Blue-Ridge Two-lined Salamander

F

2094 ***Flectonotus***
Anura - Hylidae
e Serra de Macae Tree Frogs

2095 ***Flectonotus fissilis***
Anura - Hylidae
e Serra de Macae Tree Frog; Split-backed Frog

2096 ***Flectonotus fitzgeraldi***
Anura - Hylidae
e Mount Tucutche Tree Frog

2097 ***Flectonotus goeldi***
Anura - Hylidae
e Colonia Tree Frog; Goeldi's Frog
d Schüsselrückenlaubfrosch; Bromelienlaubfrosch

2098 ***Flectonotus ohausi***
Anura - Hylidae
e Petropolis Tree Frog

2099 ***Flectonotus pygmaeus***
Anura - Hylidae
e Puerto Cabello Tree Frog
d Zwergbeutelfrosch

2100 ***Frostius***
Anura - Bufonidae
e Frost's Toads

2101 ***Frostius pernambucensis***
Anura - Bufonidae
e Frost's Toad

G

2102 ***Gastrophryne***
Anura - Microhylidae
e American Narrowmouth Toads; Narrowmouth Toads (NA); Narrow-mouthed Toads; North American Narrow-mouthed Frogs
d Engmaulfrösche; Engmundfrösche

2103 ***Gastrophryne carolinensis***
Anura - Microhylidae
e Eastern Narrow-mouthed Toad
d Karolina-Microhyla; Karolina-Engmaulfrosch; Amerikanischer Engmaulfrosch; Karolina-Engmundfrosch; Amerikanischer Engmundfrosch; Carolina-Engmundfrosch

2104 ***Gastrophryne elegans***
Anura - Microhylidae
e Elegant Narrow-mouthed Toad; Elegant Narrowmouth Toad

2105 ***Gastrophryne olivacea***
Anura - Microhylidae
e Great Plains Narrow-mouthed Toad (NA); Great Plains Narrowmouth Toad; Western Narrowmouth Toad; Western Narrow-mouthed Toad

2106 ***Gastrophryne olivacea mazatlaensis***
Anura - Microhylidae
e Sinoloan Narrow-mouthed Toad; Mazatlan Narrowmouth Toad

2107 ***Gastrophryne olivacea olivacea***
Anura - Microhylidae
e Plains Narrowmouth Toad; Great Plains Narrowmouth Toad

2108 ***Gastrophryne pictiventris***
Anura - Microhylidae
e Southern Narrow-mouthed Toad; Nicaragua Narrowmouth Toad

2109 ***Gastrophryne usta***
Anura - Microhylidae
e Two-spaded Narrow-mouthed Toad; Two-spaded Narrowmouth Toad

2110 ***Gastrophrynoides***
Anura - Microhylidae
e Borneo Narrowmouth Toads

2111 ***Gastrophrynoides borneensis***
Anura - Microhylidae
e Borneo Narrowmouth Toad

2112 ***Gastrotheca***
Anura - Hylidae
e Marsupial Frogs; Pouched Frogs
d Beutelfrösche; Taschenfrösche
f Rainettes marsupiales; Rainettes à bourse

2113 ***Gastrotheca abdita***
Anura - Hylidae
e Cordillera Colan Marsupial Frog; Cordillera Colan Pouched Frog

2114 ***Gastrotheca andaquiensis***
Anura - Hylidae
e Andes Marsupial Frog; Andes Pouched Frog

2115 ***Gastrotheca angustifrons***
Anura - Hylidae
e Pacific Marsupial Frog; Pacific Pouched Frog

2116 ***Gastrotheca argenteovirens***
Anura - Hylidae
e Popayan Marsupial Frog; Popayan Pouched Frog

2117 ***Gastrotheca aureomaculata***
Anura - Hylidae
e Gold-spotted Marsupial Frog; Gold-spotted Pouched Frog

2118 ***Gastrotheca bufona***
Anura - Hylidae
e Antioquia Marsupial Frog; Antioquia Pouched Frog

2119 ***Gastrotheca christiani***
Anura - Hylidae
e Christian's Marsupial Frog; Christian's Pouched Frog

2120 ***Gastrotheca chrysosticta***
Anura - Hylidae
e Salta Marsupial Frog; Salta Pouched Frog

2121 ***Gastrotheca cornuta***
Anura - Hylidae
e Horned Marsupial Frog; Horned Pouched Frog
d Gehörnter Beutelfrosch

2122 ***Gastrotheca dendronastes***
Anura - Hylidae
e Rio Calima Marsupial Frog; Rio Calima Pouched Frog

2123 ***Gastrotheca dunni***
Anura - Hylidae
e Dunn's Marsupial Frog; Dunn's Pouched Frog

2124 ***Gastrotheca espleletia***
Anura - Hylidae
e North Shore Marsupial Frog; North Shore Pouched Frog

2125 ***Gastrotheca excubitor***
Anura - Hylidae
e Abracanacu Marsupial Frog; Abracanacu Pouched Frog

2126 ***Gastrotheca fissipes***
Anura - Hylidae
e Igaracu Marsupial Frog; Igaracu Pouched Frog

2127 ***Gastrotheca galeata***
Anura - Hylidae
e Helmeted Marsupial Frog; Helmeted Pouched Frog

2128 ***Gastrotheca gracilis***
Anura - Hylidae
e La Banderita Marsupial Frog; La Banderita Pouched Frog

2129 ***Gastrotheca griswoldi***
Anura - Hylidae
e Griswold's Marsupial Frog; Griswold's Pouched Frog

2130 ***Gastrotheca guentheri***
Anura - Hylidae
e Günther's Marsupial Frog; Günther's Pouched Frog

2131 ***Gastrotheca helenae***
Anura - Hylidae
e Helena's Marsupial Frog; Helena's Pouched Frog

2132 ***Gastrotheca lateonota***
Anura - Hylidae
e El Tambo Marsupial Frog; El Tambo Pouched Frog

2133 ***Gastrotheca lauzirica***
Anura - Hylidae
e La Siberia Marsupial Frog; La Siberia Pouched Frog

2134 ***Gastrotheca litonedis***
Anura - Hylidae
e Azuay Marsupial Frog; Azuay Pouched Frog

2135 ***Gastrotheca longipes***
Anura - Hylidae
e Pastaza Marsupial Frog; Pastaza Pouched Frog

2136 ***Gastrotheca marsupiata***
Anura - Hylidae
e Common Marsupial Frog; Common Pouched Frog; Marsupial Frog; Pouched Frog
d Beutelfrosch; Taschenfrosch
f Rainette marsupiale; Rainette à bourse

2137 ***Gastrotheca microdiscus***
Anura - Hylidae
e Tiny fingered Marsupial Frog; Tiny-fingered Pouched Frog

2138 ***Gastrotheca monticola***
Anura - Hylidae
e Mountain Marsupial Frog; Mountain Pouched Frog

2139 ***Gastrotheca nicefori***
Anura - Hylidae
e Nicéforo's Marsupial Frog; Nicéforo's Pouched Frog

2140 ***Gastrotheca ochoai***
Anura - Hylidae
e Chilca Marsupial Frog; Chilca Pouched Frog

2141 ***Gastrotheca orophylax***
Anura - Hylidae
e Papallacta Marsupial Frog; Papallacta Pouched Frog

2142 ***Gastrotheca ovifera***
Anura - Hylidae
e Giant Marsupial Frog; Giant Pouched Frog
d Eierträger; Riesen-Beutelfrosch

2143 ***Gastrotheca pacchamama***
Anura - Hylidae
e Ayacucho Marsupial Frog ; Ayacucho Pouched Frog

2144 ***Gastrotheca peruana***
Anura - Hylidae
e Peru Marsupial Frog ; Peru Pouched Frog

2145 ***Gastrotheca plumbea***
Anura - Hylidae
e Silver Marsupial Frog; Silver Pouched Frog
d Bleifarbener Taschenfrosch

2146 ***Gastrotheca pseustes***
Anura - Hylidae
e San Lucas Marsupial Frog; San Lucas Pouched Frog

2147 ***Gastrotheca psychrophila***
Anura - Hylidae
e Ridge Marsupial Frog; Ridge Pouched Frog

2148 ***Gastrotheca rebeccae***
Anura - Hylidae
e Rebecca's Marsupial Frog; Rebecca's Pouched Frog

2149 ***Gastrotheca riobambae***
Anura - Hylidae
e Riobamba Marsupial Frog; Riobamba Pouched Frog
d Riobamba-Beutelfrosch

2150 ***Gastrotheca ruizi***
Anura - Hylidae
e Ruiz's Marsupial Frog; Ruiz's Pouched Frog

2151 ***Gastrotheca splendens***
Anura - Hylidae
e Schmidt's Marsupial Frog; Schmidt's Pouched Frog

2152 ***Gastrotheca testudinea***
Anura - Hylidae
e Espada's Marsupial Frog; Espada's Pouched Frog
d Schildkrötenfrosch

2153 ***Gastrotheca trachyceps***
Anura - Hylidae
e Cerro Munchique Marsupial Frog; Cerro Munchique Pouched Frog

2154 ***Gastrotheca walkeri***
Anura - Hylidae
e Walker's Marsupial Frog; Walker's Pouched Frog

2155 ***Gastrotheca weinlandi***
Anura - Hylidae
e Weinland's Marsupial Frog; Weinland's Pouched Frog

2156 ***Gastrotheca williamsoni***
Anura - Hylidae
e Williamson's Marsupial Frog; Williamson's Pouched Frog

2157 ***Gegeneophis***
Gymnophiona - Caeciliaidae
e Forest Caecilians

2158 ***Gegeneophis carnosus***
Gymnophiona - Caeciliaidae
e Periah Peak Caecilian; Blind Caecilian (ISC)

2159 ***Gegeneophis fulleri***
Gymnophiona - Caeciliaidae
e Kittal Ceacilian

2160 ***Gegeneophis ramaswamii***
Gymnophiona - Caeciliaidae
e Tenmalai Forest Caecilian

2161 ***Genyophryne***
Anura - Microhylidae
e Toothless Frogs

2162 ***Genyophryne thomsoni***
Anura - Microhylidae
e Toothless Frog; Thomson's Toothless Frog

2163 ***Geobatrachus***
Anura - Leptodactylidae
e Sierra Frogs

2164 ***Geobatrachus walkeri***
Anura - Leptodactylidae
e Sierra Frog; Walker's Sierra Frog

2165 ***Geocrinia***
Anura - Myobatrachidae
e Ground Froglets

2166 ***Geocrinia alba***
Anura - Myobatrachidae
e White Ground Froglet

2167 ***Geocrinia laevis***
Anura - Myobatrachidae
e Günther's Ground Froglet

2168 ***Geocrinia leai***
Anura - Myobatrachidae
e Fletcher's Ground Froglet

2169 ***Geocrinia lutea***
Anura - Myobatrachidae
e Main's Ground Froglet

2170 ***Geocrinia rosea***
Anura - Myobatrachidae
e Red Ground Froglet
d Rose-Zirpfrosch

2171 ***Geocrinia victoriana***
Anura - Myobatrachidae
e Victoria Ground Froglet

2172 ***Geocrinia vitellina***
Anura - Myobatrachidae

e Australian Ground Froglet

2173 ***Geotrypetes***
Gymnophiona - Caeciliaidae
e West African Caecilians
d Erdwühlen

2174 ***Geotrypetes angeli (Parker, 1936)***
Gymnophiona - Caeciliaidae
e Angel's Caecilian

2175 ***Geotrypetes angeli (Taylor, 1968)***
Gymnophiona - Caeciliaidae
e False Angel's Caecilian

2176 ***Geotrypetes seraphini***
Gymnophiona - Caeciliaidae
e Gaboon Caecilian; Gabon Caecilian
d Kamerun-Erdwühle
f Cécilie de Séraphin

2177 ***Glyphoglossus***
Anura - Microhylidae
e Balloon Frogs

2178 ***Glyphoglossus molossus***
Anura - Microhylidae
e Balloon Frog; Short-headed Frog (SEA)
d Stummelschnauzenfrosch

2179 ***Grandisonia***
Gymnophiona - Caeciliaidae
e Seychelles Caecilians

2180 ***Grandisonia alterenans***
Gymnophiona - Caeciliaidae
e Stejneger's Caecilian

2181 ***Grandisonia brevis***
Gymnophiona - Caeciliaidae
e Mahe Caecilian

2182 ***Grandisonia diminutiva***
Gymnophiona - Caeciliaidae
e Praslin Caecilian

2183 ***Grandisonia larvata***
Gymnophiona - Caeciliaidae
e Indian Ocean Caecilian

2184 ***Grandisonia sechellensis***
Gymnophiona - Caeciliaidae
e Seychelles Caecilian

2185 ***Gymnopis***
Gymnophiona - Caeciliaidae
e Wet Forest Caecilians

2186 ***Gymnopis multiplicata***
Gymnophiona - Caeciliaidae
e Varagua Caecilian

2187 ***Gymnopis syntremus***
Gymnophiona - Caeciliaidae
e Wet Forest Caecilian

2188 ***Gyrinophilus***
Caudata - Plethodontidae
e Cave Salamanders; Spring Salamanders
d Quellensalamander (pl); Porphyrsalamander (pl)
f Salamandres pourprées; Salamandres des sources; Salamandres pourprées des sources

2189 ***Gyrinophilus gulolineatus***
Caudata - Plethodontidae
e Berry Cave Salamander

2190 ***Gyrinophilus palleucus***
Caudata - Plethodontidae
e Tennessee Cave Salamander
d Tennessee-Höhlensalamander

2191 ***Gyrinophilus palleucus necturoides***
Caudata - Plethodontidae
e Big Mouth Cave Salamander

2192 ***Gyrinophilus palleucus palleucus***
Caudata - Plethodontidae
e Pale Salamander

2193 ***Gyrinophilus porphyriticus***
Caudata - Plethodontidae
e Spring Salamander; Purple Salamander (NA); Mountain Spring Salamander
d Porphyrsalamander; Purpursalamander
f Salamandre pourprée des sources; Salamandre pourprée; Salamandre pourpre (Qué)

2194 ***Gyrinophilus porphyriticus danielsi***
Caudata - Plethodontidae
e Blue Ridge Spring Salamander

2195 ***Gyrinophilus porphyriticus dunni***
Caudata - Plethodontidae
e Carolina Spring Salamander

2196 ***Gyrinophilus porphyriticus duryi***
Caudata - Plethodontidae
e Kentucky Spring Salamander

2197 ***Gyrinophilus porphyriticus porphyriticus***
Caudata - Plethodontidae
e Northern Spring Salamander

2198 ***Gyrinophilus subterraneus***
Caudata - Plethodontidae
e West Virginia Spring Salamander

H

2199 ***Haideotriton***
Caudata - Plethodontidae
e Blind Salamanders; Georgia Blind Salamanders
d Blindsalamander (pl)
f Salamandre aveugles de Georgie

2200 ***Haideotriton wallacei***
Caudata - Plethodontidae
e Blind Salamander; Georgia Blind Salamander
d Blindsalamander
f Salamandre aveugle de Georgie

2201 ***Hamptophryne***
Anura - Microhylidae
e Bleating Frogs
f Hamptophrynes

2202 ***Hamptophryne boliviana***
Anura - Microhylidae
e Bleating Frog; Bolivian Bleating Frog
f Hamptophryne bolivien

2203 ***Heleioporus***
Anura - Myobatrachidae
e Giant Burrowing Frogs
d Grabfrösche

2204 ***Heleioporus albopunctatus***
Anura - Myobatrachidae
e White-spotted Burrowing Frog

2205 ***Heleioporus australiacus***
Anura - Myobatrachidae
e Giant Burrowing Frog
d Südlicher Grabfrosch

2206 ***Heleioporus barycragus***
Anura - Myobatrachidae
e Yellow-wart Burrowing Frog

2207 ***Heleioporus eyrei***
Anura - Myobatrachidae
e Moaning Frog

2208 ***Heleioporus inornatus***
Anura - Myobatrachidae
e Swamp Burrowing Frog

2209 ***Heleioporus psamophilus***
Anura - Myobatrachidae
e Scarp Burrowing Frog

2210 ***Heleophryne***
Anura - Heleophrynidae
e Ghost Frogs; Torrent Frogs (CSA); Cape Platannas (CSA)
d Gespenstfrösche
f Grenouilles spectres

2211 ***Heleophryne hewitti***
Anura - Heleophrynidae
e Hewitt's African Ghost Frog; Hewitt's Ghost Frog (CSA)

2212 ***Heleophryne natalensis***
Anura - Heleophrynidae
e Natal Ghost Frog (CSA); Ghost Frog; Heleo Frog; Southeastern Ghost Frog
d Natal-Gespenstfrosch

2213 ***Heleophryne purcelli***
Anura - Heleophrynidae
e Purcell's African Ghost Frog; Cape Ghost Frog (CSA); Ghost Frog (CSA)
d Gespenstfrosch; Purcell's Gespenstfrosch
f Grenouille spectre

2214 ***Heleophryne purcelli depressa***
Anura - Heleophrynidae
e Hex River Torrent Frog; Hex River Ghost Frog (CSA)

2215 ***Heleophryne regis***
Anura - Heleophrynidae
e Royal Ghost Frog; Southern Ghost Frog (CSA)

2216 ***Heleophryne rosei***
Anura - Heleophrynidae
e Skeleton Gorge Ghost Frog; Thumbed Ghost Frog (CSA); Table Mountain Ghost Frog (CSA); Rose's Ghost Frog
d Roses Gespenstfrosch

2217 **Heleophrynidae**
Anura
e Ghost Frogs
d Gespenstfrösche; Heleophryniden
f Grenouilles spectres

2218 ***Hemidactylium***
Caudata - Plethodontidae
e Four-toed Salamanders
d Vierzehensalamander (pl)
f Salamandres à quatre doigts

2219 ***Hemidactylium scutatum***
Caudata - Plethodontidae
e Four-toed Salamander
d Vierzehensalamander
f Salamandre à quatre doigts; Salamandre à quatre orteils

2220 ***Hemiphractus***
Anura - Hylidae
e Horned Frogs
d Helmkopffrösche

2221 ***Hemiphractus bubalus***
Anura - Hylidae
e Ecuador Horned Frog
d Palmenfrosch

2222 ***Hemiphractus fasciatus***
Anura - Hylidae
e BandedHorned Frog

2223 ***Hemiphractus johnsoni***
Anura - Hylidae
e Johnson's Horned Frog

2224 ***Hemiphractus proboscideus***
Anura - Hylidae
e Sumaco Horned Frog

2225 ***Hemiphractus scutatus***
Anura - Hylidae
e Spix's Horned Frog

2226 **Hemisotidae**
Anura
e Shovelnose Frogs; Shovel-snouted Frogs
d Ferkelfrösche; Schaufelnasenfrösche
f Grenouilles à museau de porc

2227 ***Hemisus***
Anura - Hemisotidae
e Shovelnose Frogs; Shovel-nosed Frogs; Shovel-snouted Frogs; Shovelsnout Frogs; Pig-nosed Frogs (CSA); Burrowing Frogs (CSA)
d Ferkelfrösche; Schaufelnasenfrösche
f Grenouilles à museau de porc; Hémises

2228 ***Hemisus brachydactylus***
Anura - Hemisotidae
e Masiliwa Shovelsnout Frog; Masiliwa Shovel-nosed Frog; Masiliwa Shovel-snouted Frog; Masiliwa Shovelnose Frog

2229 ***Hemisus guineensis***
Anura - Hemisotidae
e Guinea Shovelsnout Frog; Guinea Shovel-nosed Frog (CSA); Guinea

Shovel-snouted Frog (CSA); Guinea Shovelnose Frog

2230 ***Hemisus guttatus***
Anura - Hemisotidae
e Spotted Shovelnose Frog; Spotted Shovel-nosed Frog (CSA); Spotted Shovelsnout Frog; Spotted Shovel-snouted Frog; Spotted Pig-nosed Frog (CSA); Spotted Burrowing Frog (CSA)
d Punktierter Ferkelfrosch; Punktierter Schaufelnasenfrosch

2231 ***Hemisus marmoratum***
Anura - Hemisotidae
e Marbled Shovelnose Frog; Marbled Shovel-nosed Frog; Marbled Shovelsnout Frog; Marbled Shovel-snouted Frog; Marbled Shovelsnout (CSA); Mottled Shovel-nosed Frog (CSA); Mottled Burrowing Frog; Marbled Shovelnose (CSA); Mottled Shovel-snouted Frog; Marbled Pig-nosed Frog (CSA); Shovel-nosed Frog; Pig-nosed Frog (CSA)
d Marmorierter Ferkelfrosch; Marmorierter Schaufelnasenfrosch
f Grenouille à museau de porc; Hémise marmorée

2232 ***Hemisus microscaphus***
Anura - Hemisotidae
e Lake Zwai Shovelnose Frog; Lake Zwai Shovel-nosed Frog; Lake Zwai Shovelsnout Frog; Lake Zwai Shovel-snouted Frog

2233 ***Hemisus olivaceus***
Anura - Hemisotidae
e Olive Shovelnose Frog; Olive Shovel-nosed Frog; Olive Shovelsnout Frog; Olive Shovel-snouted Frog

2234 ***Hemisus perreti***
Anura - Hemisotidae
e Perret's Shovelnose Frog; Perret's Shovel-nosed Frog; Perret's Shovelsnout Frog; Perret's Shovel-snouted Frog

2235 ***Herpele***
Gymnophiona - Caeciliaidae
e Congo Caecilians

2236 ***Herpele multiplicata***
Gymnophiona - Caeciliaidae
e Victoria Caecilian

2237 ***Herpele squalostoma***
Gymnophiona - Caeciliaidae
e Congo Caecilian
f Cécilie bouche-de-squale

2238 ***Heterixalus***
Anura - Hyperoliidae
e Madagascar Reed Frogs

2239 ***Heterixalus alboguttatus***
Anura - Hyperoliidae
e Whitebelly Reed Frog

2240 ***Heterixalus andrakata***
Anura - Hyperoliidae
e Andrakata Reed Frog

2241 ***Heterixalus betsileo***
Anura - Hyperoliidae
e Betsileo Reed Frog

2242 ***Heterixalus boettgeri***
Anura - Hyperoliidae
e Boettger's Reed Frog

2243 ***Heterixalus madagascariensis***
Anura - Hyperoliidae
e Madagascar Reed Frog

2244 ***Heterixalus rutenbergi***
Anura - Hyperoliidae
e Rutenberg's Reed Frog

2245 ***Heterixalus tricolor***
Anura - Hyperoliidae
e Three-coloured Reed Frog

2246 ***Hildebrandtia***
Anura - Ranidae
e Ornate Frogs

2247 ***Hildebrandtia macrotympanum***
Anura - Ranidae
e Somali Ornate Frog

2248 ***Hildebrandtia ornata***
Anura - Ranidae
e African Ornate Frog; Ornate Frog (CSA); Ornate Burrowing Frog (CSA); Hildebrandt's Burrowing Frog (CSA)
d Pracht-Grabfrosch

2249 ***Hildebrandtia ornatissima***
Anura - Ranidae
e Angola Ornate Frog

2250 ***Holoaden***
Anura - Leptodactylidae
e Highland Frogs

2251 ***Holoaden bradei***
Anura - Leptodactylidae
e Itatiaia Highland Frog

2252 ***Holoaden luederwaldti***
Anura - Leptodactylidae
e Lüderwaldt's Highland Frog

2253 ***Hoplobatrachus***
Anura - Ranidae
e Peters Frogs

2254 ***Hoplobatrachus crassus***
Anura - Ranidae
e Carnatic Peters Frog; Jerdon's Bull Frog (ISC)

2255 ***Hoplobatrachus demarchii***
Anura - Ranidae
e Ethiopia Peters Frog

2256 ***Hoplobatrachus occipitalis***
Anura - Ranidae
e African Peters Frog; Groove-crowned Bullfrog
f Grenouille tigrée africaine

2257 ***Hoplobatrachus rugulosus***
Anura - Ranidae
e Asian Peters Frog; Chinese Bull Frog (SEA)
d Chinesischer Spitzfrosch

2258 ***Hoplobatrachus tigerinus***
Anura - Ranidae
e Tiger Peters Frog; Tiger Frog; Indian Bull Frog; Tigrine Frog
d Tigerfrosch; Asiatischer Ochsenfrosch
f Grenouille tigrée; Grenouille tigrine; Grenouille-taureau indienne

2259 ***Hoplophryne***
Anura - Microhylidae
e African Banana Frogs

2260 ***Hoplophryne rogersi***
Anura - Microhylidae
e Tanzania Banana Frog; Usambara Banana Frog

2261 ***Hoplophryne ulugurensis***
Anura - Microhylidae
e African Banana Frog
d Uluguru-Engmaulfrosch; Uluguru-Schwarzfrosch

2262 ***Huia***
Anura - Ranidae
e Huia Frogs

2263 ***Huia cavitympanum***
Anura - Ranidae

e Sabah Huia Frog

2264 ***Huia nasica***
Anura - Ranidae
e Tonkin Huia Frog

2265 ***Huia sumatrana***
Anura - Ranidae
e Sumatra Huia Frog

2266 ***Huiaj avana***
Anura - Ranidae
e Java Huia Frog

2267 ***Hyalinobatrachium***
Anura - Centrolenidae
e Glass Frogs

2268 ***Hyalinobatrachium antistenesi***
Anura - Centrolenidae
e Aragua Glass Frog

2269 ***Hyalinobatrachium aureoguttatum***
Anura - Centrolenidae
e Atrato Glass Frog

2270 ***Hyalinobatrachium bergeri***
Anura - Centrolenidae
e Berger's Glass Frog

2271 ***Hyalinobatrachium chirripoi***
Anura - Centrolenidae
e Suretka Glass Frog

2272 ***Hyalinobatrachium colymbiphyllum***
Anura - Centrolenidae
e Plantation Glass Frog
d Nacktbauch-Glasfrosch

2273 ***Hyalinobatrachium duranti***
Anura - Centrolenidae
e Dunn's Glass Frog

2274 ***Hyalinobatrachium eurygnathum***
Anura - Centrolenidae
e Rio Glass Frog

2275 ***Hyalinobatrachium flavidigitatum***
Anura - Centrolenidae
e Yellow-fingered Glass Frog

2276 ***Hyalinobatrachium fleischmanni***
Anura - Centrolenidae
e Fleischmann's Glass Frog
d Glasfrosch

2277 ***Hyalinobatrachium fragile***
Anura - Centrolenidae
e Fragile Glass Frog

2278 ***Hyalinobatrachium iaspidiense***
Anura - Centrolenidae
e Yuruani Glass Frog

2279 ***Hyalinobatrachium lemur***
Anura - Centrolenidae
e Ghost Glass Frog

2280 ***Hyalinobatrachium loreocarinatum***
Anura - Centrolenidae
e La Mucuy Glass Frog

2281 ***Hyalinobatrachium munozorum***
Anura - Centrolenidae
e Upper Amazon Glass Frog

2282 ***Hyalinobatrachium orientale***
Anura - Centrolenidae
e Eastern Glass Frog

2283 ***Hyalinobatrachium ostracodermoides***
Anura - Centrolenidae
e Rivero's Glass Frog

2284 ***Hyalinobatrachium pallidum***
Anura - Centrolenidae
e Guacharaquita Glass Frog

2285 ***Hyalinobatrachium parvulum***
Anura - Centrolenidae
e Lages Glass Frog

2286 ***Hyalinobatrachium pellucidum***
Anura - Centrolenidae
e Rio Azuela Glass Frog

2287 ***Hyalinobatrachium pleurolineatum***
Anura - Centrolenidae
e El Chorotol Glass Frog

2288 ***Hyalinobatrachium pulveratum***
Anura - Centrolenidae
e Chiriqui Glass Frog

2289 ***Hyalinobatrachium revocatum***
Anura - Centrolenidae
e Colonia Tovar Glass Frog

2290 ***Hyalinobatrachium talamacae***
Anura - Centrolenidae
e Talamanca Glass Frog

2291 ***Hyalinobatrachium taylori***
Anura - Centrolenidae
e Taylor's Glass Frog
f Centronelle de Taylor

2292 ***Hyalinobatrachium uranoscopum***
Anura - Centrolenidae
e Humboldt Glass Frog

2293 ***Hyalinobatrachium valeroi***
Anura - Centrolenidae
e La Palma Glass Frog
d La Palma-Armfrosch

2294 ***Hyalinobatrachium vireovittatum***
Anura - Centrolenidae
e Starrett's Glass Frog

2295 ***Hydrolaetare***
Anura - Leptodactylidae
e Schmidt's Forest Frogs
f Hydrolaetares de Schmidt

2296 ***Hydrolaetare schmidti***
Anura - Leptodactylidae
e Schmidt's Forest Frog
f Hydrolaetare de Schmidt

2297 ***Hydromantes***
Caudata - Plethodontidae
e Web-toed Salamanders
d Höhlensalamander (pl); Schleuderzungensalamander; Amerikanischen Höhlensalamander
f Salamandres cavernicoles; Spélerpés; Hydromantes

2298 ***Hydromantes brunus***
Caudata - Plethodontidae
e Limestone Salamander
d Kalkstein-Höhlensalamander

2299 ***Hydromantes platycephalus***
Caudata - Plethodontidae
e Mount Lyell Salamander
d Mount-Lyell-Höhlensalamander; Kalifornischer Höhlensalamander

2300 ***Hydromantes shastae***
Caudata - Plethodontidae
e Shasta Salamander
d Shasta-Höhlensalamander

2301 ***Hyla***
Anura - Hylidae
e Tree Frogs; Common Tree Frogs
d Echte Laubfrösche; Laubfrösche
f Rainettes; Grenouilles arboricoles
i Raganelle

2302 ***Hyla acreana***
Anura - Hylidae
e Acre Tree Frog

2303 ***Hyla albofrenata***
Anura - Hylidae
e White-fronted Tree Frog

2304 ***Hyla alboguttata***
Anura - Hylidae
e Whitebelly Tree Frog

2305 ***Hyla albolineata***
Anura - Hylidae
e White-lined Tree Frog

2306 ***Hyla albomarginata***
Anura - Hylidae
e White-edged Tree Frog
f Rainette à bandes blanches; Rainette bordée de blanc

2307 ***Hyla albonigra***
Anura - Hylidae
e White-Black Tree Frog

2308 ***Hyla albopunctata***
Anura - Hylidae
e Spotted Tree Frog
d Bananenfrosch; Bananenlaubfrosch

2309 ***Hyla albopunctulata***
Anura - Hylidae
e White-speckled Tree Frog

2310 ***Hyla albosignata***
Anura - Hylidae
e Lutz's Tree Frog

2311 ***Hyla albovittata***
Anura - Hylidae
e White-banded Tree Frog

2312 ***Hyla alemani***
Anura - Hylidae
e Cagua Tree Frog

2313 ***Hyla allenorum***
Anura - Hylidae
e Cuzco Reserve Tree Frog

2314 ***Hyla altipotens***
Anura - Hylidae
e Yellowbelly Voiceless Tree Frog

2315 ***Hyla alvarengai***
Anura - Hylidae
e Santa Barbara Tree Frog

2316 ***Hyla alytotylax***
Anura - Hylidae
e Tandapi Tree Frog

2317 ***Hyla americana***
Anura - Hylidae
e American Tree Frog

2318 ***Hyla anataliasiasi***
Anura - Hylidae
e Goias Tree Frog

2319 ***Hyla anceps***
Anura - Hylidae
e Estrela Tree Frog

2320 ***Hyla andersonii***
Anura - Hylidae
e Pine Barrens Tree Frog; Anderson's Tree Frog
d Andersons Laubfrosch

2321 ***Hyla angustilineata***
Anura - Hylidae
e Narrow-lined Tree Frog

2322 ***Hyla annectans***
Anura - Hylidae
e Assam Tree Frog; Green Tree Frog; Indian Leaf Frog (SEA)
d Indischer Laubfrosch

2323 ***Hyla aperomea***
Anura - Hylidae
e Balzapata Tree Frog

2324 ***Hyla arborea***
Anura - Hylidae
e European Tree Frog ; Tree Frog; Common Tree Frog; Green Tree Frog; Common Hyla; Hyla
d Gemeiner Laubfrosch; Europäischer Laubfrosch; Laubfrosch; Baumfrosch
f Rainette verte; Rainette d'Europe; Rainette arboricole; Rainette européenne des arbres
i Raganella comune; Raganella

2325 ***Hyla arborescandens***
Anura - Hylidae
e Lesser Bromeliad Tree Frog

2326 ***Hyla arenicolor***
Anura - Hylidae
e Canyon Tree Frog
d Canon-Laubfrosch
f Rainette des canyons

2327 ***Hyla arianae***
Anura - Hylidae
e Santa Catarina Tree Frog

2328 ***Hyla arildae***
Anura - Hylidae
e Teresopolis Tree Frog

2329 ***Hyla armata***
Anura - Hylidae
e Armed Tree Frog

2330 ***Hyla astartea***
Anura - Hylidae
e Paranapiacaba Tree Frog

2331 ***Hyla auraria***
Anura - Hylidae
e Peters's Tree Frog

2332 ***Hyla avicova avicova***
Anura - Hylidae
e Western Bird-voiced Tree Frog

2333 ***Hyla avivoca***
Anura - Hylidae
e Bird-voiced Tree Frog
f Rainette à voix d'oiseau

2334 ***Hyla avivoca ogechiensis***
Anura - Hylidae
e Eastern Bird-voiced Tree Frog

2335 ***Hyla baileyi***
Anura - Hylidae
e Bailey's Brazilian Tree Frog

2336 ***Hyla balzani***
Anura - Hylidae
e Yungas Tree Frog

2337 ***Hyla batterbyi***
Anura - Hylidae
e Batterby's Caracas Tree Frog

2338 ***Hyla benitezi***
Anura - Hylidae
e Benitez'sTree Frog

2339 ***Hyla berthalutzae***
Anura - Hylidae
e Bertha's Tree Frog

2340 ***Hyla bifurca***
Anura - Hylidae
e Upper Amazon Tree Frog

2341 ***Hyla biobebe***
Anura - Hylidae
e Usina Tree Frog

2342 ***Hyla bipunctata***
Anura - Hylidae
e Two-spotted Tree Frog

f Rainette naine

2343 ***Hyla bischoffi***
Anura - Hylidae
e Bishoff's Tree Frog

2344 ***Hyla bistincta***
Anura - Hylidae
e Mexican Fringe-limbed Tree Frog

2345 ***Hyla boans***
Anura - Hylidae
e Rusty Tree Frog; Giant Tree Frog
d Riesenlaubfrosch; Großer Laubfrosch
f Rainette beuglante; Rainette patte-d'oie; Rainette géante

2346 ***Hyla bocourti***
Anura - Hylidae
e Bocourt's Tree Frog

2347 ***Hyla bogertae***
Anura - Hylidae
e Bogert's Aquatic Tree Frog

2348 ***Hyla bogotensis***
Anura - Hylidae
e Bogota Tree Frog

2349 ***Hyla bokermanni***
Anura - Hylidae
e Bokermann's Tarauaca Tree Frog

2350 ***Hyla brevifrons***
Anura - Hylidae
e Crump's Tree Frog
f Rainette camuse

2351 ***Hyla bromeliacia***
Anura - Hylidae
e Bromeliad Tree Frog

2352 ***Hyla caingua***
Anura - Hylidae
e Striped Tree Frog

2353 ***Hyla calcarata***
Anura - Hylidae
e Troschel's Tree Frog
f Rainette de Troschel

2354 ***Hyla callipeza***
Anura - Hylidae
e Sardinata Tree Frog

2355 ***Hyla callipleura***
Anura - Hylidae
e Charuplaya Tree Frog

2356 ***Hyla callipygia***
Anura - Hylidae
e Bocaina Tree Frog

2357 ***Hyla carinata***
Anura - Hylidae
e Tarraco Tree Frog

2358 ***Hyla carnifex***
Anura - Hylidae
e Executioner's Tree Frog

2359 ***Hyla carvalhoi***
Anura - Hylidae
e Carvalho'sTree Frog

2360 ***Hyla catracha***
Anura - Hylidae
e Zacate Blanco Tree Frog

2361 ***Hyla caucana***
Anura - Hylidae
e Cauca Tree Frog

2362 ***Hyla cavicola***
Anura - Hylidae
e Cruz's Tree Frog

2363 ***Hyla cembra***
Anura - Hylidae
e Southern Sierra Madre Tree Frog

2364 ***Hyla chaneque***
Anura - Hylidae
e Fairy Tree Frog

2365 ***Hyla charadricola***
Anura - Hylidae
e Puebla Tree Frog

2366 ***Hyla charazani***
Anura - Hylidae
e Charazani Tree Frog

2367 ***Hyla chinensis***
Anura - Hylidae
e Common ChineseTree Frog; Chinese Tree Frog; Green Tree Toad
d Chinesischer Laubfrosch
f Rainette de Chine

2368 ***Hyla chlorostea***
Anura - Hylidae
e Parjacti Tree Frog

2369 ***Hyla chryses***
Anura - Hylidae
e Golden Tree Frog

2370 ***Hyla chrysoscellis***
Anura - Hylidae
e Cope's Grey Tree Frog; Cope's Diploid; Diploid Grey Tree Frog; Southern Gray Tree Frog (NA)
f Rainette criarde

2371 ***Hyla cinerea***
Anura - Hylidae
e Green Tree Frog; Carolina Tree Frog; Rain Frog
d Grüner Laubfrosch; Gestreifter Laubfrosch; Karolina-Laubfrosch; Grüner Karolina-Laubfrosch; Cinerea-Laubfrosch; Amerikanischer Laubfrosch
f Rainette verte de la Caroline; Rainette à flancs rayés

2372 ***Hyla cinerea evittata***
Anura - Hylidae
e Northern Green Tree Frog

2373 ***Hyla circumdata***
Anura - Hylidae
e Espirito Santo Tree Frog

2374 ***Hyla claresignata***
Anura - Hylidae
e Rio Tree Frog
d Augenflecken-Laubfrosch

2375 ***Hyla clepsydra***
Anura - Hylidae
e Sao Paulo Tree Frog; Sao Paolo Tree Frog

2376 ***Hyla columbiana***
Anura - Hylidae
e Boettger's Colombian Tree Frog

2377 ***Hyla colymba***
Anura - Hylidae
e La Loma Tree Frog

2378 ***Hyla crassa***
Anura - Hylidae
e Aquatic Tree Frog

2379 ***Hyla crepitans***
Anura - Hylidae
e Emerald-eyed Tree Frog; Rattle-voiced Tree Frog
d Ruderer
f Rainette feuille-morte; Rainette crépitante

2380 ***Hyla cyanomma***
Anura - Hylidae

e Blue-eyed Tree Frog; Blue-eyed Aquatic Tree Frog

2381 ***Hyla cymbalum***
Anura - Hylidae
e Campo GrandeTree Frog

2382 ***Hyla debilis***
Anura - Hylidae
e Isla BonitaTree Frog

2383 ***Hyla decipiens***
Anura - Hylidae
e Brazilian Coastal Tree Frog

2384 ***Hyla dendroscarta***
Anura - Hylidae
e Greater Bromeliad Tree Frog

2385 ***Hyla dentei***
Anura - Hylidae
e AmapaTree Frog
f Rainette de Dente

2386 ***Hyla denticulenta***
Anura - Hylidae
e Charta Tree Frog

2387 ***Hyla dolloi***
Anura - Hylidae
e Werner's Brazilian Tree Frog

2388 ***Hyla ebraccata***
Anura - Hylidae
e Hourglass Tree Frog
d Bromelienlaubfrosch; Mittelamerikanischer Baumfrosch

2389 ***Hyla echinata***
Anura - Hylidae
e Oaxacan Fringe-limbed Tree Frog

2390 ***Hyla elegans***
Anura - Hylidae
e Elegant Forest Tree Frog

2391 ***Hyla elongata***
Anura - Hylidae
e Elongate Tree Frog

2392 ***Hyla euphorbiacea***
Anura - Hylidae
e Southern Highlands Tree Frog

2393 ***Hyla eximia***
Anura - Hylidae
e Mountain Tree Frog; Arizona Tree Frog
d Mexikanischer Laubfrosch

2394 ***Hyla eximia eximia***
Anura - Hylidae
e Mountain Tree Frog

2395 ***Hyla eximia wrightorum***
Anura - Hylidae
e Wright's Mountain Tree Frog

2396 ***Hyla faber***
Anura - Hylidae
e Blacksmith Tree Frog; Blacksmith
d Schmied; Großer Kolbenfuß; Kolbenfinger
f Rainette forgeron

2397 ***Hyla fasciata***
Anura - Hylidae
e Günther's Banded Tree Frog
f Rainette de Günther

2398 ***Hyla femoralis***
Anura - Hylidae
e Pine Woods Tree Frog
f Rainette fémorale

2399 ***Hyla fernandoi***
Anura - Hylidae
e Fernando's Tree Frog

2400 ***Hyla fimbrimembra***
Anura - Hylidae
e Heredia Tree Frog

2401 ***Hyla fluminea***
Anura - Hylidae
e Brazilian River Tree Frog

2402 ***Hyla fuentei***
Anura - Hylidae
e Fuente's Powakka Tree Frog

2403 ***Hyla fusca***
Anura - Hylidae
e Brown Tree Frog

2404 ***Hyla garagoensis***
Anura - Hylidae
e Garagoa Tree Frog

2405 ***Hyla gaycheri***
Anura - Hylidae
f Rainette de Gaucher

2406 ***Hyla geographica***
Anura - Hylidae
e Map Tree Frog
f Rainette de Spyx

2407 ***Hyla giesleri***
Anura - Hylidae
e Giesler's Tree Frog

2408 ***Hyla godmani***
Anura - Hylidae
e Godman's Tree Frog

2409 ***Hyla gouveai***
Anura - Hylidae
e Itiaia Tree Frog

2410 ***Hyla graceae***
Anura - Hylidae
e Continental Divide Tree Frog

2411 ***Hyla grandisonae***
Anura - Hylidae
e MazaruniTree Frog

2412 ***Hyla granosa***
Anura - Hylidae
e Demerara Falls Tree Frog

2413 ***Hyla gratiosa***
Anura - Hylidae
e Barking Tree Frog; Barking Frog (NA)

2414 ***Hyla gryllata***
Anura - Hylidae
e Pacific Lowland Tree Frog

2415 ***Hyla guentheri***
Anura - Hylidae
e Günther's Brazilian Tree Frog

2416 ***Hyla hadroceps***
Anura - Hylidae
e New River Tree Frog
f Rainette marteau

2417 ***Hyla hallowellii***
Anura - Hylidae
e Hallowell's Tree Frog

2418 ***Hyla haraldschultzi***
Anura - Hylidae
e Harald's Tree Frog

2419 ***Hyla hazelae***
Anura - Hylidae
e Hazel's Tree Frog

2420 ***Hyla heilprini***
Anura - Hylidae
e Los Bracitos Tree Frog

2421 ***Hyla helenae***
Anura - Hylidae

e Helena'sTree Frog

2422 ***Hyla hobbsi***
Anura - Hylidae
e Hobb's Tree Frog

2423 ***Hyla hutchinsi***
Anura - Hylidae
e Hutchin'sTree Frog

2424 ***Hyla hylax***
Anura - Hylidae
e Atlantic Forest Tree Frog

2425 ***Hyla hypselops***
Anura - Hylidae
e Pebas Tree Frog

2426 ***Hyla ibitiguara***
Anura - Hylidae
e Fazenda Salto Tree Frog

2427 ***Hyla ibitipoca***
Anura - Hylidae
e Ibitipoca Tree Frog

2428 ***Hyla imitator***
Anura - Hylidae
e Mimic Tree Frog

2429 ***Hyla inframaculata***
Anura - Hylidae
e Santarem Tree Frog

2430 ***Hyla intermedia***
Anura - Hylidae
e Italian Tree Frog
i Ragenella italiana; Raganella; Ranela

2431 ***Hyla intermixta***
Anura - Hylidae
e Lost Tree Frog

2432 ***Hyla izecksohni***
Anura - Hylidae
e Izecksohn's Tree Frog

2433 ***Hyla jahni***
Anura - Hylidae
e Jahn's Tree Frog

2434 ***Hyla japonica***
Anura - Hylidae
e Japanese Tree Frog; Japanese Tree Toad
d Kurzschnäuziger Laubfrosch; Japanischer Laubfrosch
f Rainette du Japon

2435 ***Hyla juanitae***
Anura - Hylidae
e Juanita's Earless Tree Frog

2436 ***Hyla kanaima***
Anura - Hylidae
e Kanaima Tree Frog

2437 ***Hyla karennaneae***
Anura - Hylidae
e Vaupes Tree Frog

2438 ***Hyla koechlini***
Anura - Hylidae
e Koechlin's Tree Frog

2439 ***Hyla labialis***
Anura - Hylidae
e Green Dotted Tree Frog

2440 ***Hyla lancasteri***
Anura - Hylidae
e Lancaster's Tree Frog

2441 ***Hyla lanciformis***
Anura - Hylidae
e Basin Tree Frog

2442 ***Hyla langei***
Anura - Hylidae
e Marumbi Tree Frog

2443 ***Hyla larinopypgion***
Anura - Hylidae
e Cordillera Central Tree Frog

2444 ***Hyla lascinia***
Anura - Hylidae
e Tabor's Tree Frog

2445 ***Hyla leali***
Anura - Hylidae
e Leal's Tree Frog

2446 ***Hyla lemai***
Anura - Hylidae
e Leman Tree Frog

2447 ***Hyla leptolineata***
Anura - Hylidae
e Fine-lined Tree Frog

2448 ***Hyla leucophyllata***
Anura - Hylidae
e Beireis's Tree Frog
d Eingefasster Laubkleber; Laubkleber
f Rainette leucophyle; Rainette à bandeau

2449 ***Hyla leucopygia***
Anura - Hylidae
e Guinle Tree Frog

2450 ***Hyla limai***
Anura - Hylidae
e Sao Vicente Tree Frog

2451 ***Hyla lindae***
Anura - Hylidae
e Linda's Tree Frog

2452 ***Hyla loquax***
Anura - Hylidae
e Majogany Tree Frog; Mahogany Tree Frog

2453 ***Hyla loveridgi***
Anura - Hylidae
e Loveridge's Tree Frog

2454 ***Hyla luctuosa***
Anura - Hylidae
e Reservoir Tree Frog

2455 ***Hyla luteocellata***
Anura - Hylidae
e El Mene Tree Frog

2456 ***Hyla lynchi***
Anura - Hylidae
e Lynch's Colombian Tree Frog

2457 ***Hyla marginata***
Anura - Hylidae
e Mundo Novo Tree Frog

2458 ***Hyla marianae***
Anura - Hylidae
e Spalding's Tree Frog

2459 ***Hyla marianitae***
Anura - Hylidae
e Salta Tree Frog

2460 ***Hyla marmorata***
Anura - Hylidae
e Marbled Tree Frog
f Rainette marbrée

2461 ***Hyla martinsi***
Anura - Hylidae
e Martin's Tree Frog

2462 ***Hyla mathiassoni***
Anura - Hylidae
e Mathiasson's Tree Frog

2463 ***Hyla melanogyrea***
Anura - Hylidae
e Interior Tree Frog

2464 ***Hyla melanomma***
Anura - Hylidae
e Black-eyed Tree Frog; Blackeye Tree Frog

2465 ***Hyla melanomma bivocata***
Anura - Hylidae
e Chiapan Highlands Tree Frog

2466 ***Hyla melanopleura***
Anura - Hylidae
e Lower Andes Tree Frog

2467 ***Hyla melanorhabota***
Anura - Hylidae
e Schneider's Tree Frog

2468 ***Hyla meridensis***
Anura - Hylidae
e Merida Tree Frog

2469 ***Hyla meridionalis***
Anura - Hylidae
e Mediterranenan Tree Frog; Stripeless Tree Frog
d Mittelmeer-Laubfrosch
f Rainette méditerranéenne; Rainette méridionale; Rainette du Midi
i Raganella mediterranea; Raganella meridionale; Raganella baritono

2470 ***Hyla microcephala***
Anura - Hylidae
e Yellow Tree Frog

2471 ***Hyla microcephala underwoodi***
Anura - Hylidae
e Underwood's Yellow Tree Frog

2472 ***Hyla microderma***
Anura - Hylidae
e Smallskin Tree Frog

2473 ***Hyla microps***
Anura - Hylidae
e Nova Fiburgo Tree Frog

2474 ***Hyla miliaria***
Anura - Hylidae
e Cope's Brown Tree Frog

2475 ***Hyla minera***
Anura - Hylidae
e Guatemala Tree Frog

2476 ***Hyla minima***
Anura - Hylidae
e Taperinha'sTree Frog

2477 ***Hyla miniscula***
Anura - Hylidae
e Rivero's Tiny Tree Frog

2478 ***Hyla minuta***
Anura - Hylidae
e Lesser Tree Frog
f Hyla menue

2479 ***Hyla miotympanum***
Anura - Hylidae
e Small-eared Tree Frog; Green Tree Frog

2480 ***Hyla mixe***
Anura - Hylidae
e Mixe Tree Frog

2481 ***Hyla mixomaculata***
Anura - Hylidae
e Variegated Tree Frog

2482 ***Hyla miyatai***
Anura - Hylidae
e Hosteria La Selva Tree Frog

2483 ***Hyla molitor***
Anura - Hylidae
e Schmidt's Bolivian Tree Frog

2484 ***Hyla multifasciata***
Anura - Hylidae
e Many-banded Tree Frog
f Rainette de Daudin

2485 ***Hyla musica***
Anura - Hylidae
e Musical Tree Frog

2486 ***Hyla mykter***
Anura - Hylidae
e Keelsnout Tree Frog

2487 ***Hyla nahdereri***
Anura - Hylidae
e Estrada Saraiva Tree Frog

2488 ***Hyla nana***
Anura - Hylidae
e Dwarf Clawed Frog
f Rainette naine

2489 ***Hyla nanuzae***
Anura - Hylidae
e Jaboticatubas Tree Frog

2490 ***Hyla novaisi***
Anura - Hylidae
e Bokerman's Bahia Tree Frog

2491 ***Hyla nubicola***
Anura - Hylidae
e Cloud-forest Tree Frog

2492 ***Hyla ocapia***
Anura - Hylidae
e Paku Khara Tree Frog

2493 ***Hyla oliveirai***
Anura - Hylidae
e Xeric Tree Frog

2494 ***Hyla ornatissima***
Anura - Hylidae
e Ornate Tree Frog
f Rainette très ornée

2495 ***Hyla pacha***
Anura - Hylidae
e Morona-Santiago Tree Frog

2496 ***Hyla pachyderma***
Anura - Hylidae
e Semiaquatic Tree Frog

2497 ***Hyla palliata***
Anura - Hylidae
e Oaxacan Tree Frog

2498 ***Hyla palmeri***
Anura - Hylidae
e Palmer's Tree Frog

2499 ***Hyla pantosticta***
Anura - Hylidae
e Rio Chingual Valley Tree Frog

2500 ***Hyla pardalis***
Anura - Hylidae
e Leopard Tree Frog
d Pantherlaubfrosch
f Rainette panthérine

2501 ***Hyla parviceps***
Anura - Hylidae
e Sarayacu Tree Frog

2502 ***Hyla pauiniensis***
Anura - Hylidae
e Pauini Tree Frog

2503 ***Hyla pelidna***
Anura - Hylidae
e Betania Tree Frog

2504 ***Hyla pelitta***
Anura - Hylidae
e Oaxacan Yellow Tree Frog

2505 ***Hyla pelluscens***
Anura - Hylidae
e Palmar Tree Frog

2506 ***Hyla pentheter***
Anura - Hylidae
e Mourning Tree Frog

2507 ***Hyla perkinsi***
Anura - Hylidae
e Perkin's Tree Frog

2508 ***Hyla phlebodes***
Anura - Hylidae
e San Carlos Tree Frog

2509 ***Hyla phyllognatha***
Anura - Hylidae
e Roque Tree Frog

2510 ***Hyla picadoi***
Anura - Hylidae
e Volcan Barba Tree Frog

2511 ***Hyla piceigularis***
Anura - Hylidae
e Rio Luisito Tree Frog

2512 ***Hyla picta***
Anura - Hylidae
e Painted Tree Frog

2513 ***Hyla pictipes***
Anura - Hylidae
e Pico Blanco Tree Frog

2514 ***Hyla picturata***
Anura - Hylidae
e Imbabura Tree Frog

2515 ***Hyla pinima***
Anura - Hylidae
e Sazima's Tree Frog

2516 ***Hyla pinorum***
Anura - Hylidae
e Mexican Pine Woods Tree Frog; Pine Woods Tree Frog

2517 ***Hyla platydactyla***
Anura - Hylidae
e Merida Andes Tree Frog

2518 ***Hyla plicata***
Anura - Hylidae
e Ridged Tree Frog

2519 ***Hyla polytaenia***
Anura - Hylidae
e Cope's Eastern Paraguay Tree Frog

2520 ***Hyla praestans***
Anura - Hylidae
e San Augustin Tree Frog

2521 ***Hyla prasina***
Anura - Hylidae
e Burmeister's Tree Frog

2522 ***Hyla psarolaima***
Anura - Hylidae
e Papallacta Tree Frog

2523 ***Hyla pseudopseudis***
Anura - Hylidae
e Veadeiros Tree Frog

2524 ***Hyla pseudopuma***
Anura - Hylidae
e Günther's Costa Rican Tree Frog

2525 ***Hyla ptychodactyla***
Anura - Hylidae
e Pilalo Tree Frog

2526 ***Hyla pugnax***
Anura - Hylidae
e Chiriqui-Flusse Tree Frog

2527 ***Hyla pulchella***
Anura - Hylidae
e Montevideo Tree Frog
f Rainette gentille

2528 ***Hyla pulchrilineata***
Anura - Hylidae
e Common Tree Frog (NA); Tree Frog

2529 ***Hyla punctata***
Anura - Hylidae
e Polkadot Tree Frog
f Rainette ponctuée

2530 ***Hyla quadrilineata***
Anura - Hylidae
e Four-lined Tree Frog

2531 ***Hyla raniceps***
Anura - Hylidae
e Chaco Tree Frog
f Rainette de Cope

2532 ***Hyla rhodopepla***
Anura - Hylidae
e Red-skirted Tree Frog

2533 ***Hyla riveroi***
Anura - Hylidae
e Rivero's Amazon Tree Frog

2534 ***Hyla rivularis***
Anura - Hylidae
e American Cinchona Plantation Tree Frog

2535 ***Hyla robertmertensi***
Anura - Hylidae
e Mertens's Yellow Tree Frog

2536 ***Hyla robertsorum***
Anura - Hylidae
e Roberts's Tree Frog

2537 ***Hyla roeschmanni***
Anura - Hylidae
e Roeschmann's Tree Frog

2538 ***Hyla roraima***
Anura - Hylidae
e Hoogmoed's Tree Frog

2539 ***Hyla rosenbergi***
Anura - Hylidae
e Rosenberg's Tree Frog
d Rosenbergs Laubfrosch
f Rainette de Rosenberg

2540 ***Hyla rossalleni***
Anura - Hylidae
e Ross Allen's Tree Frog

2541 ***Hyla rubicudula***
Anura - Hylidae
e Lagoa Santa Tree Frog

2542 ***Hyla rubracyla***
Anura - Hylidae
e Valle Tree Frog

2543 ***Hyla rufitela***
Anura - Hylidae
e Canal Zone Tree Frog

2544 ***Hyla ruschii***
Anura - Hylidae
e Rusch's Tree Frog

2545 ***Hyla sabrina***
Anura - Hylidae
e Sierra Juarez Tree Frog

2546 ***Hyla salvaje***
Anura - Hylidae

e Copan Tree Frog

2547 ***Hyla sanborni***
Anura - Hylidae
e Sanborn's Tree Frog

2548 ***Hyla sanchiangensis***
Anura - Hylidae
e San Chiang Tree Frog

2549 ***Hyla sarampiona***
Anura - Hylidae
e Western Andes Tree Frog

2550 ***Hyla sarayacuensis***
Anura - Hylidae
e Shreve's Sarayacu Tree Frog

2551 ***Hyla sarda***
Anura - Hylidae
e Sardinia Tree Frog; Tyrrhenian Tree Frog; Sardinian Tree Frog
d Tyrrhenischer Laubfrosch; Sardischer Laubfrosch
f Rainette sarde
i Raganella sarda; Raganella tirrenica

2552 ***Hyla sartori***
Anura - Hylidae
e Taylor's Yellow Tree Frog

2553 ***Hyla savignyi***
Anura - Hylidae
e Savigny's Tree Frog; Yellow-lemon Tree Frog
d Südosteuropäischer Laubfrosch; Kleinasiatischer Laubfrosch
f Rainette de Savigny
i Raganella di Savigny

2554 ***Hyla sazimai***
Anura - Hylidae
e Zagaia Tree Frog

2555 ***Hyla schuberti***
Anura - Hylidae
e Schubert's Rondonia Tree Frog

2556 ***Hyla secendens***
Anura - Hylidae
e Barro Branco Tree Frog

2557 ***Hyla semigutatta***
Anura - Hylidae
e Speckled Tree Frog

2558 ***Hyla senicula***
Anura - Hylidae
e Corcovado Tree Frog

2559 ***Hyla siblezi***
Anura - Hylidae
e La Escalera Tree Frog

2560 ***Hyla simmonsi***
Anura - Hylidae
e Simmon's Tree Frog

2561 ***Hyla simplex***
Anura - Hylidae
e Annam Tree Frog

2562 ***Hyla siopela***
Anura - Hylidae
e Voiceless Tree Frog

2563 ***Hyla smaragdina***
Anura - Hylidae
e Emerald Tree Frog

2564 ***Hyla smithii***
Anura - Hylidae
e Dwarf Mexican Tree Frog

2565 ***Hyla soaresi***
Anura - Hylidae
e Picos Tree Frog

2566 ***Hyla squirella***
Anura - Hylidae
e Squirrel Tree Frog; House Frog (NA); Southern Tree Frog (NA); Scraper Frog (NA); Rain Frog (NA)
d Eichhörnchen-Laubfrosch
f Rainette écureille; Rainette squirelle

2567 ***Hyla staufferorum***
Anura - Hylidae
e Jondachi Tree Frog

2568 ***Hyla stingi***
Anura - Hylidae
e Kaplan's Gargoa Tree Frog

2569 ***Hyla subocularis***
Anura - Hylidae
e Rio Turquesa Tree Frog

2570 ***Hyla sumichrasti***
Anura - Hylidae
e Sumichrast's Tree Frog

2571 ***Hyla surinamensis***
Anura - Hylidae
e Daudin's Surinam Tree Frog
f Rainette de Surinam

2572 ***Hyla suweonensis***
Anura - Hylidae
e Suweon Tree Frog

2573 ***Hyla taeniopus***
Anura - Hylidae
e Porthole Tree Frog

2574 ***Hyla thorectes***
Anura - Hylidae
e Mexican Hyla; Adler's Mottled Tree Frog

2575 ***Hyla thysanota***
Anura - Hylidae
e Cerro Mali Tree Frog

2576 ***Hyla tica***
Anura - Hylidae
e Starrett's Tree Frog

2577 ***Hyla timbeba***
Anura - Hylidae
e Cardoso's Tree Frog

2578 ***Hyla tintinnabulum***
Anura - Hylidae
e Rio Uaupes Tree Frog

2579 ***Hyla torrenticola***
Anura - Hylidae
e El Pepino Tree Frog

2580 ***Hyla triangulatum***
Anura - Hylidae
e Triangle Tree Frog

2581 ***Hyla tritaeniata***
Anura - Hylidae
e Three-banded Tree Frog

2582 ***Hyla truncata***
Anura - Hylidae
e Izecksohn's Brazilian Tree Frog

2583 ***Hyla trux***
Anura - Hylidae
e Spine-fingered Tree Frog; Spinefinger Tree Frog

2584 ***Hyla tsinglingensis***
Anura - Hylidae
e Shensi Tree Frog

2585 ***Hyla tubercolosa***
Anura - Hylidae
e Canelos Tree Frog

2586 ***Hyla uruguaya***
Anura - Hylidae
e Schmidt's Uruguay Tree Frog

2587 ***Hyla valancifer***
Anura - Hylidae
e San Martin Fringe-limbed Tree Frog

2588 ***Hyla varelae***
Anura - Hylidae
e Carrizo's Tree Frog

2589 ***Hyla vasta***
Anura - Hylidae
e Hispaniola Tree Frog; Haitian Giant Tree Frog
d Riesenlaubfrosch; Haiti-Riesenfrosch
f Rainette géante de Haiti

2590 ***Hyla versicolor***
Anura - Hylidae
e Grey Tree Frog; Common Tetraploid Grey Tree Frog; Tetraploid Grey Tree Frog; Gray Tree Frog (NA); Tree Frog (NA)
d Grauer Laubfrosch; Farbenwechselnder Laubfrosch
f Rainette grise; Rainette versicolore

2591 ***Hyla versicolor***
Anura - Hylidae
e Eastern Gray Tree Frog (NA); Eastern Grey Tree Frog

2592 ***Hyla versicolor sandersi***
Anura - Hylidae
e Central Texas Tree Frog

2593 ***Hyla vigilans***
Anura - Hylidae
e Maracaibo Basin Tree Frog

2594 ***Hyla viridifusca***
Anura - Hylidae
e Laurenti'sTree Frog

2595 ***Hyla walkeri***
Anura - Hylidae
e Walker's Tree Frog

2596 ***Hyla warreni***
Anura - Hylidae
e Warren's Tree Frog

2597 ***Hyla wavrini***
Anura - Hylidae
e Upper Orinoco Tree Frog

2598 ***Hyla weygoldti***
Anura - Hylidae
e Weygoldt's Tree Frog

2599 ***Hyla wilderi***
Anura - Hylidae
e Wilder'sTree Frog

2600 ***Hyla xanthsticta***
Anura - Hylidae
e South Fork Tree Frog

2601 ***Hyla xapuriensis***
Anura - Hylidae
e Xapuri Tree Frog

2602 ***Hyla zeteki***
Anura - Hylidae
e Zetek Tree Frog

2603 ***Hyla zhaopingensis***
Anura - Hylidae
e Zhaoping Tree Frog

2604 **Hylidae**
Anura
e Tree Frogs; Hylids; Tree Toads
d Laubfrösche; Baumfrösche; Hylide Baumfrösche; Hyliden
f Rainettes; Grenouilles arboricoles; Hylidés

2605 ***Hylodes***
Anura - Leptodactylidae
e Tree Toads
d Brasilianische Südfrösche

2606 ***Hylodes asper***
Anura - Leptodactylidae
e Warty Tree Toad

2607 ***Hylodes babax***
Anura - Leptodactylidae
e Caparao Mountains Tree Toad

2608 ***Hylodes charadranaetes***
Anura - Leptodactylidae
e Rio Tree Toad

2609 ***Hylodes glaber***
Anura - Leptodactylidae
e Itatiaia Tree Toad

2610 ***Hylodes lateristrigatus***
Anura - Leptodactylidae
e Baumann's Tree Toad

2611 ***Hylodes magalhaesi***
Anura - Leptodactylidae
e Sao Paulo Tree Toad

2612 ***Hylodes meridionalis***
Anura - Leptodactylidae
e Rio Grande Tree Toad

2613 ***Hylodes mertensi***
Anura - Leptodactylidae
e Mertens's Tree Toad

2614 ***Hylodes nasus***
Anura - Leptodactylidae
e Santa Catarina Tree Toad

2615 ***Hylodes ornatus***
Anura - Leptodactylidae
e Ornate Tree Toad

2616 ***Hylodes otavioi***
Anura - Leptodactylidae
e Otavio's Tree Toad

2617 ***Hylodes perplicatus***
Anura - Leptodactylidae
e Humboldt Tree Toad

2618 ***Hylodes phllodes***
Anura - Leptodactylidae
e Boraceia Tree Toad

2619 ***Hylodes regius***
Anura - Leptodactylidae
e Royal Tree Toad

2620 ***Hylodes vanzolinii***
Anura - Leptodactylidae
e Vanzolini's Tree Toad

2621 ***Hylomantis***
Anura - Hylidae
e Rough Leaf Frogs

2622 ***Hylomantis aspera***
Anura - Hylidae
e Rough Leaf Frog

2623 ***Hylomantis granulosa***
Anura - Hylidae
e Granular Leaf Frog

2624 ***Hylophorbus***
Anura - Microhylidae
e Mawatta Frogs

2625 ***Hylophorbus rufescens***
Anura - Microhylidae
e Mawatta Frog; Red Mawatta Frog

2626 ***Hylorina***
Anura - Leptodactylidae
e Emerald Forest Frogs

2627 ***Hylorina sylvatica***
Anura - Leptodactylidae
e Emerald Forest Frog

2628 ***Hymenochirus***
Anura - Pipidae
e Dwarf Clawed Frogs
d Zwergkrallenfrösche

2629 ***Hymenochirus boettgeri***
Anura - Pipidae
e Zaire Dwarf Clawed Frog

2630 ***Hymenochirus boulengeri***
Anura - Pipidae
e Eastern Dwarf Clawed Frog

2631 ***Hymenochirus curtipes***
Anura - Pipidae
e Western Dwarf Clawed Frog

2632 ***Hymenochirus feae***
Anura - Pipidae
e Gaboon Dwarf Clawed Frog; Gabon Dwarf Clawed Frog

2633 **Hynobiidae**
Anura
e Asiatic Salamanders; Asian Salamanders; Oriental Salamanders
d Winkelzahnmolche; Asiatische Salamander (pl)
f Hynobiidés

2634 ***Hynobius***
Caudata - Hynobiidae
e Asian Salamanders; Asiatic Salamanders
d Echte Winkelzahnmolche

2635 ***Hynobius abei***
Caudata - Hynobiidae
e Sato's Oriental Salamander
d Abés Winkelzahnmolch
f Salamandre d'Abé

2636 ***Hynobius boulengeri***
Caudata - Hynobiidae
e Boulenger's Oriental Salamander

2637 ***Hynobius chinensis***
Caudata - Hynobiidae
e Chinese Salamander
d Chinesischer Winkelzahnmolch
f Salamandre de Chine

2638 ***Hynobius dunni***
Caudata - Hynobiidae
e Dunn's Oriental Salamander
f Salamandre d'Oita

2639 ***Hynobius formosanus***
Caudata - Hynobiidae
e Taiwan Salamander

2640 ***Hynobius hidamontaus***
Caudata - Hynobiidae
e Ochikura Swamp Salamander

2641 ***Hynobius kimurae***
Caudata - Hynobiidae
e Hondo Salamander
d Kimuras Winkelzahnmolch
f Salamandre d'Hida

2642 ***Hynobius leechii***
Caudata - Hynobiidae
e Gensan Salamander; Chinese Salamander; South Manchurian Salamander
f Salamandre de Corée

2643 ***Hynobius litchenatus***
Caudata - Hynobiidae
e Aomoori Salamander
d Flechten-Winkelzahnmolch
f Salamandre du Nord-Est du Japon

2644 ***Hynobius mantschuriensis***
Caudata - Hynobiidae
e Liaoning Salamander

2645 ***Hynobius naevius***
Caudata - Hynobiidae

e Sagami Salamander; Japanese Salamander
d Punktierter Winkelzahnmolch
f Salamandre tac heté du Japon

2646 ***Hynobius nebulosus***
Caudata - Hynobiidae
e Mitsjama Salamander; Clouded Salamander; Dusky Oriental Salamander
d Nebel-Winkelzahnmolch; Japanischer Winkelzahnmolch
f Salamandre nébuleuse

2647 ***Hynobius nigrescens***
Caudata - Hynobiidae
e Sendai Salamander
d Dunkler Winkelzahnmolch
f Salamandre noire du Japon

2648 ***Hynobius okiensis***
Caudata - Hynobiidae
e Oki Island Salamander
f Salamandre d'Oki

2649 ***Hynobius retardatus***
Caudata - Hynobiidae
e Noboribetsu Salamander; Yeso Salamander
d Nördlicher Winkelzahnmolch; Hokkaido-Winkelzahnmolch
f Salamandre de Yéso

2650 ***Hynobius sonani***
Caudata - Hynobiidae
e Taichu Salamander
f Salamandre de Formose

2651 ***Hynobius stejnegeri***
Caudata - Hynobiidae
e Stejneger's Oriental Salamander
f Salamandre ambrée

2652 ***Hynobius takedai***
Caudata - Hynobiidae
e Honchu Salamander

2653 ***Hynobius tenuis***
Caudata - Hynobiidae
e Arimine Salamander

2654 ***Hynobius tokyoensis***
Caudata - Hynobiidae
e Tokyo Salamander

2655 ***Hynobius tsuensis***
Caudata - Hynobiidae
e Tsushima Salamander
f Salamandre de Tsushima

2656 ***Hynobius turkestanicus***
Caudata - Hynobiidae
e Turkestan Salamander

2657 ***Hyophryne***
Anura - Microhylidae
e Yellow Frogs

2658 ***Hyophryne histrio***
Anura - Microhylidae
e Yellow Frog; Bahia Yellow Frog

2659 **Hyperoliidae**
Anura
e African Reed Frogs
d Riedfrösche; Hyperoliiden
f Hyperoliidés

2660 ***Hyperolius***
Anura - Hyperoliidae
e African Reed Frogs; Reed Frogs (CSA); Lily Frogs (CSA)
d Riedfrösche

2661 ***Hyperolius acutirostris***
Anura - Hyperoliidae
e Sharpsnout Reed Frog

2662 ***Hyperolius adametzi***
Anura - Hyperoliidae
e Bamenda Reed Frog

2663 ***Hyperolius albifrons***
Anura - Hyperoliidae
e Ahl's African Reed Frog

2664 ***Hyperolius albofrenatus***
Anura - Hyperoliidae
e White-fronted Reed Frog

2665 ***Hyperolius albolabris***
Anura - Hyperoliidae
e White-lipped Reed Frog

2666 ***Hyperolius alticola***
Anura - Hyperoliidae
e Ruwenzori Reed Frog

2667 ***Hyperolius angolanus***
Anura - Hyperoliidae
e Angoloa Reed Frog

2668 ***Hyperolius argus***
Anura - Hyperoliidae
e Boror Reed Frog; Argus Reed Frog (CSA); Yellow-spotted Reed Frog (CSA); Argus-spotted Sedge Frog; Argus-eyed Sedge Frog; Golden Sedge Frog (CSA); Coastal Reed Frog; Broad-striped Reed Frog

2669 ***Hyperolius atrigularis***
Anura - Hyperoliidae
e Kasiki Reed Frog

2670 ***Hyperolius balfouri***
Anura - Hyperoliidae
e Balfour's Reed Frog

2671 ***Hyperolius baumanni***
Anura - Hyperoliidae
e Baumann's Reed Frog

2672 ***Hyperolius benguellensis***
Anura - Hyperoliidae
e Benguella Reed Frog

2673 ***Hyperolius bicolor***
Anura - Hyperoliidae
e Two-coloured Reed Frog

2674 ***Hyperolius bobirensis***
Anura - Hyperoliidae
e Bobiri Reed Frog

2675 ***Hyperolius bocagei***
Anura - Hyperoliidae
e Bocage's Reed Frog

2676 ***Hyperolius bolifambae***
Anura - Hyperoliidae
e Bolifamba Reed Frog

2677 ***Hyperolius bopeleti***
Anura - Hyperoliidae
e Dizangue Reed Frog

2678 ***Hyperolius brachifasciatus***
Anura - Hyperoliidae
e Ngoto Reed Frog

2679 ***Hyperolius castaneus***
Anura - Hyperoliidae
e Ahl's Reed Frog

2680 ***Hyperolius chabanaudi***
Anura - Hyperoliidae
e Chabanaud's Reed Frog

2681 ***Hyperolius chlorosteus***
Anura - Hyperoliidae
e Sierra Leone Reed Frog

2682 ***Hyperolius chrysogaster***
Anura - Hyperoliidae
e Goldbelly Reed Frog

2683 ***Hyperolius cinereus***
Anura - Hyperoliidae
e Monard's Reed Frog

2684 ***Hyperolius cinnamomeoventris***
Anura - Hyperoliidae
e Braganca Reed Frog

2685 ***Hyperolius concolor***
Anura - Hyperoliidae
e Variable Reed Frog; Hallowell's Sedge Frog; Green Reed Frog (CSA)
d Spitzkopfriedfrosch; Concolor-Riedfrosch

2686 ***Hyperolius cystocandicans***
Anura - Hyperoliidae
e Tigoni Reed Frog

2687 ***Hyperolius destafanii***
Anura - Hyperoliidae
e Nargi Reed Frog

2688 ***Hyperolius diaphanus***
Anura - Hyperoliidae
e Kigulubi Reed Frog

2689 ***Hyperolius discodactylus***
Anura - Hyperoliidae
e Disc-fingered Reed Frog

2690 ***Hyperolius endjami***
Anura - Hyperoliidae
e Yaounde Reed Frog

2691 ***Hyperolius erythromelanus***
Anura - Hyperoliidae
e Sangeve Reed Frog

2692 ***Hyperolius erythropus***
Anura - Hyperoliidae
e Medje Reed Frog

2693 ***Hyperolius fasciatus***
Anura - Hyperoliidae
e Banded Reed Frog

2694 ***Hyperolius ferreirai***
Anura - Hyperoliidae
e Rio Luiha Reed Frog

2695 ***Hyperolius ferrugineus***
Anura - Hyperoliidae
e Ituri Reed Frog

2696 ***Hyperolius fimbricatus***
Anura - Hyperoliidae
e Lambarene Reed Frog

2697 ***Hyperolius frontalis***
Anura - Hyperoliidae
e Bushoho Reed Frog

2698 ***Hyperolius fuscigula***
Anura - Hyperoliidae
e Brown-throated Reed Frog

2699 ***Hyperolius fusciventris***
Anura - Hyperoliidae
e Lime Reed Frog
d Dunkelbäuchiger Riedfrosch
f Rainette des roseaux

2700 ***Hyperolius ghesquieri***
Anura - Hyperoliidae
e Central Zaire Reed Frog

2701 ***Hyperolius gularis***
Anura - Hyperoliidae
e Loanda Reed Frog

2702 ***Hyperolius guttulatus***
Anura - Hyperoliidae
e Dotted Reed Frog; Spotted

2703 ***Hyperolius horstockii***
Anura - Hyperoliidae
e Arum Reed Frog; Arum Frog (CSA);

Arum Lily Frog (CSA); Yellow-striped Reed Frog; Arum Lily Reed Frog (CSA); Horstock's Arum Frog (CSA)
d Arumfrosch; Aronstabfrosch
f Grenouille des arums; Grenouille rieuse

2704 ***Hyperolius houyi***
Anura - Hyperoliidae
e Ussagara Reed Frog

2705 ***Hyperolius inornatus***
Anura - Hyperoliidae
e Congo Reed Frog

2706 ***Hyperolius kachalolae***
Anura - Hyperoliidae
e Kachalola Reed Frog

2707 ***Hyperolius kilbarae***
Anura - Hyperoliidae
e Shaba Reed Frog

2708 ***Hyperolius kivuensis***
Anura - Hyperoliidae
e Lake Kivu Reed Frog

2709 ***Hyperolius kuligae***
Anura - Hyperoliidae
e Camp Reed Frog

2710 ***Hyperolius lamottei***
Anura - Hyperoliidae
e Ivory Coast Reed Frog

2711 ***Hyperolius lateralis***
Anura - Hyperoliidae
e Secteur Munsenene Reed Frog

2712 ***Hyperolius laticeps***
Anura - Hyperoliidae
e Tofo Reed Frog

2713 ***Hyperolius laurenti***
Anura - Hyperoliidae
e Schiotz's Reed Frog

2714 ***Hyperolius leleupi***
Anura - Hyperoliidae
e Luvubu Reed Frog

2715 ***Hyperolius leucotaenius***
Anura - Hyperoliidae
e White-striped Reed Frog

2716 ***Hyperolius lucani***
Anura - Hyperoliidae
e Landana Reed Frog

2717 ***Hyperolius maestus***
Anura - Hyperoliidae
e Cabinda Reed Frog

2718 ***Hyperolius marmoratus***
Anura - Hyperoliidae
e Marbled Reed Frog; Painted Reed Frog (CSA); Marbled Rush Frog; Variegated Rush Frog; Spotted Tree Frog; Black-and-white-striped Sedge Frog; Spotted Tree Frog (CSA); Marbled Bush Frog (CSA)
d Marmorierter Riedfrosch
f Hyperolius marbré

2719 ***Hyperolius minutissimus***
Anura - Hyperoliidae
e Tiny Reed Frog

2720 ***Hyperolius mitchelli***
Anura - Hyperoliidae
e Mitchell's Reed Frog

2721 ***Hyperolius molleri***
Anura - Hyperoliidae
e Bedriaga Reed Frog

2722 ***Hyperolius montanus***
Anura - Hyperoliidae
e Mountain Reed Frog

2723 ***Hyperolius mosaicus***
Anura - Hyperoliidae
e South Cameroon Reed Frog

2724 ***Hyperolius nasutus***
Anura - Hyperoliidae
e Longnose Reed Frog; Long-nosed Reed Frog; Sharp-nosed Reed Frog; Long Reed Frog (CSA); Green Long Reed Frog
d Nasenriedfrosch

2725 ***Hyperolius obscurus***
Anura - Hyperoliidae
e Kasai Reed Frog

2726 ***Hyperolius occidentalis***
Anura - Hyperoliidae
e Western Reed Frog

2727 ***Hyperolius ocellatus***
Anura - Hyperoliidae
e Ocellated Reed Frog

2728 ***Hyperolius pardalis***
Anura - Hyperoliidae
e Leopard Reed Frog

2729 ***Hyperolius parkeri***
Anura - Hyperoliidae
e Parker's Reed Frog; Trilling Reed Frog (CSA)

2730 ***Hyperolius phantasticus***
Anura - Hyperoliidae
e Fantastic Reed Frog

2731 ***Hyperolius pickersgillii***
Anura - Hyperoliidae
e Avoca Reed Frog; Pickersgill's Reed Frog (CSA)

2732 ***Hyperolius picturatus***
Anura - Hyperoliidae
e Tanzania Reed Frog

2733 ***Hyperolius pictus***
Anura - Hyperoliidae
e Ahl's Reed Frog; Variable Montane Sedge Frog

2734 ***Hyperolius platyceps***
Anura - Hyperoliidae
e Benito River Reed Frog

2735 ***Hyperolius polli***
Anura - Hyperoliidae
e Tshimbulu Reed Frog

2736 ***Hyperolius polystictus***
Anura - Hyperoliidae
e Fall's Reed Frog

2737 ***Hyperolius protchei***
Anura - Hyperoliidae
e Rochebrune's Reed Frog

2738 ***Hyperolius punctulatus***
Anura - Hyperoliidae
e Quanza Reed Frog

2739 ***Hyperolius pusillus***
Anura - Hyperoliidae
e Dwarf Reed Frog; Waterlily Frog; Transparent Pygmy Sedge Frog; Transluscent Tree Frog (CSA); Waterlily Reed Frog (CSA); Lily-pad Frog (CSA)
d Wasserlilienfrosch; Seerosen-Riedfrosch

2740 ***Hyperolius pustulifer***
Anura - Hyperoliidae
e Luvungi Reed Frog

2741 ***Hyperolius quadratomaculatus***
Anura - Hyperoliidae
e Four-spotted Reed Frog

2742 ***Hyperolius quinquevittatus***
Anura - Hyperoliidae

e Tropical Reed Frog; Five-lined Sedge Frog

2743 ***Hyperolius raveni***
Anura - Hyperoliidae
e Volcano Reed Frog

2744 ***Hyperolius reesi***
Anura - Hyperoliidae
e Magambero Forest Reed Frog

2745 ***Hyperolius riggenbachi***
Anura - Hyperoliidae
e Riggenbach's Reed Frog

2746 ***Hyperolius rizophilus***
Anura - Hyperoliidae
e African Reed Frog

2747 ***Hyperolius robustus***
Anura - Hyperoliidae
e Gemba Reed Frog

2748 ***Hyperolius rubrovermiculatus***
Anura - Hyperoliidae
e Shimba Hills Reed Frog

2749 ***Hyperolius sankuruensis***
Anura - Hyperoliidae
e Omanilundu Reed Frog

2750 ***Hyperolius schoutedeni***
Anura - Hyperoliidae
e Kunungu

2751 ***Hyperolius seabrai***
Anura - Hyperoliidae
e Quillombo Reed Frog

2752 ***Hyperolius semidiscus***
Anura - Hyperoliidae
e Hewitt's Reed Frog; Yellow-striped Reed Frog (CSA); Yellow-flanked Reed Frog (CSA)

2753 ***Hyperolius sheldricki***
Anura - Hyperoliidae
e Aruba Dam Reed Frog

2754 ***Hyperolius soror***
Anura - Hyperoliidae
e Soror Reed Frog

2755 ***Hyperolius spinigularis***
Anura - Hyperoliidae
e Mulanje Reed Frog

2756 ***Hyperolius steindachneri***
Anura - Hyperoliidae
e Steindachner's Reed Frog

2757 ***Hyperolius stenodactylus***
Anura - Hyperoliidae
e Bipinidi Reed Frog

2758 ***Hyperolius sylvaticus***
Anura - Hyperoliidae
e Bobiri Forest Reed Frog

2759 ***Hyperolius tanneri***
Anura - Hyperoliidae
e Tanner's Reed Frog

2760 ***Hyperolius thoractuberculatus***
Anura - Hyperoliidae
e Warty Reed Frog

2761 ***Hyperolius tornieri***
Anura - Hyperoliidae
e Tornier's Reed Frog

2762 ***Hyperolius torrentis***
Anura - Hyperoliidae
e Ukami Reed Frog

2763 ***Hyperolius tristis***
Anura - Hyperoliidae
e Duque Reed Frog

2764 ***Hyperolius tuberculatus***
Anura - Hyperoliidae
e Rainforest Reed Frog

2765 ***Hyperolius tuberlinguis***
Anura - Hyperoliidae
e Smith's Reed Frog; Green Reed Frog (CSA); Yellow-green Reed Frog (CSA); Tinker Reed Frog (CSA)
d Grüner Riedfrosch

2766 ***Hyperolius vilhenai***
Anura - Hyperoliidae
e Luita River Reed Frog

2767 ***Hyperolius viridiflavus***
Anura - Hyperoliidae
e Kenyan Reed Frog; Green-yellow Leaf Frog (CSA); Green-yellow Tree Frog (CSA); Common Brown Tree Frog; Painted Reed Frog (CSA)
d Fleckenriedfrosch; Kreideriedfrosch; Gestreifter Riedfrosch

2768 ***Hyperolius viridigulosus***
Anura - Hyperoliidae
e Stream Reed Frog

2769 ***Hyperolius viridis***
Anura - Hyperoliidae
e Green Reed Frog

2770 ***Hyperolius wermuthi***
Anura - Hyperoliidae
e Wermuth's Reed Frog

2771 ***Hyperolius xenorhinus***
Anura - Hyperoliidae
e Teye Reed Frog

2772 ***Hyperolius zavattarii***
Anura - Hyperoliidae
e Ethiopia Reed Frog

2773 ***Hyperolius zonatus***
Anura - Hyperoliidae
e Nimba Reed Frog

2774 ***Hypogeophis***
Gymnophiona - Caeciliaidae
e Frigate Island Caecilians
d Erdwühlen

2775 ***Hypogeophis rostratus***
Gymnophiona - Caeciliaidae
e Frigate Island Caecilian
d Erdwühle
f Cécilie museau-étroit

2776 ***Hypopachus***
Anura - Microhylidae
e Sheep Frogs

2777 ***Hypopachus barberi***
Anura - Microhylidae
e Barber's Sheep Frog; Sheep Frog

2778 ***Hypopachus variolosus***
Anura - Microhylidae
e Sheep Frog

I

2779 **Ichthyophiidae**
Gymnophiona
e Fish Caecilians
d Fischwühlen
f Ichthyophidés

2780 ***Ichthyophis***
Gymnophiona - Ichthyophiidae
e Asian Caecilians; Ichthyophises (ISC)
d Fischwühle

2781 ***Ichthyophis acuminatus***
Gymnophiona - Ichthyophiidae
e Me Wang Valley Caecilian; Pointed-snouted Caecilian (SEA)
d Spitzkopf-Fischwühle

2782 ***Ichthyophis atricollaris***
Gymnophiona - Ichthyophiidae
e Long Bloee Caecilian

2783 ***Ichthyophis bannanicus***
Gymnophiona - Ichthyophiidae
e Mengla County Caecilian

2784 ***Ichthyophis beddomei***
Gymnophiona - Ichthyophiidae
e Nilgherries Caecilian; Beddome's Ichthyophis

2785 ***Ichthyophis bernisi***
Gymnophiona - Ichthyophiidae
e Indonesia Caecilian

2786 ***Ichthyophis biangularis***
Gymnophiona - Ichthyophiidae
e Metang Caecilian

2787 ***Ichthyophis billitonensis***
Gymnophiona - Ichthyophiidae
e Billiton Island Caecilian

2788 ***Ichthyophis bombayensis***
Gymnophiona - Ichthyophiidae
e Bombay Caecilian

2789 ***Ichthyophis dulitensis***
Gymnophiona - Ichthyophiidae
e Mount Dulit Caecilian

2790 ***Ichthyophis elongatus***
Gymnophiona - Ichthyophiidae
e Elongated Caecilian

2791 ***Ichthyophis glandulosus***
Gymnophiona - Ichthyophiidae
e Abungabung Caecilian

2792 ***Ichthyophis glutinosus***
Gymnophiona - Ichthyophiidae
e Ceylon Caecilian; Sticky Caecilian; Sri Lankan Striped Caecilian
d Ceylon-Wülhle
f Cécilie glutineuse

2793 ***Ichthyophis humphreyi***
Gymnophiona - Ichthyophiidae
e Humphrey's Caecilian

2794 ***Ichthyophis hypocyaneus***
Gymnophiona - Ichthyophiidae
e Marsh Caecilian

2795 ***Ichthyophis javanicus***
Gymnophiona - Ichthyophiidae
e Java Caecilian

2796 ***Ichthyophis kohtaoensis***
Gymnophiona - Ichthyophiidae
e Koh Tao Island Caecilian; Koh Tao

Caecilian (SEA)
d Koh-Tao-Fischwühle

2797 ***Ichthyophis laosensis***
Gymnophiona - Ichthyophiidae
e Upper Laos Caecilian

2798 ***Ichthyophis longicephalus***
Gymnophiona - Ichthyophiidae
e Long-headed Caecilian

2799 ***Ichthyophis malabarensis***
Gymnophiona - Ichthyophiidae
e Travancore Caecilian

2800 ***Ichthyophis mindanaoensis***
Gymnophiona - Ichthyophiidae
e Todaya Caecilian

2801 ***Ichthyophis monochrous***
Gymnophiona - Ichthyophiidae
e Western Borneo Caecilian

2802 ***Ichthyophis orthoplicatus***
Gymnophiona - Ichthyophiidae
e Pattipola Caecilian

2803 ***Ichthyophis paucisulcus***
Gymnophiona - Ichthyophiidae
e Siantar Caecilian

2804 ***Ichthyophis peninsularis***
Gymnophiona - Ichthyophiidae
e Peninsular Caecilian

2805 ***Ichthyophis pseudangularis***
Gymnophiona - Ichthyophiidae
e Taylor's Caecilian

2806 ***Ichthyophis sikkimensis***
Gymnophiona - Ichthyophiidae
e Darjeeling Caecilian

2807 ***Ichthyophis singaporensis***
Gymnophiona - Ichthyophiidae
e Singapore Caecilian

2808 ***Ichthyophis subterrestris***
Gymnophiona - Ichthyophiidae
e Cochin Caecilian

2809 ***Ichthyophis sumatranus***
Gymnophiona - Ichthyophiidae
e Sumatra Caecilian

2810 ***Ichthyophis supachai***
Gymnophiona - Ichthyophiidae
e Nakon Si Thammarat Caecilian; Supacha's Caecilian (SEA)
d Supachas Fischwühle

2811 ***Ichthyophis tricolor***
Gymnophiona - Ichthyophiidae
e Maddathorai Caecilian

2812 ***Ichthyophis youngorum***
Gymnophiona - Ichthyophiidae
e Chiang Mai Caecilian; Youngor's Caecilian (SEA)
d Yungors Fischwühle

2813 ***Idiocranium***
Gymnophiona - Caeciliaidae
e Makamunu Assumbo Caecilians

2814 ***Idiocranium russeli***
Gymnophiona - Caeciliaidae
e Makamunu Assumbo Caecilian

2815 ***Indirana***
Anura - Ranidae
e Indian Frogs

2816 ***Indirana beddomii***
Anura - Ranidae
e Beddome's Indian Frog; Beddome's Frog (ISC)

2817 ***Indirana brachytarsus***
Anura - Ranidae
e Anamallais Indian Frog

2818 ***Indirana diplosticta***
Anura - Ranidae
e Malabar Indian Frog

2819 ***Indirana gundia***
Anura - Ranidae
e Gundia Indian Frog

2820 ***Indirana leithii***
Anura - Ranidae
e Matheran Indian Frog; Leith's Frog (ISC)

2821 ***Indirana leptodactyla***
Anura - Ranidae
e Boulenger's Indian Frog

2822 ***Indirana phrynoderma***
Anura - Ranidae
e Kerala Indian Frog

2823 ***Indirana semipalmata***
Anura - Ranidae
e South Indian Frog

2824 ***Indirana tenuilingua***
Anura - Ranidae
e Rao's Indian Frog

2825 ***Indotyphlus***
Gymnophiona - Caeciliaidae
e Battersby's Caecilians; Khandala Caecilians (ISC)

2826 ***Indotyphlus battersbyi***
Gymnophiona - Caeciliaidae
e Battersby's Caecilian; Khandala Caecilian (ISC)

2827 ***Ingerana***
Anura - Ranidae
e Eastern Frogs

2828 ***Ingerana baluensis***
Anura - Ranidae
e Balu Eastern Frog

2829 ***Ingerana liui***
Anura - Ranidae
e Menglan Eastern Frog

2830 ***Ingerana mariae***
Anura - Ranidae
e Palawan Eastern Frog

2831 ***Ingerana reticulata***
Anura - Ranidae
e Reticulate Eastern Frog

2832 ***Ingerana sariba***
Anura - Ranidae
e Saribau Eastern Frog

2833 ***Ingerana tasanae***
Anura - Ranidae
e Tasan Eastern Frog; Tasan Frog (SEA)
d Tasanfrosch

2834 ***Ingerana tenasserimensis***
Anura - Ranidae
e Tenasserim Eastern Frog; Tenasserim Frog (SEA)
d Tenasserim-Frosch

2835 ***Ingerana xigangensis***
Anura - Ranidae
e Xizang Eastern Frog

2836 ***Insuetophrynus***
Anura - Leptodactylidae
e Barrio's Frogs

2837 ***Insuetophrynus arcarpicus***
Anura - Leptodactylidae

e Barrio's Frog

2838 ***Ischnocnema***
Anura - Leptodactylidae
e Big-headed Frogs

2839 ***Ischnocnema quixensis***
Anura - Leptodactylidae
e Common Big-headed Frog

2840 ***Ischnocnema saxatilis***
Anura - Leptodactylidae
e Tarapoto Big-headed Frog

2841 ***Ischnocnema simmonsi***
Anura - Leptodactylidae
e Simmons's Big-headed Frog

2842 ***Ischnocnema verrucosa***
Anura - Leptodactylidae
e Brazilian Big-headed Frog

2843 ***Ixalotriton***
Caudata - Plethodontidae
e Jumping Salamanders

2844 ***Ixalotriton niger***
Caudata - Plethodontidae
e Jumping Salamander; Black Jumping Salamander

K

2845 ***Kalophrynus***
Anura - Microhylidae
e Grainy Frogs

2846 ***Kalophrynus baluensis***
Anura - Microhylidae
e Malaysian Grainy Frog

2847 ***Kalophrynus bunguranus***
Anura - Microhylidae
e Indonesian Grainy Frog

2848 ***Kalophrynus heterochirus***
Anura - Microhylidae
e Bornepo Grainy Frog

2849 ***Kalophrynus intermedius***
Anura - Microhylidae
e Sarawak Grainy Frog

2850 ***Kalophrynus menglienicus***
Anura - Microhylidae
e Menglien Grainy Frog

2851 ***Kalophrynus palmatissimus***
Anura - Microhylidae
e Lowland Grainy Frog; Black-spotted Narrow-mouthed Frog
d Regenklebfrosch

2852 ***Kalophrynus pleurostigma***
Anura - Microhylidae
e Sumatra Grainy Frog; Groin-spot Frog; Streaked Narrowmouth Frog; Streked Narrow-mouthed Frog; Groin-spotted Narrow-mouthed Frog (SEA)
d Semmelkröte; Kleberkröte; Leistenflecken-Klebfrosch

2853 ***Kalophrynus punctatus***
Anura - Microhylidae
e Dotted Grainy Frog

2854 ***Kalophrynus robinsoni***
Anura - Microhylidae
e Robinson's Grainy Frog; Robinson's Narrow-mouthed Frog (SEA)
d Robinsons Klebfrosch

2855 ***Kalophrynus subterrestris***
Anura - Microhylidae
e Burrowing Grainy Frog

2856 ***Kaloula***
Anura - Microhylidae
e Asian Narrowmouth Toads; Kaloulas
d Asiatische Ochsenfrösche

2857 ***Kaloula baleata***
Anura - Microhylidae
e Müller's Narrowmouth Toad; Javanese Bullfrog SEA)
d Javanischer Ochsenfrosch

2858 ***Kaloula borealis***
Anura - Microhylidae
e Manchurian Narrowmouth Toad; Digging Frog; Rain Frog

2859 ***Kaloula conjuncta***
Anura - Microhylidae
e Philippine Narrowmouth Toad

2860 ***Kaloula kokacii***
Anura - Microhylidae
e Catanduanes Island Narrowmouth Toad

2861 ***Kaloula mediolineata***
Anura - Microhylidae
e Siam Narrowmouth Toad; Stripe-backed Bullfrog (SEA)

d Gestreifter Ochsenfrosch

2862 ***Kaloula picta***
Anura - Microhylidae
e Painted Narrowmouth Toad; Painted Kaloula

2863 ***Kaloula pulchra***
Anura - Microhylidae
e Malaysian Narrowmouth Toad; Malayan Bullfrog; Ceylon Kaloula; Digging Frog; Asiatic Painted Frog; Malaysian Narrow-mouthed Toad; Painted Chubby Frog; Malaysian Painted Frog; Common Bullfrog (SEA); Indian Bullfrog (SEA)
d Indischer Ochsenfrosch
f Crapaud buffle; Grenouille peinte d'Asie; Discoglosse asiatique

2864 ***Kaloula rigida***
Anura - Microhylidae
e Luzon Narrowmouth Toad

2865 ***Kaloula rugifera***
Anura - Microhylidae
e Szechwan Narrowmouth Toad

2866 ***Kaloula verrucosa***
Anura - Microhylidae
e Chinese Narrowmouth Toad

2867 ***Kassina***
Anura - Hyperoliidae
e Running Frogs; Kassinas (CSA); Kassin's Frogs (CSA)
d Streifenfrösche; Kassina; Rennfrösche
f Kassines

2868 ***Kassina arboricola***
Anura - Hyperoliidae
e Ivory Coast Running Frog

2869 ***Kassina cassinoides***
Anura - Hyperoliidae
e Silver Running Frog

2870 ***Kassina cochranae***
Anura - Hyperoliidae
e Cochran's Running Frog

2871 ***Kassina decorata***
Anura - Hyperoliidae
e Decorated Running Frog

2872 ***Kassina fusca***
Anura - Hyperoliidae
e Brown Running Frog

2873 ***Kassina kuvangensis***
Anura - Hyperoliidae
e Kuvangu Running Frog

2874 ***Kassina lamottei***
Anura - Hyperoliidae
e Rainforest Running Frog
d Lamottes Kassina

2875 ***Kassina maculata***
Anura - Hyperoliidae
e Spotted Running Frog; Red-blotched Black Frog; Red-legged Kassina (CSA); Vlei Frog (CSA); Brown-spotted Tree Frog (CSA); Red-legged Pan Frog (CSA); Spotted Hylambates
d Rotschenkelfrosch; Fleckenfrosch

2876 ***Kassina maculosa***
Anura - Hyperoliidae
e Marbled Running Frog

2877 ***Kassina mertensi***
Anura - Hyperoliidae
e Mertens's Running Frog

2878 ***Kassina parkeri***
Anura - Hyperoliidae
e Parker's Running Frog

2879 ***Kassina senegalensis***
Anura - Hyperoliidae
e Senegal Running Frog; Running Frog (CSA); Bubbling Kassina (CSA); Senegal Kassina (CSA); Short-toed Kassina (CSA)
d Senegal-Kassina; Senegal-Streifenfrosch
f Grenouille du Sénégal; Kassine de Sénégal

2880 ***Kassina somalica***
Anura - Hyperoliidae
e Somali Running Frog

2881 ***Kassina wittei***
Anura - Hyperoliidae
e De Witte's Running Frog

2882 ***Kassinula***
Anura - Hemisotidae
e Kansenia Frogs

2883 ***Kassinula wittei***
Anura - Hemisotidae
e Kansenia Frog; De Witte's Shovelnose Frog

2884 ***Kyarranus***
Anura - Myobatrachidae
e Sphagnum Frogs

2885 ***Kyarranus kundagungan***
Anura - Myobatrachidae
e Red Tunnel Frog

2886 ***Kyarranus loveridgei***
Anura - Myobatrachidae
e Loveridge's Frog

2887 ***Kyarranus sphagnicolus***
Anura - Myobatrachidae
e Sphagnum Frog

L

2888 ***Lanzarana***
Anura - Ranidae
e Lanza's Frogs

2889 ***Lanzarana largeni***
Anura - Ranidae
e Lanza's Frog

2890 ***Laurentomantis***
Anura - Ranidae
e Grainy Frogs

2891 ***Laurentomantis horrida***
Anura - Ranidae
e Boettger's Grainy Frog

2892 ***Laurentomantis malagasia***
Anura - Ranidae
e Malagasy Grainy Frog

2893 ***Laurentomantis ventrimaculata***
Anura - Ranidae
e Brown Grainy Frog

2894 ***Laurentophryne***
Anura - Bufonidae
e Parker's Tree Toads

2895 ***Laurentophryne parkeri***
Anura - Bufonidae
e Parker's Tree Toad

2896 ***Lechriodus***
Anura - Myobatrachidae
e Cannibal Frogs

2897 ***Lechriodus fletcheri***
Anura - Myobatrachidae
e Fletcher's Cannibal Frog; Fletcher's Frog (ANZ); Fletcher's Southern Frog; Fletcher's Forest Frog

2898 ***Lechriodus intergerivus***
Anura - Myobatrachidae
e Tyler's Cannibal Frog

2899 ***Lechriodus melanopyga***
Anura - Myobatrachidae
e Wiokan Cannibal Frog; Black-rumped Southern Frog

2900 ***Lechriodus platyceps***
Anura - Myobatrachidae
e Arfak Cannibal Frog; Broad-headed Southern Frog

2901 ***Leiopelma***
Anura - Leiopelmatidae
e New Zealand Frogs; Leiopelmas
d Neuseeländischer Frösche; Neuseeländische Urfrösche

2902 ***Leiopelma archeyi***
Anura - Leiopelmatidae
e Coromandel New Zealand Frog; Coromandel Leiopelma; Archey's Frog
d Archeys Frosch; Archey-Frosch
f Grenouille d'Archey

2903 ***Leiopelma hamiltonii***
Anura - Leiopelmatidae
e Brown New Zealand Frog; Hamilton's Frog; Hamilton's Leiopelma; Stephen Island Leiopelma
d Hamiltons Frosch; Hamilton-Frosch
f Grenouille de Hamilton

2904 ***Leiopelma hochstetteri***
Anura - Leiopelmatidae
e Hochstetter's New Zealand Frog;

North Island Leiopelma; Hochstetter's Frog
d Hochstetters Frosch
f Grenouille de Hochstetter

2905 **Leiopelmatidae**
Anura
e New Zealand Frogs; Tailed Frogs; Tail-muscled Frogs
d Urfrösche; Schwanzfrösche; Leiopalmatiden
f Leiopelmatidés

2906 ***Lepidobatrachus***
Anura - Leptodactylidae
e Paraguay Horned Frogs

2907 ***Lepidobatrachus asper***
Anura - Leptodactylidae
e Paraguay Horned Frog; Budgett's Frog
d Chaco-Pfeiffrosch
f Grenouille de Budgett

2908 ***Lepidobatrachus laevis***
Anura - Leptodactylidae
e Budgett's Horned Frog

2909 ***Lepidobatrachus llanensis***
Anura - Leptodactylidae
e Llanos Frog

2910 ***Leptobrachella***
Anura - Megophryidae
e Borneo Frogs; Slim-armed Frogs; Slender-armed Frogs

2911 ***Leptobrachella brevicrus***
Anura - Megophryidae
e Dring's Borneo Frog

2912 ***Leptobrachella mjobergi***
Anura - Megophryidae
e Mount Gadin Borneo Frog

2913 ***Leptobrachella naluensis***
Anura - Megophryidae
e Kamborangah Borneo Frog

2914 ***Leptobrachella natunae***
Anura - Megophryidae
e Natuna Borneo Frog

2915 ***Leptobrachella palmata***
Anura - Megophryidae
e Palm Borneo Frog

2916 ***Leptobrachella parva***
Anura - Megophryidae
e Gunnung Mulu Borneo Frog

2917 ***Leptobrachella serasanae***
Anura - Megophryidae
e Searasan Borneo Frog

2918 ***Leptobrachium***
Anura - Megophryidae
e Eastern Spadefoot Toads; Litter Frogs (SEA)
d Schlankarmfrösche

2919 ***Leptobrachium chapaense***
Anura - Megophryidae
e Chapa Spadefoot Toad; White-eyed Litter Frog (SEA)
d Weißaugen-Schlankarmfrosch

2920 ***Leptobrachium hasseltii***
Anura - Megophryidae
e Java Spadefoot Toad
d Van Hasselterscher Krötenfrosch

2921 ***Leptobrachium hendricksoni***
Anura - Megophryidae
e Thai Spadefoot Toad; Spotted Litter Frog (SEA)
d Hendricksons Schlankarmfrosch

2922 ***Leptobrachium montanum***
Anura - Megophryidae

e Mountain Spadefoot Toad

2923 ***Leptobrachium nigrops***
Anura - Megophryidae
e Singapore Spadefoot Toad; Black-eyed Litter Frog (SEA)
d Schwarzaugen-Schlankarmfrosch

2924 ***Leptobrachium pullum***
Anura - Megophryidae
e Vietnam Spadefoot Toad; Dark Litter Frog (SEA)
d Dunkler Schlankarmfrosch

2925 **Leptodactylidae**
Anura
e Leptodactylids; Tropical Frogs; Southern Frogs
d Südfrösche; Pfeiffrösche; Leptodactyliden
f Leptodactylidés

2926 ***Leptodactylodon***
Anura - Arthroleptidae
e Egg Frogs

2927 ***Leptodactylodon albiventris***
Anura - Arthroleptidae
e Whitebelly Egg Frog

2928 ***Leptodactylodon axillaris***
Anura - Arthroleptidae
e Mountain Egg Frog

2929 ***Leptodactylodon bicolor***
Anura - Arthroleptidae
e Two-coloured Egg Frog

2930 ***Leptodactylodon boulengeri***
Anura - Arthroleptidae
e Boulenger's Egg Frog

2931 ***Leptodactylodon erythrogastser***
Anura - Arthroleptidae
e Redbelly Egg Frog

2932 ***Leptodactylodon mertensi***
Anura - Arthroleptidae
e Mertens's Egg Frog

2933 ***Leptodactylodon ornatus***
Anura - Arthroleptidae
e Ornate Egg Frog

2934 ***Leptodactylodon ovatus***
Anura - Arthroleptidae
e Cameroon Egg Frog

2935 ***Leptodactylodon perreti***
Anura - Arthroleptidae
e Perret's Egg Frog

2936 ***Leptodactylodon polycanthus***
Anura - Arthroleptidae
e African Egg Frog

2937 ***Leptodactylodon ventrimarmoratus***
Anura - Arthroleptidae
e Speckled Egg Frog

2938 ***Leptodactylus***
Anura - Leptodactylidae
e White-lipped Frogs; Tropical Frogs; Nest-building Frogs
d Pfeiffrösche; Echte Pfeiffrösche; Ladenbläser; Südfrösche
f Leptodactyles

2939 ***Leptodactylus albilabris***
Anura - Leptodactylidae
e Günther's White-lipped Frog; White-lipped Frog (WI)

2940 ***Leptodactylus bolivianus***
Anura - Leptodactylidae
e Bolivian White-lipped Frog
d Bolivianischer Pfeiffrosch
f Leptodactyle bolivien

2941 ***Leptodactylus bufonius***
Anura - Leptodactylidae

e Vizcacheras's Frog

2942 ***Leptodactylus camaquara***
Anura - Leptodactylidae
e Jaboticatubas Frog

2943 ***Leptodactylus chaquensis***
Anura - Leptodactylidae
e Cei's White-lipped Frog

2944 ***Leptodactylus cunicularius***
Anura - Leptodactylidae
e Sazima's White-lipped Frog

2945 ***Leptodactylus dantasi***
Anura - Leptodactylidae
e Feijo White-lipped Frog

2946 ***Leptodactylus elenae***
Anura - Leptodactylidae
e Marbled Frog

2947 ***Leptodactylus fallax***
Anura - Leptodactylidae
e Dominican White-lipped Frog; Mountain Chicken Frog
f Leptodactyle des Antilles

2948 ***Leptodactylus flavopictus***
Anura - Leptodactylidae
e Yellow Painted Frog

2949 ***Leptodactylus furnarius***
Anura - Leptodactylidae
e Campo Grande Frog

2950 ***Leptodactylus fuscus***
Anura - Leptodactylidae
e Rufous Frog
f Grenouille galonnée; Leptodactyle galonné

2951 ***Leptodactylus geminus***
Anura - Leptodactylidae
e Grey-striped Frog

2952 ***Leptodactylus gracilis***
Anura - Leptodactylidae
e Dumeril's Striped Frog

2953 ***Leptodactylus hallowelli***
Anura - Leptodactylidae
e Hallowell's White-lipped Frog

2954 ***Leptodactylus insularum***
Anura - Leptodactylidae
e San Miguel Island Frog

2955 ***Leptodactylus jolyi***
Anura - Leptodactylidae
e Joly's Frog

2956 ***Leptodactylus knudseni***
Anura - Leptodactylidae
e Knudsen's Frog
f Leptodactyle de Knudsen

2957 ***Leptodactylus labialis***
Anura - Leptodactylidae
e Mexican White-lipped Frog; Cope's White-lipped Frog

2958 ***Leptodactylus labrosus***
Anura - Leptodactylidae
e Pimocha White-lipped Frog

2959 ***Leptodactylus labyrinthicus***
Anura - Leptodactylidae
e Labyrinth Frog

2960 ***Leptodactylus laticeps***
Anura - Leptodactylidae
e Santa Fe Frog

2961 ***Leptodactylus latinasus***
Anura - Leptodactylidae
e Oven Frog

2962 ***Leptodactylus longirostris***
Anura - Leptodactylidae
e Longnose Santarem Frog
f Leptodactyle à long museau

2963 ***Leptodactylus macrosternum***
Anura - Leptodactylidae
e Miranda's White-lipped Frog

2964 ***Leptodactylus marambaiae***
Anura - Leptodactylidae
e Marambaia White-lipped Frog

2965 ***Leptodactylus melanonotus***
Anura - Leptodactylidae
e Sabinal Frog; Mexican River Frog
d Schwarzrückiger Pfeiffrosch

2966 ***Leptodactylus myersi***
Anura - Leptodactylidae
f Leptodactyle de Myers

2967 ***Leptodactylus mystaceus***
Anura - Leptodactylidae
e Basin White-lipped Frog
f Leptodactyle à moustache

2968 ***Leptodactylus mystacinus***
Anura - Leptodactylidae
e Moustached Frog
d Schnurbartpfeifer

2969 ***Leptodactylus notoaktites***
Anura - Leptodactylidae
e Iporanga White-lipped Frog

2970 ***Leptodactylus ocellatus***
Anura - Leptodactylidae
e Criolla Frog
d Augenpfeifer
f Grenouille ocellée

2971 ***Leptodactylus pentadactylus***
Anura - Leptodactylidae
e South American Bullfrog; Smoky Jungle Frog; Slender-fingered Bladder Frog
d Südamerikanischer Ochsenfrosch; Fünffingeriger Pfeiffrosch
f Grenouille taureau; Grenouille taureau d'Amérique du Sud; Leptodactyle géant

2972 ***Leptodactylus plaumanni***
Anura - Leptodactylidae
e Nova Teutonia White-lipped Frog

2973 ***Leptodactylus podicipinus***
Anura - Leptodactylidae
e Pointedbelly Frog

2974 ***Leptodactylus poecilochilus***
Anura - Leptodactylidae
e Turbo White-lipped Frog

2975 ***Leptodactylus pustulatus***
Anura - Leptodactylidae
e Ceara White-lipped Frog; Pasture Frog

2976 ***Leptodactylus rhodomystax***
Anura - Leptodactylidae
e Loreto White-lipped Frog
f Leptodactyle rougeâtre

2977 ***Leptodactylus rhodonotus***
Anura - Leptodactylidae
e Peru White-lipped Frog

2978 ***Leptodactylus rhodostima***
Anura - Leptodactylidae
e Nauta White-lipped Frog

2979 ***Leptodactylus riveroi***
Anura - Leptodactylidae
e Rivero's White-lipped Frog

2980 ***Leptodactylus rugosus***
Anura - Leptodactylidae

e Guyana White-lipped Frog

2981 ***Leptodactylus silvanimbus***
Anura - Leptodactylidae
e Honduras White-lipped Frog

2982 ***Leptodactylus spixii***
Anura - Leptodactylidae
e Spix's White-lipped Frog

2983 ***Leptodactylus stenodema***
Anura - Leptodactylidae
e San Jose White-lipped Frog
f Leptodactyle étroit

2984 ***Leptodactylus syphax***
Anura - Leptodactylidae
e Matto Grosso White-lipped Frog

2985 ***Leptodactylus tapiti***
Anura - Leptodactylidae
e Goias White-lipped Frog

2986 ***Leptodactylus troglodytes***
Anura - Leptodactylidae
e Pernambuco White-lipped Frog

2987 ***Leptodactylus ventrimaculatus***
Anura - Leptodactylidae
e Bulun White-lipped Frog

2988 ***Leptodactylus viridis***
Anura - Leptodactylidae
e Jim's White-lipped Frog

2989 ***Leptodactylus wagneri***
Anura - Leptodactylidae
e Wagner's White-lipped Frog

2990 ***Leptolalax***
Anura - Megophryidae
e Asian Toads; Slender Frogs (SEA)
d Schlankfrösche

2991 ***Leptolalax alpinus***
Anura - Megophryidae
e Alpine Asian Toad

2992 ***Leptolalax bourreti***
Anura - Megophryidae
e Bourret's Asian Toad

2993 ***Leptolalax dringi***
Anura - Megophryidae
e Dring's Asian Toad

2994 ***Leptolalax gracilis***
Anura - Megophryidae
e Matang Asian Toad

2995 ***Leptolalax heteropus***
Anura - Megophryidae
e Malaysian Asian Toad; Variableslender Frog (SEA)
d Variabler Schlankfrosch

2996 ***Leptolalax liui***
Anura - Megophryidae
e Fujian Asian Toad

2997 ***Leptolalax pelodytoides***
Anura - Megophryidae
e Thao Asian Toad; Mudslender Frog (SEA)
d Schlamm-Schlankfrosch

2998 ***Leptolalax ventripunctatus***
Anura - Megophryidae
e Yunnan Asian Toad

2999 ***Leptopelis***
Anura - Hyperoliidae
e Forest Tree Frogs; Tree Frogs (CSA); Leaf Frogs (CSA)
d Waldsteiger; Wsaldsteigerfrösche

3000 ***Leptopelis anchiella***
Anura - Hyperoliidae
e Huila Forest Tree Frog

3001 ***Leptopelis argenteus***
Anura - Hyperoliidae
e Bagamoyo Forest Tree Frog; Glade Tree Frog (CSA); Silver Tree Frog (CSA)

3002 ***Leptopelis aubryi***
Anura - Hyperoliidae
e Gaboon Forest Tree Frog; Gabon Forest Tree Frog

3003 ***Leptopelis barbouri***
Anura - Hyperoliidae
e Barbour's Forest Tree Frog

3004 ***Leptopelis bequaerti***
Anura - Hyperoliidae
e Gbanga Forest Tree Frog

3005 ***Leptopelis bocagei***
Anura - Hyperoliidae
e Horseshoe Forest Tree Frog; Bocages's Frog; Bocages Tree Frog (CSA); Bocage's Burrowing Frog

3006 ***Leptopelis boulengeri***
Anura - Hyperoliidae
e Victoria Forest Tree Frog

3007 ***Leptopelis brevipes***
Anura - Hyperoliidae
e Musole Forest Tree Frog

3008 ***Leptopelis brevirostris***
Anura - Hyperoliidae
e Cameroon Forest Tree Frog
d Kurzschschnäuziger Waldsteiger

3009 ***Leptopelis broadleyi***
Anura - Hyperoliidae
e Broadley's Forest Tree Frog

3010 ***Leptopelis bufonides***
Anura - Hyperoliidae
e Savannah Forest Tree Frog

3011 ***Leptopelis calcaratus***
Anura - Hyperoliidae
e Efulen Forest Tree Frog

3012 ***Leptopelis christyi***
Anura - Hyperoliidae
e Christy's Forest Tree Frog

3013 ***Leptopelis cinnamomeus***
Anura - Hyperoliidae
e Angola Forest Tree Frog; Brown-backed Tree Frog (CSA); Brown-backed eaf Frog (CSA)
d Braunrückiger Waldsteigerfrosch

3014 ***Leptopelis concolor***
Anura - Hyperoliidae
e Witu Forest Tree Frog

3015 ***Leptopelis fenestratus***
Anura - Hyperoliidae
e Zaire Forest Tree Frog

3016 ***Leptopelis fiziensis***
Anura - Hyperoliidae
e Mokanga Forest Tree Frog

3017 ***Leptopelis flavomaculatus***
Anura - Hyperoliidae
e Brown Forest Tree Frog; Ornate Tree Frog (CSA); Spotted Tree Frog (CSA)

3018 ***Leptopelis gramineus***
Anura - Hyperoliidae
e Badditu Forest Tree Frog

3019 ***Leptopelis hyloides***
Anura - Hyperoliidae
e Bolama Forest Tree Frog

3020 ***Leptopelis jordani***
Anura - Hyperoliidae
e Congulu Forest Tree Frog

3021 ***Leptopelis karissimbensis***
Anura - Hyperoliidae
e Karrissimbi Forest Tree Frog

3022 ***Leptopelis kivuensis***
Anura - Hyperoliidae
e Kisenyi Forest Tree Frog

3023 ***Leptopelis lebeaui***
Anura - Hyperoliidae
e Nyonga Forest Tree Frog

3024 ***Leptopelis macrotis***
Anura - Hyperoliidae
e Big-eyed Forest Tree Frog

3025 ***Leptopelis marginatus***
Anura - Hyperoliidae
e Quissanga Forest Tree Frog

3026 ***Leptopelis millsoni***
Anura - Hyperoliidae
e Niger Forest Tree Frog

3027 ***Leptopelis modestus***
Anura - Hyperoliidae
e Modest Forest Tree Frog; African Tree Frog

3028 ***Leptopelis mossambicus***
Anura - Hyperoliidae
e Mozambique Forest Tree Frog; Brown-backed Tree Frog (CSA)

3029 ***Leptopelis natalensis***
Anura - Hyperoliidae
e Natal Forest Tree Frog; Forest Tree Frog (CSA); Raucous Tree Frog (CSA); Natal Leaf Frog (CSA)
d Natal-Waldsteiger

3030 ***Leptopelis nordequatorialis***
Anura - Hyperoliidae
e West Cameroon Forest Tree Frog

3031 ***Leptopelis notatus***
Anura - Hyperoliidae
e Common Forest Tree Frog

3032 ***Leptopelis occidentalis***
Anura - Hyperoliidae
e Tai Forest Tree Frog

3033 ***Leptopelis ocellatus***
Anura - Hyperoliidae
e Ocellated Forest Tree Frog

3034 ***Leptopelis omissus***
Anura - Hyperoliidae
e Kala Forest Tree Frog

3035 ***Leptopelis oryi***
Anura - Hyperoliidae
e Garamba Forest Tree Frog

3036 ***Leptopelis palmatus***
Anura - Hyperoliidae
e Palm Forest Tree Frog

3037 ***Leptopelis parbocagii***
Anura - Hyperoliidae
e Lake Upemba Forest Tree Frog

3038 ***Leptopelis parkeri***
Anura - Hyperoliidae
e Parker's Forest Tree Frog

3039 ***Leptopelis parvus***
Anura - Hyperoliidae
e Kanole Forest Tree Frog

3040 ***Leptopelis ragazzii***
Anura - Hyperoliidae
e Shoa Forest Tree Frog

3041 ***Leptopelis rufus***
Anura - Hyperoliidae
e Red Tree Frog

3042 ***Leptopelis susanae***
Anura - Hyperoliidae
e Susana's Forest Tree Frog

3043 ***Leptopelis ulungurensis***
Anura - Hyperoliidae
e Uluguru Forest Tree Frog

3044 ***Leptopelis vannutellii***
Anura - Hyperoliidae
e Dime Forest Tree Frog

3045 ***Leptopelis vermiculatus***
Anura - Hyperoliidae
e Amani Forest Tree Frog

3046 ***Leptopelis viridis***
Anura - Hyperoliidae
e Rusty Forest Tree Frog

3047 ***Leptopelis xenodactylus***
Anura - Hyperoliidae
e Weza Forest Tree Frog; Long-toed Tree Frog (CSA); Long-toed Leaf Frog (CSA); Long-toed Burrowing Tree Frog (CSA); Long-toed Burrowing Leaf Frog (CSA)

3048 ***Leptopelis yaldeni***
Anura - Hyperoliidae
e Grassland Forest Tree Frog

3049 ***Leptophryne***
Anura - Bufonidae
e Indonesia Tree Toads

3050 ***Leptophryne borbonica***
Anura - Bufonidae
e Bourbon Toad; Java Tree Toad; Long-legged Slender Toad (SEA)
d Urwaldkröte; Langbein-Schlangkröte

3051 ***Leptophryne cruentata***
Anura - Bufonidae
e Indonesia Tree Toad
f Crapaud ensanglanté

3052 ***Limnodynastes***
Anura - Myobatrachidae
e Australian Swamp Frogs
d Australische Sumpffrösche; Sumpffrösche

3053 ***Limnodynastes covexiusculus***
Anura - Myobatrachidae
e Australian Marbled Frog; Marbled Southern Frog; Convex Southern Frog; Marbled Frog (ANZ)

3054 ***Limnodynastes depressus***
Anura - Myobatrachidae
e Lake Argyle Frog

3055 ***Limnodynastes dorsalis***
Anura - Myobatrachidae
e Western Banjo Frog; Pobblebonk (ANZ); Sandfrog; Bullfrog
d Banjo-Frosch; Sandfrosch

3056 ***Limnodynastes dumerilii***
Anura - Myobatrachidae
e Eastern Banjo Frog; Bullfrog

3057 ***Limnodynastes fletcheri***
Anura - Myobatrachidae
e Long-thumb Frog; Long-thumbed Frog (ANZ); Barking Frog

3058 ***Limnodynastes interioris***
Anura - Myobatrachidae
e Giant Banjo Frog; Giant Bullfrog

3059 ***Limnodynastes ornatus***
Anura - Myobatrachidae
e Ornate Burrowing Frog

3060 ***Limnodynastes peronii***
Anura - Myobatrachidae
e Brown-striped Frog (ANZ); Brown Frog; Striped Marsh Frog

d Brauner Australischer Sumpffrosch; Langbeiniger Sumpffrosch

3061 ***Limnodynastes salmini***
Anura - Myobatrachidae
e Salmon-striped Frog; Steindachner's Frog

3062 ***Limnodynastes spenceri***
Anura - Myobatrachidae
e Sandy Burrowing Frog; Spencer's Nurrowing Frog

3063 ***Limnodynastes tasmaniensis***
Anura - Myobatrachidae
e Spotted Grass Frog; Spotted Marsh Frog
d Tasmanischer Sumpffrosch

3064 ***Limnodynastes terraereginae***
Anura - Myobatrachidae
e Northern Banjo Frog; Northern Bullfrog (ANZ)

3065 ***Limnomedusa***
Anura - Leptodactylidae
e Rapids Frogs

3066 ***Limnomedusa macroglossa***
Anura - Leptodactylidae
e Rapids Frog

3067 ***Limnonectes***
Anura - Ranidae
e Wart Frogs

3068 ***Limnonectes acanthi***
Anura - Ranidae
e Busuanga Wart Frog

3069 ***Limnonectes andamanensis***
Anura - Ranidae
e Andaman Wart Frog

3070 ***Limnonectes arathooni***
Anura - Ranidae
e Djikoro Wart Frog

3071 ***Limnonectes blythii***
Anura - Ranidae
e Blyth's Wart Frog; Blyth's Giant Frog (SEA)
d Blyths Riesenfrosch

3072 ***Limnonectes brevipalmatus***
Anura - Ranidae
e Pegu Wart Frog

3073 ***Limnonectes cancrivorus***
Anura - Ranidae
e Java Wart Frog; Mangrove Frog; Crab-eating Frog
d Philippinen-Frosch; Krabbenfresser

3074 ***Limnonectes corrugatus***
Anura - Ranidae
e Sri Lanka Wart Frog

3075 ***Limnonectes dabanus***
Anura - Ranidae
e Annam Wart Frog

3076 ***Limnonectes dammeremani***
Anura - Ranidae
e Dammerman's Wart Frog

3077 ***Limnonectes diuatus***
Anura - Ranidae
e Tagibo River Wart Frog

3078 ***Limnonectes doriae***
Anura - Ranidae
e Burmese Wart Frog; Dori's Frog (SEA)
d Doris Frosch

3079 ***Limnonectes finchi***
Anura - Ranidae
e Finch's Wart Frog

3080 ***Limnonectes fragilis***
Anura - Ranidae
e Fragile Wart Frog

3081 ***Limnonectes greenii***
Anura - Ranidae
e Hainan Wart Frog

3082 ***Limnonectes grunniens***
Anura - Ranidae
e Amboine Wart Frog
f Grenouille grognante

3083 ***Limnonectes heinrichi***
Anura - Ranidae
e Heinrich's Wart Frog

3084 ***Limnonectes ibanorum***
Anura - Ranidae
e Tributary Wart Frog

3085 ***Limnonectes ingeri***
Anura - Ranidae
e Inger's Wart Frog

3086 ***Limnonectes kenepaiensis***
Anura - Ranidae
e Kenepai Wart Frog

3087 ***Limnonectes keralensis***
Anura - Ranidae
e Kerala Wart Frog

3088 ***Limnonectes khammonensis***
Anura - Ranidae
e Khammon Wart Frog

3089 ***Limnonectes khasianus***
Anura - Ranidae
e Khasi Wart Frog

3090 ***Limnonectes kohchangae***
Anura - Ranidae
e Koh Chang Wart Frog; Koh Chang Frog (SEA)
d Koh-Chang-Frosch

3091 ***Limnonectes kuhlii***
Anura - Ranidae
e Kuhl's Wart Frog; Kuhl's Stream Frog (SEA)
d Kuhls Bachfrosch

3092 ***Limnonectes laticeps***
Anura - Ranidae
e Khasi Wart Frog; Broad-headed Frog (SEA)
d Breitkopffrosch

3093 ***Limnonectes leytensis***
Anura - Ranidae
e Leyte Wart Frog
d Leytescher Bergfrosch

3094 ***Limnonectes limnocharis***
Anura - Ranidae
e Boie's Wart Frog; Rice Frog; Indian Cricket Frog (ISC); Paddyfield Frog; Paddy Frog; Rice-paddy Frog
d Reisfrosch; Südostasiatischer Reisfrosch

3095 ***Limnonectes macrocephalus***
Anura - Ranidae
e Big-headed Wart Frog

3096 ***Limnonectes macrodo***
Anura - Ranidae
e Malaya Wart Frog

3097 ***Limnonectes macrognathus***
Anura - Ranidae
e Karin Hills Wart Frog; Big-headed Frog (SEA)
d Großkopffrosch

3098 ***Limnonectes magnus***
Anura - Ranidae
e Mount Apo Wart Frog

3099 ***Limnonectes malesianus***
Anura - Ranidae
e Singpore Wart Frog; Malayan Giant Frog (SEA)
d Malaiischer Riesenfrosch

3100 ***Limnonectes mawlyndipi***
Anura - Ranidae
e Mawlyndipi Wart Frog

3101 ***Limnonectes mawphlangensis***
Anura - Ranidae
e Mawphlang Wart Frog

3102 ***Limnonectes micrixalus***
Anura - Ranidae
e Abungabung Wart Frog

3103 ***Limnonectes microdiscus***
Anura - Ranidae
e Indonesia Wart Frog
d Bergfrosch

3104 ***Limnonectes microtympanum***
Anura - Ranidae
e Sulawesi Wart Frog

3105 ***Limnonectes modestus***
Anura - Ranidae
e Moluccas Wart Frog

3106 ***Limnonectes murthii***
Anura - Ranidae
e Ghats Wart Frog

3107 ***Limnonectes namiyei***
Anura - Ranidae
e Okinawa Wart Frog

3108 ***Limnonectes nepalensis***
Anura - Ranidae
e Nepal Wart Frog

3109 ***Limnonectes nilagiricus***
Anura - Ranidae
e Nilgiri Wart Frog

3110 ***Limnonectes nitidus***
Anura - Ranidae
e Tanah Rata Wart Frog; Brightness Frog (SEA)
d Glanzfrosch

3111 ***Limnonectes palavanensis***
Anura - Ranidae
e Palawan Wart Frog; Guardian Frog

3112 ***Limnonectes paramacrodon***
Anura - Ranidae
e Tawau Wart Frog; Masked Swamp Frog (SEA)
d Maskierter Sumpffrosch

3113 ***Limnonectes parambikulamana***
Anura - Ranidae
e Parabikulam Wart Frog

3114 ***Limnonectes parvus***
Anura - Ranidae
e Bunawan Wart Frog

3115 ***Limnonectes pierrei***
Anura - Ranidae
e Pierre's Wart Frog

3116 ***Limnonectes pileatus***
Anura - Ranidae
e Thai Wart Frog; Caped Frog (SEA)
d Kappenfrosch

3117 ***Limnonectes plicatellus***
Anura - Ranidae
e Penang Wart Frog; Rhinocerus Frog (SEA)
d Höckerfrosch

3118 ***Limnonectes raja***
Anura - Ranidae
e Pattani Wart Frog; Rajah Frog (SEA)

d Rajah-Frosch

3119 ***Limnonectes rufescens***
Anura - Ranidae
e Malabar Wart Frog

3120 ***Limnonectes sauriceps***
Anura - Ranidae
e Mysore Wart Frog

3121 ***Limnonectes syhadrensis***
Anura - Ranidae
e Bombay Wart Frog

3122 ***Limnonectes teraiensis***
Anura - Ranidae
e Terai Wart Frog

3123 ***Limnonectes timorensis***
Anura - Ranidae
e Timor Wart Frog

3124 ***Limnonectes toumanoffi***
Anura - Ranidae
e Toumanoff's Wart Frog; Tweedie's Frog (SEA)
d Tweedies Frosch

3125 ***Limnonectes tweediei***
Anura - Ranidae
e Tweedie's Wart Frog

3126 ***Limnonectes verruculosus***
Anura - Ranidae
e Sundas Wart Frog

3127 ***Limnonectes visavanus***
Anura - Ranidae
e Philippine Wart Frog

3128 ***Limnonectes vittiger***
Anura - Ranidae
e Luzon Wart Frog

3129 ***Limnonectes woodworthi***
Anura - Ranidae
e Woodworth's Wart Frog; Woodworth's Frog

3130 ***Lineatriton***
Caudata - Plethodontidae
e Lined Salamanders; Veracruz Worm Salamanders
d Veracruz-Salamander (pl)

3131 ***Lineatriton lineolus***
Caudata - Plethodontidae
e Lined Salamander; Veracruz Worm Salamander
d Veracruz-Salamander

3132 ***Lithodytes***
Anura - Leptodactylidae
e Gold-striped Frogs
f Lithodytes

3133 ***Lithodytes lineatus***
Anura - Leptodactylidae
e Gold-striped Frog
f Lithodyte rayé

3134 ***Litoria***
Anura - Hylidae
e Australasian Tree Frogs
d Australische Laubfrösche
f Litories

3135 ***Litoria adelaidensis***
Anura - Hylidae
e Slender Tree Frog

3136 ***Litoria alboguttata***
Anura - Hylidae
e Australian Striped Tree Frog; Striped Burrowing Frog

3137 ***Litoria albolabris***
Anura - Hylidae
e Wandolleck's White-lipped Tree Frog

3138 ***Litoria amboinensis***
Anura - Hylidae
e Horst's Tree Frog

3139 ***Litoria angiana***
Anura - Hylidae
e Angiana Tree Frog

3140 ***Litoria arfakiana***
Anura - Hylidae
e Arfakiana Tree Frog

3141 ***Litoria aruensis***
Anura - Hylidae
e Aruens Tree Frog

3142 ***Litoria aurea***
Anura - Hylidae
e Green-and-golden Bell Frog; Golden Bell Frog; Bell Frog
d Australischer Goldlaubfrosch; Goldlaubfrosch
f Rainette d'or; Litorie d'or

3143 ***Litoria becki***
Anura - Hylidae
e Beck's Tree Frog

3144 ***Litoria bicolor***
Anura - Hylidae
e Northern Dwarf Tree Frog

3145 ***Litoria booroolongensis***
Anura - Hylidae
e Booroolong Tree Frog; Booroolong Frog (ANZ)

3146 ***Litoria brevipalmata***
Anura - Hylidae
e Green-thighed Tree Frog; Gold-thighed Frog (ANZ)

3147 ***Litoria brongersmai***
Anura - Hylidae
e Brongersma's Tree Frog

3148 ***Litoria bulmeri***
Anura - Hylidae
e Bulmer's Tree Frog

3149 ***Litoria burrowsae***
Anura - Hylidae
e Tasmanian Tree Frog (ANZ); Burrow's Tree Frog

3150 ***Litoria capitula***
Anura - Hylidae
e Samlakki Tree Frog

3151 ***Litoria castanea***
Anura - Hylidae
e Cream-spotted Tree Frog

3152 ***Litoria chloris***
Anura - Hylidae
e Australian Red-eyed Tree Frog; Red-eyed Tree Frog (ANZ); North Coast Tree Frog; Green Forest Frog
d Rotäugiger Baumfrosch

3153 ***Litoria chloronota***
Anura - Hylidae
e Arfak Mountains Tree Frog

3154 ***Litoria citropa***
Anura - Hylidae
e Blue Mountains Tree Frog
f Rainette à pates orangées; Litorie à pates orangées

3155 ***Litoria congenita***
Anura - Hylidae
e Yule Island Tree Frog

3156 ***Litoria contrastens***
Anura - Hylidae
e Barabuna Tree Frog

3157 ***Litoria coolcoolensis***
Anura - Hylidae
e Cooloola Tree Frog

3158 ***Litoria coplandi***
Anura - Hylidae
e Saxicolina Tree Frog; Saxicoline Tree Frog (ANZ); Copland's Tree Frog

3159 ***Litoria cyclorhynchus***
Anura - Hylidae
e Spotted-thighed Tree Frog; Green-and-golden Bell Frog; Spotted-thighed Frog (ANZ)

3160 ***Litoria dahlii***
Anura - Hylidae
e Dahl's Olive Tree Frog

3161 ***Litoria darlingtoni***
Anura - Hylidae
e Darlington's Madang Tree Frog

3162 ***Litoria dentata***
Anura - Hylidae
e Bleating Tree Frog (ANZ); Keferstein's Tree Frog
d Blökender Baumfrosch

3163 ***Litoria dorsalis***
Anura - Hylidae
e Dwarf Rocket Frog

3164 ***Litoria dorsivena***
Anura - Hylidae
e Eastern Mountains Tree Frog

3165 ***Litoria electrica***
Anura - Hylidae
e Electric Tree Frog

3166 ***Litoria eucnemis***
Anura - Hylidae
e Lonnberg's Tree Frog; Green-eyed Frog

3167 ***Litoria everetti***
Anura - Hylidae
e Everett's Timor Tree Frog

3168 ***Litoria ewingii***
Anura - Hylidae
e Australian Brown Tree Frog; Ewing's Frog; Brown Tree Frog (ANZ); Whistling Frog
f Rainette d'Ewing; Litorie d'Ewing

3169 ***Litoria exophthalmia***
Anura - Hylidae
e Big-eyed Tree Frog

3170 ***Litoria fallax***
Anura - Hylidae
e Eastern Dwarf Tree Frog (ANZ); Green Reed Frog

3171 ***Litoria flavipunctata***
Anura - Hylidae
e Yellow-speckled Tree Frog

3172 ***Litoria freycineti***
Anura - Hylidae
e Freycinet's Tree Frog; Freycinet's Frog (ANZ)
f Rainette de Freycet; Litorie de Freycet

3173 ***Litoria genimaculata***
Anura - Hylidae
e Brown-spotted Tree Frog

3174 ***Litoria gracilenta***
Anura - Hylidae
e Dainty Green Tree Frog; Slender Gree Tree Frog

3175 ***Litoria graminea***
Anura - Hylidae
e Northern New Guinea Tree Frog

3176 ***Litoria impura***
Anura - Hylidae
e Southern New Guinea Tree Frog

3177 ***Litoria inermis***
Anura - Hylidae
e Fleck-lipped Tree Frog

3178 ***Litoria infrafrenata***
Anura - Hylidae
e Australian Giant Tree Frog; Giant Tree Frog; New Guinea Tree Frog
d Riesenlaubfrosch; Papuanischer Riesenlaubfrosch; Australischer Riesenlaubfrosch
f Rainette géante; Rainette géante de Nouvelle-Guinée; Litorie géante; Litorie géante de Nouvelle-Guinée

3179 ***Litoria iris***
Anura - Hylidae
e Western Highland Tree Frog

3180 ***Litoria jenolaensis***
Anura - Hylidae
e Jenolan Caves Tree Frog

3181 ***Litoria jervisiensis***
Anura - Hylidae
e Jervis Bay Tree Frog
f Rainette de Jervis; Litorie de Jervis

3182 ***Litoria jeudii***
Anura - Hylidae
e Garman New Guinea Tree Frog

3183 ***Litoria kinghorni***
Anura - Hylidae
e Kinghorn's Tree Frog

3184 ***Litoria latopalmata***
Anura - Hylidae
e Australian Variable Tree Frog; Broad-palmed Tree Frog

3185 ***Litoria lesuerrii***
Anura - Hylidae
e Lesueur's Tree Frog; Lesueur's Frog (ANZ)
f Rainette de Lesueur; Litorie de Lesueur

3186 ***Litoria leucova***
Anura - Hylidae
e West Sepik Tree Frog

3187 ***Litoria longicrus***
Anura - Hylidae
e Wendessi Tree Frog

3188 ***Litoria longirostris***
Anura - Hylidae
e Longnose Tree Frog

3189 ***Litoria lorica***
Anura - Hylidae
e Alexandra Creek Tree Frog

3190 ***Litoria louisiadensis***
Anura - Hylidae
e Rossel Island Tree Frog

3191 ***Litoria lutea***
Anura - Hylidae
e Faro Island Tree Frog

3192 ***Litoria meirana***
Anura - Hylidae
e Australian Cross-banded Tree Frog

3193 ***Litoria microbelos***
Anura - Hylidae
e Cairns Tree Frog

3194 ***Litoria micromembrana***
Anura - Hylidae
e Nondugl Tree Frog

3195 ***Litoria modica***
Anura - Hylidae
e Oruge Tree Frog

3196 ***Litoria moorei***
Anura - Hylidae

e Western Green-and-golden Bell Frog (ANZ); Bull Frog

3197 ***Litoria multiplica***
Anura - Hylidae
e Kassam Tree Frog

3198 ***Litoria mystax***
Anura - Hylidae
e Moalf Tree Frog

3199 ***Litoria nannotis***
Anura - Hylidae
e Torrent Tree Frog; Waterfall Frog

3200 ***Litoria napae***
Anura - Hylidae
e Snow Mountains Tree Frog

3201 ***Litoria nasuta***
Anura - Hylidae
e Australian Rocket Frog; Streamlined Rocket Frog; Rocket Frog (ANZ)

3202 ***Litoria nigrofreneta***
Anura - Hylidae
e Black-striped Tree Frog

3203 ***Litoria nigropunctata***
Anura - Hylidae
e Black-dotted Tree Frog

3204 ***Litoria nyakalensis***
Anura - Hylidae
e Henrietta Creek Tree Frog

3205 ***Litoria obtusirostris***
Anura - Hylidae
e Jobi Tree Frog

3206 ***Litoria oenicolen***
Anura - Hylidae
e Trauna River Tree Frog

3207 ***Litoria olangburensis***
Anura - Hylidae
e Pepper-throated Tree Frog

3208 ***Litoria pallida***
Anura - Hylidae
e Coastal Floodplains Tree Frog

3209 ***Litoria paraewingi***
Anura - Hylidae
e Red Cryptic Tree Frog

3210 ***Litoria pearsonianana***
Anura - Hylidae
e Cedar Creek Tree Frog; Pearson's Green Tree Frog

3211 ***Litoria peronii***
Anura - Hylidae
e Peron's Tree Frog (ANZ); Peron's Emerald-spotted Tree Frog
f Rainette de Peron

3212 ***Litoria personata***
Anura - Hylidae
e Sandstone Tree Frog

3213 ***Litoria phyllochroa***
Anura - Hylidae
e Leaf-green Tree Frog
d Australischer Grüner Laubfrosch

3214 ***Litoria piperata***
Anura - Hylidae
e Back Creek Tree Frog

3215 ***Litoria pratti***
Anura - Hylidae
e Pratt's Tree Frog

3216 ***Litoria prora***
Anura - Hylidae
e Efogi Tree Frog

3217 ***Litoria pygmea***
Anura - Hylidae
e Geelvink Pygmy Tree Frog

3218 ***Litoria quadrilineata***
Anura - Hylidae
e Lined Tree Frog

3219 ***Litoria raniformis***
Anura - Hylidae
e Blue-thighed Tree Frog; Green-and-golden Bell Frog

3220 ***Litoria reveltata***
Anura - Hylidae
e Orange-thighed Tree Frog

3221 ***Litoria rheocola***
Anura - Hylidae
e Atherton Tableland Tree Frog

3222 ***Litoria rothii***
Anura - Hylidae
e Rust-eyed Tree Frog; Roth's Tree Frog

3223 ***Litoria rubella***
Anura - Hylidae
e Desert Tree Frog

3224 ***Litoria sanguinolenta***
Anura - Hylidae
e Sabang Tree Frog

3225 ***Litoria spenceri***
Anura - Hylidae
e Spencer's Tree Frog

3226 ***Litoria spinifera***
Anura - Hylidae
e Spiny Tree Frog

3227 ***Litoria subglandulosa***
Anura - Hylidae
e Gold-striped Tree Frog

3228 ***Litoria thesaurensis***
Anura - Hylidae
e Treasury Island Tree Frog

3229 ***Litoria timida***
Anura - Hylidae
e Menemsorae Tree Frog

3230 ***Litoria tornieri***
Anura - Hylidae
e Tornier's Australian Tree Frog

3231 ***Litoria tyleri***
Anura - Hylidae
e Tyler's Australian Tree Frog

3232 ***Litoria umbonata***
Anura - Hylidae
e Balievalley Tree Frog

3233 ***Litoria vagabonda***
Anura - Hylidae
e Wahal Tree Frog

3234 ***Litoria vereauxii***
Anura - Hylidae
e Fawn Tree Frog; Verreaux's Tree Frog
d Verrauxs Laubfrosch
f Rainette de Verreaux; Litorie de Verreaux

3235 ***Litoria vinosa***
Anura - Hylidae
e Ithaca Creek Tree Frog

3236 ***Litoria vocivincens***
Anura - Hylidae
e Brown River Tree Frog

3237 ***Litoria watjulumensis***
Anura - Hylidae

e Watjulum Mission Tree Frog

3238 ***Litoria wisselensis***
Anura - Hylidae
e Wissel Lakes Tree Frog

3239 ***Litoria wollastoni***
Anura - Hylidae
e Highland Tree Frog

3240 ***Litoria xanthomera***
Anura - Hylidae
e Lime Tree Frog

3241 ***Liua***
Caudata - Hynobiidae
e Sichuan Salamanders
d Krallentragende Winkelzahnmolche

3242 ***Liua shihi***
Caudata - Hynobiidae
e Sichuan Salamander
d Krallentragender Winkelzahnmolch; Krallen-Winkelzahnmolch; Chinesischer Froschzahnmolch

3243 ***Luetkenotyphlus***
Gymnophiona - Caeciliaidae
e Sao Paolo Caecilians; Sao Paulo Caecilians

3244 ***Luetkenotyphlus brasiliensis***
Gymnophiona - Caeciliaidae
e Sao Paolo Caecilian; Sao Paulo Caecilian

3245 ***Lynchophrys***
Anura - Leptodactylidae
e Amable Maria Frogs

3246 ***Lynchophrys brachydactyla***
Anura - Leptodactylidae
e Amable Maria Frog

3247 ***Lysapsus***
Anura - Pseudidae
e Harlequin Frogs

3248 ***Lysapsus caraya***
Anura - Pseudidae
e Mato Verde Harlequin Frog

3249 ***Lysapsus laevis***
Anura - Pseudidae
e Guyana Harlequin Frog

3250 ***Lysapsus limellus***
Anura - Pseudidae
e Uruguay Harlequin Frog
d Zwerg-Harlekinfrosch

M

3251 ***Macrogenioglottus***
Anura - Leptodactylidae
e Bahia Forest Frogs

3252 ***Macrogenioglottus alipioi***
Anura - Leptodactylidae
e Bahia Forest Frog

3253 ***Madecassophryne***
Anura - Microhylidae
e Madagascar Tree Frogs

3254 ***Madecassophryne truebae***
Anura - Microhylidae
e Madagascar Tree Frog; Trueb's Madagascar Tree Frog

3255 ***Mannophryne***
Anura - Dendrobatidae
e Fingered Poison Frogs; Fingered Poison-arrow Frogs

3256 ***Mannophryne collaris***
Anura - Dendrobatidae
e Collared Poison Frog; Collared Poison-arrow Frog

3257 ***Mannophryne herminae***
Anura - Dendrobatidae
e Hermina's Poison Frog; Hermina's Poison-arrow Frog

3258 ***Mannophryne neblina***
Anura - Dendrobatidae
e Aragua Poison Frog; Aragua Poison-arrow Frog

3259 ***Mannophryne obliterata***
Anura - Dendrobatidae
e St. Teresa Poison Frog; St. Teresa Poison-arrow Frog

3260 ***Mannophryne olmonae***
Anura - Dendrobatidae
e Bloody Bay Poison Frog; Bloody Bay Poison-arrow Frog

3261 ***Mannophryne riveroi***
Anura - Dendrobatidae
e Rivero's Poison Frog; Rivero's Poison-arrow Frog

3262 ***Mannophryne trinitatis***
Anura - Dendrobatidae
e Trinidad Poison Frog ; Trinidad Poison-arrow Frog
d Venezuela-Baumsteiger; Halsbandfrosch; Gelbkehl-Colostethus

3263 ***Mannophryne yustizi***
Anura - Dendrobatidae
e Yacambu Poison Frog; Yacambu Poison-arrow Frog

3264 ***Mantella***
Anura - Ranidae
e Golden Frogs; Mantellas
d Goldfröschchen (pl); Madagaskar-Buntfrösche
f Grenouilles dorées

3265 ***Mantella aurantiaca***
Anura - Ranidae
e Golden Frog; Golden Mantella; Ginger Tree Frog
d Goldfröschchen
f Grenouille rouge; Grenouille dorée; Grenouille dorée de Madagascar

3266 ***Mantella betsileo***
Anura - Ranidae
e Betsileo Golden Frog

3267 ***Mantella cowani***
Anura - Ranidae
e Black Golden Frog

3268 ***Mantella crocea***
Anura - Ranidae
e Eastern Golden Frog

3269 ***Mantella expectata***
Anura - Ranidae
e Tular Golden Frog

3270 ***Mantella laevigata***
Anura - Ranidae
e Folohy Golden Frog

3271 ***Mantella madagascariensis***
Anura - Ranidae
e Madagascar Golden Frog

3272 ***Mantella pulchra***
Anura - Ranidae
e Parker's Golden Frog

3273 ***Mantella viridis***
Anura - Ranidae
e Green Golden Frog; Green Mantella

3274 ***Mantidactylus***
Anura - Ranidae
e Madagascar Frogs

3275 ***Mantidactylus acuticeps***
Anura - Ranidae
e Sharp Madagascar Frog

3276 ***Mantidactylus aerumnalis***
Anura - Ranidae
e Adrangoloaka Madagascar Frog

3277 ***Mantidactylus aglavei***
Anura - Ranidae
e Anamalozoatra Madagascar Frog

3278 ***Mantidactylus albofrenatus***
Anura - Ranidae
e Eastern Madagascar Frog

3279 ***Mantidactylus albolineatus***
Anura - Ranidae
e White-lined Madagascar Frog

3280 ***Mantidactylus alutus***
Anura - Ranidae
e Mountain Madagascar Frog

3281 ***Mantidactylus ambohimitombi***
Anura - Ranidae
e Abohimitombi Madagascar Frog

3282 ***Mantidactylus argenteus***
Anura - Ranidae
e Folohy Madagascar Frog

3283 ***Mantidactylus asper***
Anura - Ranidae
e East Betsileo Madagascar Frog

3284 ***Mantidactylus bertini***
Anura - Ranidae
e Isaka-Ivondro Madagascar Frog

3285 ***Mantidactylus betsileanus***
Anura - Ranidae
e Betsileo Madagascar Frog

3286 ***Mantidactylus bicalcaratus***
Anura - Ranidae
e Sainte Marie Madagascar Frog

3287 ***Mantidactylus biporus***
Anura - Ranidae
e Two-pore Madagascar Frog

3288 ***Mantidactylus blommersae***
Anura - Ranidae
e Moramanga Madagascar Frog

3289 ***Mantidactylus boulengeri***
Anura - Ranidae
e Boulenger's Madagascar Frog

3290 ***Mantidactylus cornutus***
Anura - Ranidae
e Horned Madagascar Frog

3291 ***Mantidactylus curtus***
Anura - Ranidae
e Ankafana Madagascar Frog

3292 ***Mantidactylus decaryi***
Anura - Ranidae
e Decary's Madagascar Frog

3293 ***Mantidactylus depressiceps***
Anura - Ranidae
e Pulsing Madagascar Frog

3294 ***Mantidactylus domerguei***
Anura - Ranidae
e Domergue's Madagascar Frog

3295 ***Mantidactylus eiselti***
Anura - Ranidae
e Eselt's Madagascar Frog

3296 ***Mantidactylus elegans***
Anura - Ranidae
e Elegnt Madagascar Frog

3297 ***Mantidactylus femoralis***
Anura - Ranidae
e Fort Madagascar Frog

3298 ***Mantidactylus flavobrunneus***
Anura - Ranidae
e Blommers's Madagascar Frog

3299 ***Mantidactylus grandidieri***
Anura - Ranidae
e Grandidier's Madagascar Frog

3300 ***Mantidactylus grandisonae***
Anura - Ranidae
e Ambana Madagascar Frog

3301 ***Mantidactylus granulatus***
Anura - Ranidae
e Grainy Madagascar Frog

3302 ***Mantidactylus guibei***
Anura - Ranidae
e Guibe's Madagascar Frog

3303 ***Mantidactylus guttulatus***
Anura - Ranidae
e Grey Madagascar Frog

3304 ***Mantidactylus klemmeri***
Anura - Ranidae
e Klemmer's Madagascar Frog

3305 ***Mantidactylus leucomaculatus***
Anura - Ranidae
e White-spotted Madagascar Frog

3306 ***Mantidactylus liber***
Anura - Ranidae
e Free Madagascar Frog

3307 ***Mantidactylus lugubris***
Anura - Ranidae
e Dumerils Madagascar Frog

3308 ***Mantidactylus luteus***
Anura - Ranidae
e White Madagascar Frog

3309 ***Mantidactylus madecassus***
Anura - Ranidae
e Andringitra Madagascar Frog

3310 ***Mantidactylus majori***
Anura - Ranidae
e Ivohimanita Madagascar Frog

3311 ***Mantidactylus microtympanum***
Anura - Ranidae
e Angel's Madagascar Frog

3312 ***Mantidactylus mocquardi***
Anura - Ranidae
e Mocquard's Madagascar Frog

3313 ***Mantidactylus opparis***
Anura - Ranidae
e Central Madagascar Frog

3314 ***Mantidactylus perracae***
Anura - Ranidae
e Peracca's Madagascar Frog

3315 ***Mantidactylus plicifer***
Anura - Ranidae
e Madagascar Frog; Common Madagascar Frog

3316 ***Mantidactylus pseudoasper***
Anura - Ranidae
e Massif Madagascar Frog

3317 ***Mantidactylus pulcher***
Anura - Ranidae
e Tsarafidy Madagascar Frog

3318 ***Mantidactylus punctatus***
Anura - Ranidae
e Dotted Madagascar Frog

3319 ***Mantidactylus spinifer***
Anura - Ranidae
e Greater Spiny Madagascar Frog

3320 ***Mantidactylus spiniferus***
Anura - Ranidae
e Lesser Spiny Madagascar Frog

3321 ***Mantidactylus ulcerosus***
Anura - Ranidae
e Warty Madagascar Frog

3322 ***Mantidactylus webbi***
Anura - Ranidae
e Webb's Madagascar Frog

3323 ***Mantidactylus wittei***
Anura - Ranidae
e De Witte's Madagascar Frog

3324 ***Mantophryne***
Anura - Microhylidae
e Archipelago Frogs

3325 ***Mantophryne infulata***
Anura - Microhylidae
e Archipelago Frog

3326 ***Mantophryne lateralis***
Anura - Microhylidae
e Archipelago Frog

3327 ***Mantophryne louisiadensis***
Anura - Microhylidae
e Archipelago Frog

3328 ***Megaelosia***
Anura - Leptodactylidae
e Big-tooth Frogs

3329 ***Megaelosia bocainensis***
Anura - Leptodactylidae
e Bocaina Big-tooth Frog

3330 ***Megaelosia goeldii***
Anura - Leptodactylidae
e Rio Big-tooth Frog
d Goeldis Frosch

3331 ***Megaelosia lutzae***
Anura - Leptodactylidae
e Lutz's Big-tooth Frog

3332 ***Megaelosia massarti***
Anura - Leptodactylidae
e Brazilian Big-tooth Frog

3333 ***Megistolotis***
Anura - Myobatrachidae
e Flat-headed Toadlets; Woodworker Frogs

3334 ***Megistolotis lignarius***
Anura - Myobatrachidae
e Flat-headed Toadlet; Woodworker Frog

3335 ***Megophrys***
Anura - Megophryidae
e Asian Spadefoot Toads; Horned Frogs (SEA)
d Zipfelfrösche; Asiatischer Zipfelfrösche; Krötenfrösche; Zipfelfrösche
f Grenouilles cornues; Grenouilles cornues asiatiques

3336 ***Megophrys aceras***
Anura - Megophryidae
e Perak Spadefoot Toad; Malayan Horned Frog (SEA)
d Malaiischer Zipfelfrosch

3337 ***Megophrys baluensis***
Anura - Megophryidae
e Balu Spadefoot Toad

3338 ***Megophrys boettgeri***
Anura - Megophryidae
e Boettger's Spadefoot Toad
d Boettgers Krötenfrosch

3339 ***Megophrys brachykolos***
Anura - Megophryidae
e Peak Spadefoot Toad; Short-legged Toad
d Kurzbeiniger Krötenfrosch

3340 ***Megophrys edwardinae***
Anura - Megophryidae
e Edwardina's Spadefoot Toad

3341 ***Megophrys feae***
Anura - Megophryidae
e Kakhien Hills Spadefoot Toad; Fea's Horned Frog (SEA)
d Feas Zipfelfrosch

3342 ***Megophrys gigantica***
Anura - Megophryidae
e Giant Asian Spadefoot Toad

3343 ***Megophrys glandulosa***
Anura - Megophryidae
e Jingdong Spadefoot Toad

3344 ***Megophrys intermedia***
Anura - Megophryidae
e Annam Spadefoot Toad

3345 ***Megophrys kempii***
Anura - Megophryidae
e Kemp's Spadefoot Toad

3346 ***Megophrys kuatunensis***
Anura - Megophryidae
e Kuatun Spadefoot Toad

3347 ***Megophrys lateralis***
Anura - Megophryidae
e Anderson's Spadefoot Toad; Large Spadefoot Toad (ISC); Greater Stream Horned Frog (SEA)
d Großer Bach-Zipfelfrosch

3348 ***Megophrys longipes***
Anura - Megophryidae
e Malacca Spadefoot Toad; Long-legged Horned Frog (SEA)
d Langbeiniger Krötenfrosch; Langbein-Zipfelfrosch

3349 ***Megophrys mangshanensis***
Anura - Megophryidae
e Mangshan Spadefoot Toad

3350 ***Megophrys minor***
Anura - Megophryidae

e Kwangshien Spadefoot Toad; Dwarf Horned Frog (SEA)
d Chinesischer Krötenfrosch; Zwerg-Zipfelfrosch

3351 ***Megophrys montana***
Anura - Megophryidae
e Asian Spadefoot Toad
d Zipfelfrosch; Asiatischer Zipfelfrosch; Zipfelkröte; Zipfelkrötenfrosch
f Grenouille cornue; Grenouille cornue asiatique

3352 ***Megophrys nankiangensis***
Anura - Megophryidae
e Nankiang Spadefoot Toad

3353 ***Megophrys nasuta***
Anura - Megophryidae
e Bornean Horned Toad; Long-nosed Horned Frog (SEA)
d Nasenzipfelfrosch

3354 ***Megophrys omeimontis***
Anura - Megophryidae
e Mount Omei Spadefoot Toad
d Omeizi-Krötenfrosch

3355 ***Megophrys pachyproctus***
Anura - Megophryidae
e Gelin Spadefoot Toad

3356 ***Megophrys palpebralespinosa***
Anura - Megophryidae
e Tonkin Spadefoot Toad

3357 ***Megophrys parva***
Anura - Megophryidae
e Burmese Spadefoot Toad; Little Spadefoot Toad (ISC); Small Horned Toad; Lesser Stream Horned Frog (SEA)
d Kleiner Bach-Zipfelfrosch

3358 ***Megophrys robusta***
Anura - Megophryidae
e Bengal Spadefoot Toad

3359 ***Megophrys spinata***
Anura - Megophryidae
e Spiny Spadefoot Toad

3360 ***Melanobatrachus***
Anura - Microhylidae
e Malaysian Tree Frogs

3361 ***Melanobatrachus indicus***
Anura - Microhylidae
e Kerala Hills Tree Frog; Black Microhylid (ISC)
d Indischer Schwarzfrosch

3362 ***Melanobatrachus pollicaris***
Anura - Microhylidae
e Malaysian Tree Frog; Treehole Frog
d Baumhöhlenfrosch

3363 ***Melanophryniscus***
Anura - Bufonidae
e South American Redbelly Toads
d Schwarzkrötchen

3364 ***Melanophryniscus atroluteus***
Anura - Bufonidae
e Uruguay Redbelly Toad

3365 ***Melanophryniscus cambaraensis***
Anura - Bufonidae
e Brazilian Redbelly Toad

3366 ***Melanophryniscus devincenzii***
Anura - Bufonidae
e Rivera Redbelly Toad

3367 ***Melanophryniscus macrogranulosus***
Anura - Bufonidae
e Torres Redbelly Toad

3368 ***Melanophryniscus montevidensis***
Anura - Bufonidae
e Montevideo Redbelly Toad

3369 ***Melanophryniscus moreirae***
Anura - Bufonidae
e Maldonado Redbelly Toad

3370 ***Melanophryniscus orejasmirandai***
Anura - Bufonidae
e Orejas-Miranda Redbelly Toad

3371 ***Melanophryniscus rubriventris***
Anura - Bufonidae
e Yungas Redbelly Toad

3372 ***Melanophryniscus sanmartini***
Anura - Bufonidae
e San Martin Redbelly Toad

3373 ***Melanophryniscus stelzneri***
Anura - Bufonidae
e Redbelly Toad
d Argentinischer Stummelfuß; Stelzners Stummelfuß; Schwarzer Stummelfuß

3374 ***Melanophryniscus tumifrons***
Anura - Bufonidae
e Swelling-fronted Redbelly Toad

3375 ***Meristogenys***
Anura - Ranidae
e Borneo Frogs

3376 ***Meristogenys amoropalamus***
Anura - Ranidae
e Mountain Borneo Frog

3377 ***Meristogenys jeroboa***
Anura - Ranidae
e Günther's Borneo Frog

3378 ***Meristogenys kinabaluensis***
Anura - Ranidae
e Kiau Borneo Frog

3379 ***Meristogenys macrophthalmus***
Anura - Ranidae
e Matsui's Borneo Frog

3380 ***Meristogenys orphnocnemis***
Anura - Ranidae
e Sabah Borneo Frog

3381 ***Meristogenys phaeomerus***
Anura - Ranidae
e Kapit Borneo Frog

3382 ***Meristogenys poecilus***
Anura - Ranidae
e Malaysian Borneo Frog

3383 ***Meristogenys whiteheadi***
Anura - Ranidae
e Whitehead's Borneo Frog

3384 ***Mertensiella***
Caudata - Salamandridae
e Caucasus Salamanders
d Schlanksalamander (pl); Kleinasiatische Salamander (pl); Transkaukasische Salamander (pl)
f Salamandres du Caucase

3385 ***Mertensiella caucasica***
Caudata - Salamandridae
e Caucasus Salamander; Caucasian Salamander
d Kaukasus-Salamander
f Salamandre du Caucase
i Mertensiella caucasica

3386 ***Mertensiella luschani***
Caudata - Salamandridae
e Luschan's Salamander; Lycian Salamander
d Lykischer Salamander; Kleinasiatischer Salamander

f Salamandre de Lycie; Salamandre de Luschan; Salamandre d'Anatolie
i Mertensiella di Luschan

3387 ***Mertensophryne***
Anura - Bufonidae
e Snouted Frogs

3388 ***Mertensophryne micranotis***
Anura - Bufonidae
e Loveridge's Snouted Frog; Bunty's Dwarf Toad (CSA)

3389 ***Mertensophryne schmidti***
Anura - Bufonidae
e Schmidt's Snouted Frog

3390 ***Metacrinia***
Anura - Myobatrachidae
e Nicholl's Toadlets

3391 ***Metacrinia nichollsi***
Anura - Myobatrachidae
e Nicholl's Toadlet

3392 ***Metaphrynella***
Anura - Microhylidae
e Borneo Tree Frogs

3393 ***Metaphrynella sundana***
Anura - Microhylidae
e Borneo Tree Frog

3394 ***Micrixalus***
Anura - Ranidae
e Tropical Frogs

3395 ***Micrixalus fuscus***
Anura - Ranidae
e Brown Tropical Frog

3396 ***Micrixalus nudis***
Anura - Ranidae
e Kerala Tropical Frog

3397 ***Micrixalus phyllophilus***
Anura - Ranidae
e Nilgiri Tropical Frog

3398 ***Micrixalus saxicola***
Anura - Ranidae
e Malabar Tropical Frog

3399 ***Micrixalus silvaticus***
Anura - Ranidae
e Southern Tropical Frog

3400 ***Micrixalus thampii***
Anura - Ranidae
e Silent Valley Tropical Frog

3401 ***Microbatrachella***
Anura - Ranidae
e Cape Flats Frogs; Micro Frogs (CSA)

3402 ***Microbatrachella capensis***
Anura - Ranidae
e Cape Flats Frog; Micro Frog (CSA)

3403 ***Microcaecilia***
Gymnophiona - Caeciliaidae
e Tiny Caecilians

3404 ***Microcaecilia albiceps***
Gymnophiona - Caeciliaidae
e Tiny White Caecilian

3405 ***Microcaecilia rabei***
Gymnophiona - Caeciliaidae
e Tiny Venezuelan Caecilian

3406 ***Microcaecilia supernumeraria***
Gymnophiona - Caeciliaidae
e Tiny Brazan Caecilian

3407 ***Microcaecilia taylori***
Gymnophiona - Caeciliaidae
e Tiny Taylor's Caecilian

3408 ***Microcaecilia unicolor***
Gymnophiona - Caeciliaidae
e Tiny Cayenne Caecilian
f Microcécilie unicolore

3409 ***Microhyla***
Anura - Microhylidae
e Rice Frogs; Narrow-mouthed Toads
d Engmaulfrösche
f Microhyles

3410 ***Microhyla achatina***
Anura - Microhylidae
e Javas Rice Frog
f Microhyle agatine

3411 ***Microhyla annamensis***
Anura - Microhylidae
e Viet Nam Rice Frog; Narrow-mouthed Frog (SEA)
d Annam-Engmaulfrosch

3412 ***Microhyla annectens***
Anura - Microhylidae
e Larut Hills Rice Frog; Mountain Narrow-mouthed Frog (SEA)
d Berg-Engmaulfrosch

3413 ***Microhyla berdmorei***
Anura - Microhylidae
e Pegu Rice Frog; Answering Frog; Colonel Berdmore's Narrow-mouthed Frog; Berdmore's Narrow-mouthed Frog (SEA)
d Gelbbauch-Engmaulfrosch

3414 ***Microhyla borneensis***
Anura - Microhylidae
e Borneo Rice Frog; Bornean Narrow-mouthed Frog (SEA)
d Borneo-Engmaulfrosch

3415 ***Microhyla butleri***
Anura - Microhylidae
e Butler's Rice Frog; Butler's Pygmy Frog; Butler's Narrow-mouthed Frog (SEA)
d Butlers Engmaulfrosch

3416 ***Microhyla chakrapanii***
Anura - Microhylidae
e Mayabunder Rice Frog

3417 ***Microhyla fowleri***
Anura - Microhylidae
e Fowler's Rice Frog

3418 ***Microhyla fusca***
Anura - Microhylidae
e Brown Rice Frog

3419 ***Microhyla heymonsi***
Anura - Microhylidae
e Taiwan Rice Frog; Heymon's Narrow-mouthed Frog (SEA)
d Heymonses Engmaulfrosch

3420 ***Microhyla maculifera***
Anura - Microhylidae
e Sabah Rice Frog

3421 ***Microhyla mixtura***
Anura - Microhylidae
e Chinese Rice Frog

3422 ***Microhyla okinavensis***
Anura - Microhylidae
e Okinawa Rice Frog

3423 ***Microhyla ornata***
Anura - Microhylidae
e Ornate Rice Frog; Ornate Narrowhead Frog; Ornate Frog; Microhylid Frog; Ant Frog; Marshy Toad; Ornate Microhylid (ISC); Ornate Narrow-mouthed Frog (SEA)
d Ornament-Engmaulfrosch

3424 ***Microhyla palmipes***
Anura - Microhylidae
e Pengalengan Rice Frog; Malayan

Narrow-mouthed Frog
d Malaiischer Engmaulfrosch

3425 ***Microhyla perparva***
Anura - Microhylidae
e Labang Forest Rice Frog

3426 ***Microhyla petrigena***
Anura - Microhylidae
e Kapit Rice Frog

3427 ***Microhyla picta***
Anura - Microhylidae
e Painted Rice Frog

3428 ***Microhyla pulchra***
Anura - Microhylidae
e Guangdong Rice Frog; Beautiful Narrow-headed Frog; Beautiful Pygmy Frog; Marbled Narrow-mouthed Frog (SEA)
d Orangeschenkel; Orangeschenkel-Frosch; Pracht-Engmaulfrosch

3429 ***Microhyla rubra***
Anura - Microhylidae
e Indian Rice Frog; Red Microhylid (ISC)

3430 ***Microhyla superciliaris***
Anura - Microhylidae
e Batu Cave Rice Frog; Horned Narrow-mouthed Frog (SEA)
d Zipfel-Engmaulfrosch

3431 ***Microhyla zeylanica***
Anura - Microhylidae
e Sri Lanka Rice Frog

3432 **Microhylidae**
Anura
e Narrowmouth Toads; Narrowmouth Frogs; Narrow-mouthed Frogs; Narrow-mouthed Toads; Microhylids
d Engmundfrösche; Engmaulfrösche; Kurzkopffrösche; Microhyliden
f Crapauds à bouche étroite; Microhylidés

3433 ***Micryletta***
Anura - Microhylidae
e Paddy Frogs

3434 ***Micryletta inornata***
Anura - Microhylidae
e Deli Paddy Frog; Brown Pygmy Frog (SEA)
d Ungemusterter Engmaulfrosch

3435 ***Micryletta steinegeri***
Anura - Microhylidae
e Stejneger's Paddy Frog

3436 ***Mimosiphonops***
Gymnophiona - Caeciliaidae
e Worm-patterned Caecilians

3437 ***Mimosiphonops reinhardti***
Gymnophiona - Caeciliaidae
e Reinhardt's Caecilian

3438 ***Mimosiphonops vermiculatus***
Gymnophiona - Caeciliaidae
e Worm-patterned Caecilian

3439 ***Minyobates***
Anura - Dendrobatidae
e Tiny Poison Frogs; Tiny Poison-arrow Frogs

3440 ***Minyobates abditus***
Anura - Dendrobatidae
e Collins's Poison Frog; Collins's Poison-arrow Frog

3441 ***Minyobates altobueyensis***
Anura - Dendrobatidae
e Alto del Buey Poison Frog; Alto del Buey Poison-arrow Frog

3442 ***Minyobates bombetes***
Anura - Dendrobatidae
e Cauca Poison Frog; Cauca Poison-arrow Frog

3443 ***Minyobates fulguritus***
Anura - Dendrobatidae
e Yellowbelly Poison Frog; Yellowbelly Poison-arrow Frog; Yellow-bellied Poison Frog

3444 ***Minyobates minutus***
Anura - Dendrobatidae
e Bluebelly Poison Frog; Bluebelly Poison-arrow Frog; Blue-bellied Poison-arrow Frog
d Zwergbaumsteiger

3445 ***Minyobates ophistomelas***
Anura - Dendrobatidae
e Andean Poison Frog; Andean Poison-arrow Frog

3446 ***Minyobates steyermarki***
Anura - Dendrobatidae
e Demonic Poison Frog; Demonic Poison-arrow Frog

3447 ***Minyobates viridis***
Anura - Dendrobatidae
e Green Poison Frog; Green Poison-arrow Frog

3448 ***Minyobates virolensis***
Anura - Dendrobatidae
e Santander Poison Frog; Santander Poison-arrow Frog

3449 ***Mixophyes***
Anura - Myobatrachidae
e Barred Frogs; Barred River Frogs; Great Barred Frogs

3450 ***Mixophyes balbus***
Anura - Myobatrachidae
e Grey Barred Frog

3451 ***Mixophyes fasciolatus***
Anura - Myobatrachidae
e Great Barred Frog

3452 ***Mixophyes fleayi***
Anura - Myobatrachidae
e Queensland Barred Frog

3453 ***Mixophyes hihihorlo***
Anura - Myobatrachidae
e Mamosado Barred Frog

3454 ***Mixophyes iteratus***
Anura - Myobatrachidae
e Giant Barred Frog

3455 ***Mixophyes schevilli***
Anura - Myobatrachidae
e Northern Barred Frog

3456 ***Myersiella***
Anura - Microhylidae
e Elongated Frogs

3457 ***Myersiella microps***
Anura - Microhylidae
e Elongated Frog; Rio Elongated Frog

3458 **Myobatrachidae**
Anura
e Australian Froglets
d Australische Südfrösche; Myobatrachiden

3459 ***Myobatrachus***
Anura - Myobatrachidae
e Turtle Frogs
d Schildkrötenfrösche

3460 ***Myobatrachus gouldii***
Anura - Myobatrachidae
e Turtle Frog
d Schildkrötenfrosch

N

3461 ***Nannophrys***
Anura - Ranidae
e Streamlined Frogs

3462 ***Nannophrys ceylonnensis***
Anura - Ranidae
e Ceylon Streamlined Frog

3463 ***Nannophrys guentheri***
Anura - Ranidae
e Günther's Streamlined Frog

3464 ***Nannophrys marmorata***
Anura - Ranidae
e Marbled Streamlined Frog

3465 ***Nanorana***
Anura - Ranidae
e Yunnan Slow Frogs

3466 ***Nanorana pleskei***
Anura - Ranidae
e Songpan Slow Frog

3467 ***Nanorana ventripunctata***
Anura - Ranidae
e Yunnan Slow Frog

3468 ***Natalobatrachus***
Anura - Ranidae
e Natal Diving Frogs; Kloof Frogs; Natal Frogs (CSA); Boneberg's Frogs
d Natal-Frösche

3469 ***Natalobatrachus bonebergi***
Anura - Ranidae
e Natal Diving Frog; Gloomy Kloof Frog; Kloof Frog (CSA); Boneberg's Frog (CSA); Natal Frog (CSA)
d Natal-Frosch

3470 ***Nectocaecilia***
Gymnophiona - Typhlonectidae
e Upper Amazon Caecilians

3471 ***Nectocaecilia cooperi***
Gymnophiona - Typhlonectidae
e Cooper's American Caecilian

3472 ***Nectocaecilia haydee***
Gymnophiona - Typhlonectidae
e Estado Zulia Caecilian

3473 ***Nectocaecilia petersii***
Gymnophiona - Typhlonectidae
e Upper Amazon Caecilian

3474 ***Nectophryne***
Anura - Bufonidae
e African Tree Toads
d Afrikanische Baumkröten; Baumkröten; Laubkröten

3475 ***Nectophryne afra***
Anura - Bufonidae
e African Tree Toad
d Laubkröte

3476 ***Nectophryne batesii***
Anura - Bufonidae
e Bates's Tree Toad

3477 ***Nectophrynoides***
Anura - Bufonidae
e African Live-bearing Toads; Viviparous Toads
d Lebendgebärige Kröten; Falsche Schwimmkröten; Falsche Baumkröten; Baumkröten

3478 ***Nectophrynoides cryptus***
Anura - Bufonidae
e Secret Tree Toad

3479 ***Nectophrynoides minutus***
Anura - Bufonidae
e Minute Tree Toad

3480 ***Nectophrynoides tornieri***
Anura - Bufonidae
e Tornier's Tree Toad; Usambara Viviparous Toad
d Torniers Baumkröte

3481 ***Nectophrynoides viviparus***
Anura - Bufonidae
e Morogoro Tree Toad; Rungwe Viviparous Toad
d Tanganjika Baumkröte

3482 ***Nectophrynoides wendyae***
Anura - Bufonidae
e Uzungwe Scarp Tree Toad

3483 ***Necturus***
Caudata - Proteidae
e Waterdogs
d Furchenmolche
f Nectures

3484 ***Necturus alabamensis***
Caudata - Proteidae
e Alabama Waterdog; Black Warrior Waterdog
d Alanama-Furchenmolch

3485 ***Necturus beyeri***
Caudata - Proteidae
e Gulf Coast Waterdog
d Golfküsten-Furchenmolch

3486 ***Necturus lewisi***
Caudata - Proteidae
e Neuse River Waterdog
d Neuse-River-Furchenmolch

3487 ***Necturus maculosus***
Caudata - Proteidae
e Mudpuppy; Waterdog
d Furchenmolch; Gefleckter Furchenmolch
f Necture tacheté
i Cane d'acqua

3488 ***Necturus maculosus louisianensis***
Caudata - Proteidae
e Red River Mudpuppy; Red River Waterdog

3489 ***Necturus maculosus maculosus***
Caudata - Proteidae
e Mudpuppy; Common Mudpuppy

3490 ***Necturus punctatus***
Caudata - Proteidae
e Dwarf Waterdog; Lake Winnebago Mudpuppy
d Zwergfurchenmolch
f Necture pygmée

3491 ***Necturus punctatus lodingi***
Caudata - Proteidae
e Mobile Dwarf Waterdog; Mobile Waterdog

3492 ***Necturus punctatus punctatus***
Caudata - Proteidae
e Dwarf Waterdog
d Zwergfurchenmolch
f Necture pygmée

3493 ***Nelsonophryne***
Anura - Microhylidae
e Nelson Frogs

3494 ***Nelsonophryne aequatorialis***
Anura - Microhylidae
e Cuenca Nelson Frog

3495 ***Nelsonophryne aterrimus***
Anura - Microhylidae

e Costa Rica Nelson Frog

3496 ***Neobatrachus***
Anura - Myobatrachidae
e Stubby Frogs

3497 ***Neobatrachus albipes***
Anura - Myobatrachidae
e White Stubby Frog

3498 ***Neobatrachus aquilonius***
Anura - Myobatrachidae
e Burrowing Rain Frog

3499 ***Neobatrachus centralis***
Anura - Myobatrachidae
e Trilling Frog

3500 ***Neobatrachus fulvus***
Anura - Myobatrachidae
e Mulga Stubby Frog

3501 ***Neobatrachus kunapalari***
Anura - Myobatrachidae
e Yellow Stubby Frog

3502 ***Neobatrachus pelobatoides***
Anura - Myobatrachidae
e Humming Frog
d Knoblauchkrötenähnlicher Frosch

3503 ***Neobatrachus pictus***
Anura - Myobatrachidae
e Painted Stubby Frog; Spotted Burrowing Frog; Meeowing Painted Frog; Spadefoot Toad; Painted Burrowing Frog

3504 ***Neobatrachus sudelli***
Anura - Myobatrachidae
e Striped Stubby Frog

3505 ***Neobatrachus sutor***
Anura - Myobatrachidae
e Shoemaker Frog

3506 ***Neobatrachus willsmorei***
Anura - Myobatrachidae
e Goldfields Bullfrog

3507 ***Nesionixalus***
Anura - Hyperoliidae
e Gulf Frogs

3508 ***Nesionixalus molleri***
Anura - Hyperoliidae
e Moller's Gulf Frog

3509 ***Nesionixalus thomensis***
Anura - Hyperoliidae
e Sao Tome Gulf Frog

3510 ***Nesomantis***
Anura - Sooglossidae
e Thomasset's Frogs; Thomasset's Seychelles Frogs
d Thomassets Seychellen-Frösche

3511 ***Nesomantis thomasseti***
Anura - Sooglossidae
e Thomasset's Frog; Thomasset's Seychelles Frog
d Thomassets Seychellen-Frosch

3512 ***Neurergus***
Caudata - Salamandridae
e Kurdistan Newts
d Salamanderartige Bergmolche; Bergbachmolche
f Tritons du Kurdistan
i Neurergi

3513 ***Neurergus crocatus***
Caudata - Salamandridae
e Lake Urmia Newt; Azerbaijan Newt
d Urmia-Molch; Van-Bergbachmolch; Salamanderartiger Bergmolch
f Triton du Kurdistan
i Neurergo croceo; Neurergo curdo

3514 ***Neurergus kaiseri***
Caudata - Salamandridae
e Luristan Newt; Lorestan Newt
d Lurischer Bergbachmolch; Kaisers Bergbachmolch

3515 ***Neurergus microspilotus***
Caudata - Salamandridae
e Kurdistan Newt

3516 ***Neurergus strauchii***
Caudata - Salamandridae
e Anatolia Newt
d Kürdischer Bergbachmolch; Strauchs Gebirgsbachmolch; Türkischer Bergbachmolch
f Triton de Strauch
i Neurergo di Strauch; Neurergo turcico

3517 ***Nimbaphrynoides***
Anura - Bufonidae
e Nimba Toads

3518 ***Nimbaphrynoides liberiensis***
Anura - Bufonidae
e Liberian Nimba Toad

3519 ***Nimbaphrynoides occidentalis***
Anura - Bufonidae
e Western Nimba Toad; Mountain Nimba Viviparous Toad
d Westliche Lebendgebärige Kröte

3520 ***Notaden***
Anura - Myobatrachidae
e Australian Spadefoot Toads

3521 ***Notaden bennettii***
Anura - Myobatrachidae
e Crucifix Toad (ANZ); Holy Cross Toad; Catholic Frog
d Katholikenfrosch
f Grenouille catholique

3522 ***Notaden melanoscaphus***
Anura - Myobatrachidae
e Northern Spadefoot Toad

3523 ***Notaden nichollsi***
Anura - Myobatrachidae
e Desert Spadefoot Toad
d Nicholls Frosch

3524 ***Notaden weigeli***
Anura - Myobatrachidae
e Grey-limbed Spadefoot Toad

3525 ***Nothophryne***
Anura - Ranidae
e Broadley's Mountain Frogs; Mongrel Frogs

3526 ***Nothophryne broadleyi***
Anura - Ranidae
e Broadley's Mountain Frog; Mongrel Frog

3527 ***Notophthalmus***
Caudata - Salamandridae
e Red-spotted Newts; Newts
d Ostamerikanische Wassermolche; Wassermolche
f Notophthalmes; Tritons

3528 ***Notophthalmus meridionalis***
Caudata - Salamandridae
e Black-spotted Newt
d Schwarzgefleckter Wassermolch; Schwarzfleckenmolch

3529 ***Notophthalmus meridionalis kallerti***
Caudata - Salamandridae
e Mexican Black-spotted Newt

3530 ***Notophthalmus meridionalis meridionalis***
Caudata - Salamandridae
e Texas Black-spotted Newt; Southern Black-spotted Newt

3531 ***Notophthalmus perstriatus***
Caudata - Salamandridae
e Striped Newt
d Gestreifter Wassermolch; Streifenmolch

3532 ***Notophthalmus viridescens***
Caudata - Salamandridae
e Eastern Newt; Eastern Red-spotted Newt (NA)
d Grünlicher Wassermolch; Rotfleckenmolch
f Notophthalme verdâtre; Triton vert; Triton vert à taches rouges

3533 ***Notophthalmus viridescens dorsalis***
Caudata - Salamandridae
e Broken-striped Newt

3534 ***Notophthalmus viridescens louisianensis***
Caudata - Salamandridae
e Central Newt

3535 ***Notophthalmus viridescens piaropicola***
Caudata - Salamandridae
e Peninsula Newt

3536 ***Notophthalmus viridescens viridescens***
Caudata - Salamandridae
e Red-spotted Newt

3537 ***Nototriton***
Caudata - Plethodontidae
e Moss Salamander

3538 ***Nototriton adelos***
Caudata - Plethodontidae
e Sierra Juarez Salamander

3539 ***Nototriton alvarezdeltoroi***
Caudata - Plethodontidae
e Alvarez del Toro Salamander

3540 ***Nototriton barbouri***
Caudata - Plethodontidae
e Yoro Salamander

3541 ***Nototriton diminitus***
Caudata - Plethodontidae
e Quebarada Salamander

3542 ***Nototriton nasalis***
Caudata - Plethodontidae
e Cortes Salamander

3543 ***Nototriton picadoi***
Caudata - Plethodontidae
e La Estrella Salamander

3544 ***Nototriton richardi***
Caudata - Plethodontidae
e Richard's Salamander

3545 ***Nototriton veraepacis***
Caudata - Plethodontidae
e Baja Verapaz Salamander

3546 ***Nyctanolis***
Caudata - Plethodontidae
e Long-limbed Salamanders

3547 ***Nyctanolis pernix***
Caudata - Plethodontidae
e Long-limbed Salamander; Nimble Long-limbed Salamander; Midland Long-tailed Salamander

3548 ***Nyctibatrachus***
Anura - Ranidae
e Night Frogs

3549 ***Nyctibatrachus aliciae***
Anura - Ranidae
e Alicia's Night Frog; Wrinkled Frog

3550 ***Nyctibatrachus beddomii***
Anura - Ranidae

e Beddome's Night Frog

3551 ***Nyctibatrachus deccanensis***
Anura - Ranidae
e Anamallai Night Frog

3552 ***Nyctibatrachus humayuni***
Anura - Ranidae
e Bombay Night Frog; Humayun's Wrinkled Frog

3553 ***Nyctibatrachus kempholeyensis***
Anura - Ranidae
e Kempholey Night Frog

3554 ***Nyctibatrachus major***
Anura - Ranidae
e Malabar Night Frog; Large Wrinkled Frog

3555 ***Nyctibatrachus minor***
Anura - Ranidae
e Kerala Night Frog

3556 ***Nyctibatrachus modestus***
Anura - Ranidae
e Shimoga Night Frog

3557 ***Nyctibatrachus sanctipalustris***
Anura - Ranidae
e Coorg Night Frog

3558 ***Nyctibatrachus sylvaticus***
Anura - Ranidae
e Forest Night Frog

3559 ***Nyctimantis***
Anura - Hylidae
e Brown-eyed Tree Frogs

3560 ***Nyctimantis rugiceps***
Anura - Hylidae
e Brown-eyed Tree Frog

3561 ***Nyctimystes***
Anura - Hylidae
e Big-eyed Tree Frogs

3562 ***Nyctimystes avocalis***
Anura - Hylidae
e Loud Big-eyed Tree Frog

3563 ***Nyctimystes cheesmani***
Anura - Hylidae
e Cheesman's Big-eyed Tree Frog

3564 ***Nyctimystes dayi***
Anura - Hylidae
e Day's Big-eyed Tree Frog; Day's Frog

3565 ***Nyctimystes daymani***
Anura - Hylidae
e Dayman's Big-eyed Tree Frog

3566 ***Nyctimystes disruptus***
Anura - Hylidae
e Madang Big-eyed Tree Frog

3567 ***Nyctimystes fluciatilis***
Anura - Hylidae
e Indonesian Big-eyed Tree Frog

3568 ***Nyctimystes foricula***
Anura - Hylidae
e Kaironk Big-eyed Tree Frog

3569 ***Nyctimystes granti***
Anura - Hylidae
e Grant's Big-eyed Tree Frog

3570 ***Nyctimystes gularis***
Anura - Hylidae
e Mondo Big-eyed Tree Frog

3571 ***Nyctimystes humeralis***
Anura - Hylidae
e Green Big-eyed Tree Frog

3572 ***Nyctimystes kubori***
Anura - Hylidae
e Sandy Big-eyed Tree Frog

3573 ***Nyctimystes montanus***
Anura - Hylidae
e Mountain Big-eyed Tree Frog

3574 ***Nyctimystes narinosus***
Anura - Hylidae
e Common Big-eyed Tree Frog; Big-eyed Tree Frog

3575 ***Nyctimystes obsoletus***
Anura - Hylidae
e Simbang Big-eyed Tree Frog

3576 ***Nyctimystes oktediensis***
Anura - Hylidae
e Richards's Big-eyed Tree Frog

3577 ***Nyctimystes papua***
Anura - Hylidae
e Papua Big-eyed Tree Frog

3578 ***Nyctimystes perimetri***
Anura - Hylidae
e Archipeligo Big-eyed Tree Frog

3579 ***Nyctimystes persimilis***
Anura - Hylidae
e Milne Bay Big-eyed Tree Frog

3580 ***Nyctimystes pulcher***
Anura - Hylidae
e Spurred Big-eyed Tree Frog

3581 ***Nyctimystes rueppelli***
Anura - Hylidae
e Rüppel's Big-eyed Tree Frog

3582 ***Nyctimystes semipalmatus***
Anura - Hylidae
e Kokoda Big-eyed Tree Frog

3583 ***Nyctimystes trachydermis***
Anura - Hylidae
e Morobe Big-eyed Tree Frog

3584 ***Nyctimystes zweifeli***
Anura - Hylidae
e Zweifel's Big-eyed Tree Frog

3585 ***Nyctixalus***
Anura - Rhacophoridae
e Indonesian Tree Frogs

3586 ***Nyctixalus anodon***
Anura - Rhacophoridae
e Sumatra Indonesian Tree Frog

3587 ***Nyctixalus flavosignatus***
Anura - Rhacophoridae
e Java Indonesian Tree Frog

3588 ***Nyctixalus moloch***
Anura - Rhacophoridae
e Assam Indonesian Tree Frog

3589 ***Nyctixalus pictus***
Anura - Rhacophoridae
e Painted Indonesian Tree Frog; White-spotted Tree Frog (SEA)
d Weißflecken-Katzenaugenfrosch

3590 ***Nyctixalus spinosus***
Anura - Rhacophoridae
e Spiny Indonesian Tree Frog

3591 ***Nyctymystes tyleri***
Anura - Hylidae
e Tyler's Big-eyed Tree Frog

O

3592 ***Occidozyga***
Anura - Ranidae
e Java Frogs
d Spitzzungenfrösche

3593 ***Occidozyga lima***
Anura - Ranidae
e Java Frog; Common Floating Frog (SEA)
d Spitzzungenfrosch
f Oxyglosse lime

3594 ***Occidozyga semipalmata***
Anura - Ranidae
e Lowah Frog

3595 ***Odontophrynus***
Anura - Leptodactylidae
e Escurzos; Escuerzos
d Amerikanische Hornfrösche

3596 ***Odontophrynus achalensis***
Anura - Leptodactylidae
e Cordoba Escurzo; Cordoba Escuerzo

3597 ***Odontophrynus americanus***
Anura - Leptodactylidae
e Common Lesser Escuerzo; Common Lesser Escurzo; Lesser Escuerzo; Lesser Escurzo

3598 ***Odontophrynus barrioi***
Anura - Leptodactylidae
e Barrio's Escurzo; Barrio's Escuerzo

3599 ***Odontophrynus carvalhoi***
Anura - Leptodactylidae
e Carvalho's Escurzo; Carvalho's Escuerzo

3600 ***Odontophrynus lavillai***
Anura - Leptodactylidae
e Cei's Escurzo; Cei's Escuerzo

3601 ***Odontophrynus moratoi***
Anura - Leptodactylidae
e Botucatu Escurzo; Botucatu Escuerzo

3602 ***Odontophrynus occidentalis***
Anura - Leptodactylidae
e Cururu Lesser Escurzo; Cururu Lesser Escuerzo

3603 ***Odontophrynusn cultripes***
Anura - Leptodactylidae
e Rio Grande Escurzo; Rio Grande Escuerzo

3604 ***Oedipina***
Caudata - Plethodontidae
e Worm Salamanders
d Tropensalamander (pl)

3605 ***Oedipina alfaroi***
Caudata - Plethodontidae
e Limon Worm Salamander
d Spitzkopf-Tropensalamander

3606 ***Oedipina altura***
Caudata - Plethodontidae
e Cartago Worm Salamander
d Gebirgs-Tropensalamander

3607 ***Oedipina carablanca***
Caudata - Plethodontidae
e Los Diamantes Worm Salamander
d Gedrungener Tropensalamander

3608 ***Oedipina collaris***
Caudata - Plethodontidae
e Collared Worm Salamander
d Natternkopf-Tropensalamander

3609 ***Oedipina complex***
Caudata - Plethodontidae
e Gamboa Worm Salamander
d Maulwurf-Tropensalamander

3610 ***Oedipina cyclocauda***
Caudata - Plethodontidae
e Costa Rica Worm Salamander
d Grauschwarzer Tropensalamander

3611 ***Oedipina elongata***
Caudata - Plethodontidae
e Central American Worm Salamander; Galliwasp
d Blaugrauer Tropensalamander

3612 ***Oedipina gephyra***
Caudata - Plethodontidae
e La Fortuna Worm Salamander

3613 ***Oedipina grandis***
Caudata - Plethodontidae
e Cerro Pando Worm Salamander

3614 ***Oedipina ignea***
Caudata - Plethodontidae
e Chimaltenango Worm Salamander
d Guatemala-Tropensalamander

3615 ***Oedipina parvipes***
Caudata - Plethodontidae
e Colombian Worm Salamander
d Wühlender Tropensalamander

3616 ***Oedipina paucidentata***
Caudata - Plethodontidae
e El Empalme Worm Salamander
d Schleichen-Tropensalamander

3617 ***Oedipina poelzi***
Caudata - Plethodontidae
e Quarry Worm Salamander
d Poelzs Tropensalamander

3618 ***Oedipina stuarti***
Caudata - Plethodontidae
e Stuart's Worm Salamander
d Stuarts Tropensalamander

3619 ***Oedipina taylori***
Caudata - Plethodontidae
e Taylor's Worm Salamander
d Taylors Tropensalamander

3620 ***Oedipina uniformis(Brame 1968)***
Caudata - Plethodontidae
e Cienega Colorado Worm Salamander
d Sumpf-Tropensalamander

3621 ***Oedipina uniformis(Keferstein 1868)***
Caudata - Plethodontidae
e Keferstein's Worm Salamander
d Kefersteins Tropensalamander; Langschwänziger Tropensalamander

3622 ***Onychodactylus***
Caudata - Hynobiidae
e Clawed Salamanders
d Krallenfingermolche; Krallensalamander (pl); Krallenmolche
f Salamandres onguiculée; Onychodactyles

3623 ***Onychodactylus fischeri***
Caudata - Hynobiidae
e Fischer's Clawed Salamander; Long-tailed Salamander
d Koreanischer Krallenfingermolch; Ussuri-Molch; Ussurischer Krallenmolch
f Salamandre onguiculée de l'Oussouri; Onychodactyle de l'Oussouri

3624 ***Onychodactylus japonicus***
Caudata - Hynobiidae
e Japanese Clawed Salamander
d Japanischer Krallenfingermolch; Japanischer Krallensalamander; Japanischer Krallenmolch

f Salamandre onguiculée du Japon; Onychodactyle du Japon; Onychodactyle japonais

3625 ***Ophryophryne***
Anura - Megophryidae
e Mountain Toads

3626 ***Ophryophryne microstoma***
Anura - Megophryidae
e Asian Mountain Toad

3627 ***Ophryophryne pachyproctus***
Anura - Megophryidae
e Zhushihe Mountain Toad

3628 ***Ophryophryne poilani***
Anura - Megophryidae
e Spiny Mountain Toad

3629 ***Opisthothylax***
Anura - Hyperoliidae
e Grey-eyed Frogs

3630 ***Opisthothylax immaculatus***
Anura - Hyperoliidae
e Grey-eyed Frog

3631 ***Oreophryne***
Anura - Microhylidae
e Cross Frogs

3632 ***Oreophryne albopunctata***
Anura - Microhylidae
e White-dotted Cross Frog

3633 ***Oreophryne annulata***
Anura - Microhylidae
e Davao Cross Frog

3634 ***Oreophryne anthonyi***
Anura - Microhylidae
e Anthony's Cross Frog

3635 ***Oreophryne biroi***
Anura - Microhylidae
e New Guinea Cross Frog

3636 ***Oreophryne brachypus***
Anura - Microhylidae
e Gazelle Cross Frog

3637 ***Oreophryne brevicrus***
Anura - Microhylidae
e Lake Habbema Cross Frog

3638 ***Oreophryne celebensis***
Anura - Microhylidae
e Celebes Cross Frog

3639 ***Oreophryne crucifer***
Anura - Microhylidae
e Common Cross Frog; Cross Frog

3640 ***Oreophryne flava***
Anura - Microhylidae
e Iran Cross Frog; Yellow Frog

3641 ***Oreophryne frontifasciata***
Anura - Microhylidae
e Horst's Cross Frog

3642 ***Oreophryne geislerorum***
Anura - Microhylidae
e Madang Cross Frog

3643 ***Oreophryne idenburgensis***
Anura - Microhylidae
e Idenburg Cross Frog

3644 ***Oreophryne inornata***
Anura - Microhylidae
e Dull Cross Frog

3645 ***Oreophryne insulana***
Anura - Microhylidae
e Goodenough Cross Frog

3646 ***Oreophryne jeffersoniana***
Anura - Microhylidae
e Komodo Cross Frog

3647 ***Oreophryne kampeni***
Anura - Microhylidae
e Moroka Cross Frog

3648 ***Oreophryne moluccensis***
Anura - Microhylidae
e Moluccan Cross Frog

3649 ***Oreophryne monticola***
Anura - Microhylidae
e Lombok Cross Frog

3650 ***Oreophryne nana***
Anura - Microhylidae
e Volcano Cross Frog

3651 ***Oreophryne parkeri***
Anura - Microhylidae
e Parker's Cross Frog

3652 ***Oreophryne rookmakeri***
Anura - Microhylidae
e Flores Cross Frog

3653 ***Oreophryne variabilis***
Anura - Microhylidae
e Variable Cross Frog

3654 ***Oreophryne wolterstorffi***
Anura - Microhylidae
e Wolterstorff's Cross Frog

3655 ***Oreophryne zimmeri***
Anura - Microhylidae
e Zimmer's Cross Frog

3656 ***Oreophrynella***
Anura - Bufonidae
e Bush Toads

3657 ***Oreophrynella hubneri***
Anura - Bufonidae
e Bolivar Bush Toad

3658 ***Oreophrynella macconnelli***
Anura - Bufonidae
e MacConnell's Bush Toad

3659 ***Oreophrynella quelchii***
Anura - Bufonidae
e Roraima Bush Toad

3660 ***Oscaecilia***
Gymnophiona - Caeciliaidae
e South American Caecilians; Panamanian Caecilians
f Oscécilies

3661 ***Oscaecilia bassleeri***
Gymnophiona - Caeciliaidae
e Pastaza River Caecilian

3662 ***Oscaecilia elongata***
Gymnophiona - Caeciliaidae
e Yavisa Caecilian

3663 ***Oscaecilia equatorialis***
Gymnophiona - Caeciliaidae
e Equatorial Caecilian

3664 ***Oscaecilia hypernumeces***
Gymnophiona - Caeciliaidae
e Joinville Caecilian

3665 ***Oscaecilia keopckeorum***
Gymnophiona - Caeciliaidae
e Quisto Cocha Caecilian

3666 ***Oscaecilia ochrocephala***
Gymnophiona - Caeciliaidae
e Yellow-headed Caecilian

3667 ***Oscaecilia ozae***
Gymnophiona - Caeciliaidae

e Airstrip Caecilian

3668 ***Oscaecilia polyzona***
Gymnophiona - Caeciliaidae
e New Granada Caecilian

3669 ***Oscaecilia zweifeli***
Gymnophiona - Caeciliaidae
e Tributary Caecilian
f Oscécilie de Zweifel

3670 ***Osornophryne***
Anura - Bufonidae
e Plump Toads

3671 ***Osornophryne antisana***
Anura - Bufonidae
e Napo Plump Toad

3672 ***Osornophryne bufoniformis***
Anura - Bufonidae
e Peracca's Plump Toad

3673 ***Osornophryne guacamayo***
Anura - Bufonidae
e Guacamayo Plump Toad

3674 ***Osornophryne percrassa***
Anura - Bufonidae
e Herveo Plump Toad

3675 ***Osornophryne talipes***
Anura - Bufonidae
e Cannatella's Plump Toad

3676 ***Osteocephalus***
Anura - Hylidae
e Slender-legged Tree Frogs

3677 ***Osteocephalus buckleyi***
Anura - Hylidae
e Buckley's Slender-legged Tree Frog
f Ostéocéphale de Buckley

3678 ***Osteocephalus ellkejungingerae***
Anura - Hylidae
e Henle's Slender-legged Tree Frog

3679 ***Osteocephalus langsdorffi***
Anura - Hylidae
e Ocellated Tree Frog
f Rainette de Langsdorff

3680 ***Osteocephalus leprieuri***
Anura - Hylidae
e Cayenne Slender-legged Tree Frog
f Rainette de Leprieur

3681 ***Osteocephalus oophagus***
Anura - Hylidae
f Ostéocéphale oophage

3682 ***Osteocephalus pearsoni***
Anura - Hylidae
e Pearson's Slender-legged Tree Frog

3683 ***Osteocephalus subtilis***
Anura - Hylidae
e Brazilian Slender-legged Tree Frog

3684 ***Osteocephalus taurinus***
Anura - Hylidae
e Manaus Slender-legged Tree Frog
f Ostéoéphale taurin

3685 ***Osteocephalus verruciger***
Anura - Hylidae
e Ecuador Slender-legged Tree Frog

3686 ***Osteopilus***
Anura - Hylidae
e Cuban Tree Frogs; West Indian Tree Frogs
d Antillen-Laubfrösche

3687 ***Osteopilus brunneus***
Anura - Hylidae
e Savanna-la-Mar Tree Frog

3688 ***Osteopilus dominicensis***
Anura - Hylidae
e Dominican Tree Frog
d Dominikanischer Laubfrosch

3689 ***Osteopilus septentrionalis***
Anura - Hylidae
e Cuban Tree Frog; Giant Tree Frog (NA)
d Kuba-Laubfrosch; Riesenlaubfrosch
f Rainette de Cuba

3690 ***Otophryne***
Anura - Microhylidae
e Pancake Frogs

3691 ***Otophryne pyburni***
Anura - Microhylidae
f Otophryne de Pyburn

3692 ***Otophryne robusta***
Anura - Microhylidae
e Pancake Frog

P

3693 ***Paa***
Anura - Ranidae
e Paa Frogs

3694 ***Paa annandalii***
Anura - Ranidae
e Annandale's Paa Frog

3695 ***Paa arnoldi***
Anura - Ranidae
e Arnold's Paa Frog

3696 ***Paa blanfordii***
Anura - Ranidae
e Blanford's Paa Frog

3697 ***Paa bourreti***
Anura - Ranidae
e Bourret's Paa Frog; Bourret's Frog (SEA)
d Bourrets Frosch

3698 ***Paa boulengeri***
Anura - Ranidae
e Boulenger's Paa Frog

3699 ***Paa chayunensis***
Anura - Ranidae
e Chayu Paa Frog

3700 ***Paa conaensis***
Anura - Ranidae
e Cona Paa Frog

3701 ***Paa ercepeae***
Anura - Ranidae
e Torrent Paa Frog

3702 ***Paa exilispinosa***
Anura - Ranidae
e Hong Kong Paa Frog; Hong Kong Spiny Frog

3703 ***Paa fasciculispina***
Anura - Ranidae
e Thai Paa Frog; Spiny-breasted Frog (SEA)
d Stachelbrust-Frosch

3704 ***Paa feae***
Anura - Ranidae
e Kakhien Paa Frog

3705 ***Paa hazarensis***
Anura - Ranidae
e Kashmir Paa Frog

3706 ***Paa jiulongensis***
Anura - Ranidae
e Jiulong Paa Frog

3707 ***Paa liebigii***
Anura - Ranidae
e Sikkim Paa Frog

3708 ***Paa liui***
Anura - Ranidae
e Ninglang Paa Frog

3709 ***Paa maculosa***
Anura - Ranidae
e Spotted Paa Frog

3710 ***Paa minica***
Anura - Ranidae
e Nepal Paa Frog

3711 ***Paa polunini***
Anura - Ranidae
e Langtang Paa Frog

3712 ***Paa rara***
Anura - Ranidae
e Rara Paa Frog

3713 ***Paa rostandi***
Anura - Ranidae
e Dubois's Paa Frog

3714 ***Paa shini***
Anura - Ranidae
e Chinese Paa Frog

3715 ***Paa sichuaensis***
Anura - Ranidae
e Sichian Paa Frog

3716 ***Paa spinosa***
Anura - Ranidae
e Spiny Paa Frog; Chinese Edible Frog; Chinese Spiny Frog

3717 ***Paa sternosignata***
Anura - Ranidae
e Malir Paa Frog; Baluch Frog; Baluch Mountain Frog

3718 ***Paa vicina***
Anura - Ranidae
e Himalaya Paa Frog

3719 ***Paa yunnanensis***
Anura - Ranidae
e Yunnan Paa Frog

3720 ***Pachyhynobius***
Caudata - Hynobiidae
e Shangcheng Salamanders
d Graubraune Winkelzahnmolche

3721 ***Pachyhynobius shangchengensis***
Caudata - Hynobiidae
e Shangcheng Salamander
d Graubrauner Winkelzahnmolch

3722 ***Pachymedusa***
Anura - Hylidae
e Mexican Leaf Frogs; Mexican Giant Tree Frogs
d Gespenstfrösche

3723 ***Pachymedusa dacnicolor***
Anura - Hylidae
e Mexican Leaf Frog
d Gespenstfrosch; Mexikanischer Gespenstfrosch

3724 ***Pachytriton***
Caudata - Salamandridae
e Chinese Newts
d Kurzfußmolche

3725 ***Pachytriton brevipes***
Caudata - Salamandridae
e Tsitou Newt
d Chinesischer Kurzfußmolch

3726 ***Pachytriton labiatus***
Caudata - Salamandridae
e Unterstein's Newt

3727 ***Palmatorappia***
Anura - Ranidae
e Solomon Islands Palm Frogs

3728 ***Palmatorappia solomonensis***
Anura - Ranidae
e Solomon Islands Palm Frog

3729 ***Paracassina***
Anura - Hyperoliidae
e Common Striped Frogs

3730 ***Paracassina kouhiensis***
Anura - Hyperoliidae
e Kounhi Valley Striped Frog

3731 ***Paracassina obscura***
Anura - Hyperoliidae

e Ethiopia Striped Frog

3732 ***Paracrinia***
Anura - Myobatrachidae
e Haswell's Frogs

3733 ***Paracrinia haswelli***
Anura - Myobatrachidae
e Haswell's Frog; Haswell's Froglet

3734 ***Paradoxhyla***
Anura - Microhylidae
e Web-foot Frogs; Web-footed Frogs

3735 ***Paradoxhyla palmata***
Anura - Microhylidae
e Web-foot Frog; Web-footed Frog

3736 ***Paramesotriton***
Caudata - Salamandridae
e Warty Newts; Wart-Salamanders
d Warzenmolche
f Tritons; Paramétritons

3737 ***Paramesotriton caudopunctatus***
Caudata - Salamandridae
e Guizhou Warty Newt
d Schwanzfleckmolch

3738 ***Paramesotriton chinensis***
Caudata - Salamandridae
e Chinese Warty Newt; Chinese Newt
d Chinesischer Warzenmolch; Chinesischer Warzensalamander
f Triton de Chine

3739 ***Paramesotriton deloustali***
Caudata - Salamandridae
e Vietnam Warty Newt; Tam Dao Newt
d Nordvietnamesischer Warzenmolch; Nordvietnamesischer Wassermolch
f Salamandre di Tonkin; Triton de Deloustal

3740 ***Paramesotriton fuzhongensis***
Caudata - Salamandridae
e Wanggao Warty Newt
d Hongkong-Warzenmolch

3741 ***Paramesotriton guanxiensis***
Caudata - Salamandridae
e Guangxi Warty Newt

3742 ***Paramesotriton hongkongensis***
Caudata - Salamandridae
e Hong Kong Warty Newt; Hong Kong Newt
d Hongkong-Molch; Hongkong-Warzenmolch
f Triton de Hong-Kong

3743 ***Paratelmatobius***
Anura - Leptodactylidae
e Coastal Frogs

3744 ***Paratelmatobius gaigeae***
Anura - Leptodactylidae
e Gaige's Coastal Frog

3745 ***Paratelmatobius lutzii***
Anura - Leptodactylidae
e Lutz's Coastal Frog

3746 ***Paratelmatobius poecilogaster***
Anura - Leptodactylidae
e San Andre Coastal Frog

3747 ***Parhopophryne***
Anura - Microhylidae
e Amari Forest Frogs

3748 ***Parhopophryne usambarica***
Anura - Microhylidae
e Amari Forest Frog

3749 ***Parvicaecilia***
Gymnophiona - Caeciliaidae
e Colombian Caecilians

3750 ***Parvicaecilia nicefori***
Gymnophiona - Caeciliaidae
e Honda Caecilian

3751 ***Parvicaecilia pricei***
Gymnophiona - Caeciliaidae
e El Centro Caecilian

3752 ***Parvimolge***
Caudata - Plethodontidae
e Dwarf Salamanders; Mexican Dwarf Salamanders
d Mexikanische Pygmäensalamander (pl)

3753 ***Parvimolge townsendi***
Caudata - Plethodontidae
e Dwarf Salamander; Townsend's Dwarf Salamander
d Dunns Pygmäensalamander

3754 ***Pedostibes***
Anura - Bufonidae
e Asian Tree Toads; Tree Toads

3755 ***Pedostibes everetti***
Anura - Bufonidae
e Everett's Asian Tree Toad

3756 ***Pedostibes hosii***
Anura - Bufonidae
e Boulenger's Asian Tree Toad; Hose's Tree Toad (SEA)
d Baumkröte; Hoses Baumkröte
f Crapaud arboricole

3757 ***Pedostibes kempi***
Anura - Bufonidae
e Kemp's Asian Tree Toad

3758 ***Pedostibes maculatus***
Anura - Bufonidae
e Spotted Asian Tree Toad

3759 ***Pedostibes rugosus***
Anura - Bufonidae
e Inger's Asian Tree Toad

3760 ***Pedostibes tuberculosus***
Anura - Bufonidae
e Warty Asian Tree Toad; Malabar Tree Toad (ISC)

3761 ***Pelobates***
Anura - Pelobatidae
e European Spadefoot Toads; Spadefoots
d Knoblauchkröten; Europäische Schaufelkröten; Schaufelkröten
f Pélobates; Crapauds à couteaux

3762 ***Pelobates cultripes***
Anura - Pelobatidae
e Wagler's Spadefoot Toad; Western Spadefoot Toad; Spanish Spadefoot; Spanish Spadefoot Toad; Western Spadefoot; Iberian Spadefoot
d Messerfuß; Südfranzösische Knoblauchkröte
f Crapaud à couteaux; Pélobate cultripède

3763 ***Pelobates fuscus***
Anura - Pelobatidae
e Common Eurasian Spadefoot Toad; Eurasian Spadefoot Toad; Common Spadefoot; Spadefooted Toad; Garlic Toad; Digging Toad; Common Spadefoot Toad
d Knoblauchkröte
f Pélobate brun; Pélobate brun d'Europe; Crapaud brun
i Pelobate foco; Rospo bruno di Cornalia

3764 ***Pelobates fuscus insubricus***
Anura - Pelobatidae
d Insubrische Knoblauchkröte
i Pelobate insubrico

3765 ***Pelobates syriacus***
Anura - Pelobatidae

e Syrian Spadefoot Toad; Eastern Spadefoot Toad; Southwest Asian Spadefoot Toad; Eastern Spadefoot; Syrian Spadefoot
d Syrische Schaufelkröte
f Pélobate syrien

3766 ***Pelobates varaldii***
Anura - Pelobatidae
e Moroccan Spadefoot Toad
f Pélobate marocain; Pélobate de Varaldi; Pélobate du Maroc

3767 **Pelobatidae**
Anura
e Spadefoots; Spadefoot Toads; Spade-footed Toads; Toad-Frogs
d Krötenfrösche; Schaufelfüße; Froschkröten; Lehmkröten; Wühlkröten; Pelobatiden
f Crapauds fournisseurs; Pélobatidés

3768 ***Pelodryas***
Anura - Hylidae
e White's Tree Frogs

3769 ***Pelodryas caerulea***
Anura - Hylidae
e White's Tree Frog; Green Tree Frog (ANZ); Emerald-green Tree Frog; Dumpy Frog
d Korallenfinger; Korallenfinger-Laubfrosch
f Rainette bleue; Rainette géante; Rainette de White

3770 ***Pelodryas cavernicola***
Anura - Hylidae
e Long Cave Tree Frog

3771 ***Pelodryas splendida***
Anura - Hylidae
e Splendid Tree Frog; Magnificent Tree Frog

3772 ***Pelodytes***
Anura - Pelodytidae
e Parsley Frogs; Mud Divers
d Schlammtaucher (pl)
f Pélodytes
i Pelodíti

3773 ***Pelodytes caucasicus***
Anura - Pelodytidae
e Caucasus Parsley Frog; Caucasian Parsley Frog; Caucasian Mud Diver
d Kaukasischer Schlammtaucher
f Pélodyte du Caucase
i Pelodite

3774 ***Pelodytes ibericus***
Anura - Pelodytidae
e Iberian Parsley Frog
f Pélodyte ibérique

3775 ***Pelodytes punctatus***
Anura - Pelodytidae
e Parsley Frog; Common Parsley Frog; Mud Diver
d Westlicher Schlammtaucher; Schlammtaucher; Punktierter Schlammtaucher
f Pélodyte ponctué; Pélodyte persillé; Grenouille persillé
i Pelodite puntato; Pelodite punteggiato

3776 **Pelodytidae**
Anura
e Pelodites; Mud Divers; Parsley Frogs
d Schlammtaucher (pl); Pelodytiden
f Pélodytidés; Pélodytes ponctués

3777 ***Pelophryne***
Anura - Bufonidae
e Flathead Toads; Dwarf Toads
d Philippinen-Kröten

3778 ***Pelophryne albolineata***
Anura - Bufonidae
e White-striped Flathead Toad

3779 ***Pelophryne brevipes***
Anura - Bufonidae
e Zamboanga Flathead Toad; St. Andrew's Cross Toadlet (SEA)
d Stummelfußkröte

3780 ***Pelophryne guentheri***
Anura - Bufonidae
e Günther's Flathead Toad

3781 ***Pelophryne lighti***
Anura - Bufonidae
e Mindanao Flathead Toad

3782 ***Pelophryne macrotis***
Anura - Bufonidae
e Boulenger's Flathead Toad

3783 ***Pelophryne misdera***
Anura - Bufonidae
e Black Flathead Toad

3784 ***Pelophryne scalptus***
Anura - Bufonidae
e Hainan Flathead Toad

3785 ***Peltophryne***
Anura - Bufonidae
e Caribbean Toads

3786 ***Peltophryne cataulicieps***
Anura - Bufonidae
e Schwartz's Caribbean Toad

3787 ***Peltophryne empusa***
Anura - Bufonidae
e Cope's Caribbean Toad; Cuban Toad
d Kuba-Kröte
f Crapaud de Cuba

3788 ***Peltophryne fluviatica***
Anura - Bufonidae
e Dominican Caribbean Toad

3789 ***Peltophryne guentheri***
Anura - Bufonidae
e Günther's Caribbean Toad

3790 ***Peltophryne gundlachi***
Anura - Bufonidae
e Gundlach's Caribbean Toad

3791 ***Peltophryne lemur***
Anura - Bufonidae
e Lowland Caribbean Toad

3792 ***Peltophryne longinasus***
Anura - Bufonidae
e Stejneger's Caribbean Toad
d Langnasenkröte

3793 ***Peltophryne peltocephala***
Anura - Bufonidae
e Tschudi's Caribbean Toad; Cuban Toad
f Crapaud peltocéphale

3794 ***Peltophryne taladai***
Anura - Bufonidae
e Cuban Caribbean Toad; Cuban Toad

3795 ***Perhohapsis***
Anura - Microhylidae
e School Frogs

3796 ***Perhohapsis menziesi***
Anura - Microhylidae
e School Frog; Iarowari School Frog

3797 ***Petropedetes***
Anura - Ranidae
e African Water Frogs

3798 ***Petropedetes cameroonensis***
Anura - Ranidae
e Cameroon Water Frog
d Kamerun-Frosch

3799 ***Petropedetes johnstoni***
Anura - Ranidae
e Johnstone's Water Frog

3800 ***Petropedetes natator***
Anura - Ranidae
e Sierra Leone Water Frog; Swimmer
d Schwimmer
f Nageur

3801 ***Petropedetes newtoni***
Anura - Ranidae
e Newton's Water Frog

3802 ***Petropedetes palmipes***
Anura - Ranidae
e Efulen Water Frog

3803 ***Petropedetes parkeri***
Anura - Ranidae
e Parker's Water Frog

3804 ***Petropedetes perreti***
Anura - Ranidae
e Perret's Water Frog

3805 ***Phaeognathus***
Caudata - Plethodontidae
e Red Hills Salamanders
d Schleichensalamander (pl)

3806 ***Phaeognathus hubrichti***
Caudata - Plethodontidae
e Red Hills Salamander
d Schleichensalamander; Hubrichts Schleichensalamander

3807 ***Phasmahyla***
Anura - Hylidae
e Shining Leaf Frogs

3808 ***Phasmahyla cochranae***
Anura - Hylidae
e Chocolatefoot Leaf Frog

3809 ***Phasmahyla exilis***
Anura - Hylidae
e Mottled Leaf Frog

3810 ***Phasmahyla guttata***
Anura - Hylidae
e Spotted Leaf Frog

3811 ***Phasmahyla jandaia***
Anura - Hylidae
e Sazima's Leaf Frog

3812 ***Philautus***
Anura - Rhacophoridae
e Bubble-nest Frogs; Bush Frogs (ISC)

3813 ***Philautus acutirostris***
Anura - Rhacophoridae
e Philippine Bubble-nest Frog

3814 ***Philautus acutus***
Anura - Rhacophoridae
e Sharp Bubble-nest Frog

3815 ***Philautus adspersus***
Anura - Rhacophoridae
e Sri Lanka Bubble-nest Frog

3816 ***Philautus albopunctatus***
Anura - Rhacophoridae
e Dotted Bubble-nest Frog

3817 ***Philautus alticola***
Anura - Rhacophoridae
e Bongao Bubble-nest Frog

3818 ***Philautus amoensis***
Anura - Rhacophoridae
e Kamboranga Bubble-nest Frog

3819 ***Philautus andersoni***
Anura - Rhacophoridae
e Anderson's Bubble-nest Frog

3820 ***Philautus annandalii***
Anura - Rhacophoridae
e Himalaya Bubble-nest Frog; East Himalayan Bush Frog (ISC); Annandale's Bush Frog (ISC)

3821 ***Philautus aurantium***
Anura - Rhacophoridae
e Medolong Bubble-nest Frog

3822 ***Philautus aurifasciatus***
Anura - Rhacophoridae
e Java Bubble-nest Frog

3823 ***Philautus banaensis***
Anura - Rhacophoridae
e Bana Bubble-nest Frog

3824 ***Philautus beddomi***
Anura - Rhacophoridae
e Beddome's Bubble-nest Frog

3825 ***Philautus bombayensis***
Anura - Rhacophoridae
e Bombay Bubble-nest Frog; Konkan Bush Frog (ISC)

3826 ***Philautus carinensis***
Anura - Rhacophoridae
e Burmese Bubble-nest Frog; Burmese Bush Frog (SEA)
d Birmesischer Buschfrosch

3827 ***Philautus cavirostris***
Anura - Rhacophoridae
e Tibetan Bubble-nest Frog

3828 ***Philautus chalazodes***
Anura - Rhacophoridae
e Chalazodes Bubble-nest Frog

3829 ***Philautus charius***
Anura - Rhacophoridae
e Karnataka Bubble-nest Frog

3830 ***Philautus cherrapunjiae***
Anura - Rhacophoridae
e Cherrapunji Bubble-nest Frog

3831 ***Philautus cornutus***
Anura - Rhacophoridae
e Horned Bubble-nest Frog

3832 ***Philautus cruri***
Anura - Rhacophoridae
e Kempholey Bubble-nest Frog

3833 ***Philautus disegrus***
Anura - Rhacophoridae
e Malysian Bubble-nest Frog

3834 ***Philautus dubius***
Anura - Rhacophoridae
e Darjeeling Bubble-nest Frog

3835 ***Philautus elegans***
Anura - Rhacophoridae
e Elegant Bubble-nest Frog

3836 ***Philautus emembranatus***
Anura - Rhacophoridae
e Mindanao Bubble-nest Frog

3837 ***Philautus femoralis***
Anura - Rhacophoridae
e Southwestern Bubble-nest Frog

3838 ***Philautus flaviventris***
Anura - Rhacophoridae
e Malabar Bubble-nest Frog

3839 ***Philautus garo***
Anura - Rhacophoridae
e Garo Hills Bubble-nest Frog

3840 ***Philautus glandulosus***
Anura - Rhacophoridae
e Southern Bubble-nest Frog

3841 ***Philautus gracilipes***
Anura - Rhacophoridae
e Chapa Bubble-nest Frog; Slender-legged Bush Frog (SEA)
d Schlankbeiniger Buschfrosch

3842 ***Philautus gryllus***
Anura - Rhacophoridae
e Langbian Bubble-nest Frog

3843 ***Philautus hassanensis***
Anura - Rhacophoridae
e Dutta's Bubble-nest Frog

3844 ***Philautus hosii***
Anura - Rhacophoridae
e Borneo Bubble-nest Frog

3845 ***Philautus ingeri***
Anura - Rhacophoridae
e Inger's Bubble-nest Frog

3846 ***Philautus jacobsoni***
Anura - Rhacophoridae
e Jacobson's Bubble-nest Frog

3847 ***Philautus jerdonii***
Anura - Rhacophoridae
e Jerdon's Bubble-nest Frog

3848 ***Philautus jinxiuensis***
Anura - Rhacophoridae
e Jinxiu Bubble-nest Frog

3849 ***Philautus kempiae***
Anura - Rhacophoridae
e Turu Bubble-nest Frog

3850 ***Philautus kerangae***
Anura - Rhacophoridae
e Kerangas Bubble-nest Frog

3851 ***Philautus kottigeharensis***
Anura - Rhacophoridae
e Kottigehar Bubble-nest Frog

3852 ***Philautus leitensis***
Anura - Rhacophoridae
e Leyte Bubble-nest Frog

3853 ***Philautus leucorhinus***
Anura - Rhacophoridae
e Whitenose Bubble-nest Frog; Marten's Bush Frog (ISC)

3854 ***Philautus lissobrachius***
Anura - Rhacophoridae
e Davao Bubble-nest Frog

3855 ***Philautus longchuanensis***
Anura - Rhacophoridae
e Longchuan Bubble-nest Frog

3856 ***Philautus longicrus***
Anura - Rhacophoridae
e Palawan Bubble-nest Frog

3857 ***Philautus maosonensis***
Anura - Rhacophoridae
e Mao-Son Bubble-nest Frog

3858 ***Philautus medogensis***
Anura - Rhacophoridae
e Medog Bubble-nest Frog

3859 ***Philautus melanensis***
Anura - Rhacophoridae
e Rao's Bubble-nest Frog

3860 ***Philautus menglaensis***
Anura - Rhacophoridae
e Zhishihe's Bubble-nest Frog

3861 ***Philautus microdiscus***
Anura - Rhacophoridae
e Kobo Bubble-nest Frog

3862 ***Philautus microtympanum***
Anura - Rhacophoridae
e Günther's Bubble-nest Frog
d Ceylonesischer Ruderfrosch
f Grenouille volante de Ceylan

3863 ***Philautus mjobergi***
Anura - Rhacophoridae
e Murud Bubble-nest Frog

3864 ***Philautus namdaphaensis***
Anura - Rhacophoridae
e Tirap Bubble-nest Frog

3865 ***Philautus narainensis***
Anura - Rhacophoridae
e Kottigehar Bubble-nest Frog

3866 ***Philautus nasutus***
Anura - Rhacophoridae
e Snouted Bubble-nest Frog

3867 ***Philautus noblei***
Anura - Rhacophoridae
e Noble's Bubble-nest Frog

3868 ***Philautus ocellatus***
Anura - Rhacophoridae
e Ocellated Bubble-nest Frog

3869 ***Philautus pallidipes***
Anura - Rhacophoridae
e Pangaerango Bubble-nest Frog

3870 ***Philautus palpebralis***
Anura - Rhacophoridae
e Vietnam Bubble-nest Frog

3871 ***Philautus parkeri***
Anura - Rhacophoridae
e Parker's Bubble-nest Frog

3872 ***Philautus parvulus***
Anura - Rhacophoridae
e Karin Bubble-nest Frog; Dwarf Busch Frog (SEA)
d Kleiner Buschfrosch

3873 ***Philautus petersi***
Anura - Rhacophoridae
e Peters's Bubble-nest Frog

3874 ***Philautus pleurostictus***
Anura - Rhacophoridae
e Indian Bubble-nest Frog

3875 ***Philautus pulcherrimus***
Anura - Rhacophoridae
e Manantoddy Bubble-nest Frog

3876 ***Philautus rhododiscus***
Anura - Rhacophoridae
e Chinese Bubble-nest Frog

3877 ***Philautus romeri***
Anura - Rhacophoridae
e Romer's Bubble-nest Frog

3878 ***Philautus schmakeri***
Anura - Rhacophoridae
e Schmaker's Bubble-nest Frog

3879 ***Philautus shillongensis***
Anura - Rhacophoridae
e Shillong Bubble-nest Frog

3880 ***Philautus shyamrupus***
Anura - Rhacophoridae
e Hornbill Bubble-nest Frog

3881 ***Philautus signatus***
Anura - Rhacophoridae
e Boulenger's Bubble-nest Frog

3882 ***Philautus similis***
Anura - Rhacophoridae
e Sumatra Bubble-nest Frog

3883 ***Philautus surdus***
Anura - Rhacophoridae
e Luzon Bubble-nest Frog

3884 ***Philautus swamianus***
Anura - Rhacophoridae
e Swami Bubble-nest Frog

3885 ***Philautus tectus***
Anura - Rhacophoridae
e Gunung Mulu Bubble-nest Frog

3886 ***Philautus travancorius***
Anura - Rhacophoridae
e Travancore Bubble-nest Frog

3887 ***Philautus tytthus***
Anura - Rhacophoridae
e Triangle Bubble-nest Frog

3888 ***Philautus umbra***
Anura - Rhacophoridae
e Pinnacle Bubble-nest Frog

3889 ***Philautus variabilis***
Anura - Rhacophoridae
e Variable Bubble-nest Frog

3890 ***Philautus vermiculatus***
Anura - Rhacophoridae
e Perak Bubble-nest Frog; Spotted Bush Frog (SEA)
d Gefleckter Buschfrosch

3891 ***Philautus vittiger***
Anura - Rhacophoridae
e Indonesian Bubble-nest Frog

3892 ***Philautus williamsi***
Anura - Rhacophoridae
e Williams's Bubble-nest Frog

3893 ***Philoria***
Anura - Myobatrachidae
e Baw Baw Frogs

3894 ***Philoria frosti***
Anura - Myobatrachidae
e Baw Baw Frog

3895 ***Phlyctimantis***
Anura - Hyperoliidae
e African Striped Frogs
d Gelbschenkelfrösche

3896 ***Phlyctimantis boulengeri***
Anura - Hyperoliidae
e Boulenger's Striped Frog

3897 ***Phlyctimantis keithae***
Anura - Hyperoliidae
e Keith's Striped Frog

3898 ***Phlyctimantis leonardi***
Anura - Hyperoliidae
e Olive Striped Frog
d Leonards Baumfrosch

3899 ***Phlyctimantis verrucosus***
Anura - Hyperoliidae
e Warty Striped Frog

3900 ***Phrynella***
Anura - Microhylidae
e Malacca Frogs

3901 ***Phrynella pulchra***
Anura - Microhylidae
e Malacca Frog; Beauty Pygmy Frog (SEA)
d Schmuck-Engmaulfrosh

3902 ***Phrynobatrachus***
Anura - Ranidae
e African River Frogs; Puddle Frogs (CSA); Toad-Frogs (CSA)
d Silberfrösche; Springfrösche; Pfützenfrösche

3903 ***Phrynobatrachus accraensis***
Anura - Ranidae
e Accra River Frog

3904 ***Phrynobatrachus acridoides***
Anura - Ranidae
e Zanzibar River Frog; Small Puddle Frog (CSA); Zanzibar Puddle Frog (CSA); East African Puddle Frog (CSA); Cope's Toad-Frog (CSA)

3905 ***Phrynobatrachus acutiro***
Anura - Ranidae
e Rugegewald River Frog

3906 ***Phrynobatrachus albolabris***
Anura - Ranidae
e White-lipped River Frog

3907 ***Phrynobatrachus albomarginatus***
Anura - Ranidae
e Mauda River Frog

3908 ***Phrynobatrachus alleni***
Anura - Ranidae
e Allen's River Frog

3909 ***Phrynobatrachus alticola***
Anura - Ranidae
e Forest River Frog

3910 ***Phrynobatrachus annulatus***
Anura - Ranidae
e Ringed River Frog

3911 ***Phrynobatrachus anotis***
Anura - Ranidae
e Usinga River Frog

3912 ***Phrynobatrachus asper***
Anura - Ranidae
e Itombwe River Frog

3913 ***Phrynobatrachus auritus***
Anura - Ranidae
e Eared River Frog

3914 ***Phrynobatrachus batesii***
Anura - Ranidae
e Bates's River Frog

3915 ***Phrynobatrachus bequaerti***
Anura - Ranidae
e Vissoke River Frog

3916 ***Phrynobatrachus bottegi***
Anura - Ranidae
e Auta River Frog

3917 ***Phrynobatrachus calcaratus***
Anura - Ranidae
e Boutry River Frog

3918 ***Phrynobatrachus congicus***
Anura - Ranidae
e Congo River Frog

3919 ***Phrynobatrachus cornutus***
Anura - Ranidae
e Horned River Frog

3920 ***Phrynobatrachus cricogaster***
Anura - Ranidae
e Nkongsamba River Frog

3921 ***Phrynobatrachus cryptotis***
Anura - Ranidae
e Cryptic River Frog

3922 ***Phrynobatrachus dalcqi***
Anura - Ranidae
e Kiva River Frog

3923 ***Phrynobatrachus dendrobates***
Anura - Ranidae
e Medje River Frog

3924 ***Phrynobatrachus dispar***
Anura - Ranidae
e Peters's River Frog

3925 ***Phrynobatrachus elberti***
Anura - Ranidae
e Elbert's River Frog

3926 ***Phrynobatrachus feae***
Anura - Ranidae
e Principe River Frog

3927 ***Phrynobatrachus fraterculus***
Anura - Ranidae
e Macenta River Frog

3928 ***Phrynobatrachus gastoni***
Anura - Ranidae
e Buta River Frog

3929 ***Phrynobatrachus ghanensis***
Anura - Ranidae
e Ghana River Frog

3930 ***Phrynobatrachus giorgii***
Anura - Ranidae
e Yambata River Frog

3931 ***Phrynobatrachus graueri***
Anura - Ranidae
e Rugege River Frog

3932 ***Phrynobatrachus guieensis***
Anura - Ranidae
e Guinea River Frog

3933 ***Phrynobatrachus gutturosus***
Anura - Ranidae
e Chabanaud's River Frog; Guttural Cricket Frog; Guttural Puddle Frog (CSA)

3934 ***Phrynobatrachus hylaios***
Anura - Ranidae
e Foulassi River Frog

3935 ***Phrynobatrachus keniensis***
Anura - Ranidae
e Kenya River Frog

3936 ***Phrynobatrachus kiangopensis***
Anura - Ranidae
e Kingapop River Frog

3937 ***Phrynobatrachus krefftii***
Anura - Ranidae
e Krefft's River Frog

3938 ***Phrynobatrachus latifrons***
Anura - Ranidae
e Ahl's River Frog

3939 ***Phrynobatrachus liberiensis***
Anura - Ranidae
e Liberia River Frog

3940 ***Phrynobatrachus mababiensis***
Anura - Ranidae
e Mababe River Frog; Dwarf Puddle Frog (CSA); Common Cricket Frog; Cricket Frog; Mamabe Toad-Frog (CSA)

3941 ***Phrynobatrachus manengoubensis***
Anura - Ranidae
e Manengouba River Frog

3942 ***Phrynobatrachus minutus***
Anura - Ranidae
e Tiny River Frog

3943 ***Phrynobatrachus nanus***
Anura - Ranidae
e Buala River Frog

3944 ***Phrynobatrachus natalensis***
Anura - Ranidae
e Natal River Frog; Snoring Puddle Frog (CSA); Natal Puddle Frog

(CSA); Smith's Frog; Natal Frog; Toad-Frog; Common Toad-Frog (CSA); Common Puddle Frog (CSA)
d Natal-Silberfrosch
f Grenouille de Smith

3945 ***Phrynobatrachus ogoensis***
Anura - Ranidae
e Ogowe River Frog

3946 ***Phrynobatrachus pakenhami***
Anura - Ranidae
e Pokenham's River Frog

3947 ***Phrynobatrachus parkeri***
Anura - Ranidae
e Parker's River Frog

3948 ***Phrynobatrachus parvulus***
Anura - Ranidae
e Loanda River Frog; Small Puddle Frog (CSA)

3949 ***Phrynobatrachus perpalmatus***
Anura - Ranidae
e Lake Mwero River Frog

3950 ***Phrynobatrachus petropeditoides***
Anura - Ranidae
e Ruwenzori River Frog

3951 ***Phrynobatrachus plicatus***
Anura - Ranidae
e Coast River Frog

3952 ***Phrynobatrachus pygmaeus***
Anura - Ranidae
e Chad River Frog

3953 ***Phrynobatrachus rouxi***
Anura - Ranidae
e Roux's River Frog

3954 ***Phrynobatrachus rungwensis***
Anura - Ranidae
e Rungwe River Frog

3955 ***Phrynobatrachus scapularis***
Anura - Ranidae
e Buta River Frog

3956 ***Phrynobatrachus sciangallarum***
Anura - Ranidae
e Murle River Frog

3957 ***Phrynobatrachus steindachneri***
Anura - Ranidae
e Steindachner's River Frog

3958 ***Phrynobatrachus sternfeldi***
Anura - Ranidae
e Sternfeld's River Frog

3959 ***Phrynobatrachus stewarti***
Anura - Ranidae
e Stewart's River Frog

3960 ***Phrynobatrachus sulfureogularis***
Anura - Ranidae
e Central River Frog

3961 ***Phrynobatrachus taiensis***
Anura - Ranidae
e Tai River Frog

3962 ***Phrynobatrachus tellinii***
Anura - Ranidae
e Peracca's River Frog

3963 ***Phrynobatrachus tokba***
Anura - Ranidae
e Tokba River Frog

3964 ***Phrynobatrachus ukingensis***
Anura - Ranidae
e Ukinga River Frog; Dwarf Puddle Frog; Common Cricket Frog (CSA); Cricket Frog (CSA); Pygmy Puddle Frog (CSA); Dwarf Toad-Frog (CSA)

3965 ***Phrynobatrachus versicolor***
Anura - Ranidae
e Rwanda River Frog

3966 ***Phrynobatrachus villiersi***
Anura - Ranidae
e Yapo River Frog

3967 ***Phrynobatrachus vogti***
Anura - Ranidae
e Vogt's River Frog

3968 ***Phrynobatrachus werneri***
Anura - Ranidae
e Werner's River Frog

3969 ***Phrynobatrachus zavattarii***
Anura - Ranidae
e Scortecci's River Frog

3970 ***Phrynodon***
Anura - Ranidae
e Hook Frogs

3971 ***Phrynodon sandersoni***
Anura - Ranidae
e Hook Frog; Sanderson's Hook Frog

3972 ***Phrynoglossus***
Anura - Ranidae
e Oriental Frogs

3973 ***Phrynoglossus baluensis***
Anura - Ranidae
e Balu Oriental Frog

3974 ***Phrynoglossus borealis***
Anura - Ranidae
e Rotung Oriental Frog

3975 ***Phrynoglossus celebensis***
Anura - Ranidae
e Celebes Oriental Frog

3976 ***Phrynoglossus diminutivus***
Anura - Ranidae
e Tiny Oriental Frog

3977 ***Phrynoglossus floresianus***
Anura - Ranidae
e Flores Oriental Frog

3978 ***Phrynoglossus laevis***
Anura - Ranidae
e Philippine Oriental Frog

3979 ***Phrynoglossus magnapustulosus***
Anura - Ranidae
e Thai Oriental Frog

3980 ***Phrynoglossus martensii***
Anura - Ranidae
e Martens Oriental Frog

3981 ***Phrynoglossus vittatus***
Anura - Ranidae
e Striped Oriental Frog

3982 ***Phrynohyas***
Anura - Hylidae
e Golden-eyed Tree Frogs; Marbled Tree Frogs
d Krötenlaubfrösche

3983 ***Phrynohyas coriacea***
Anura - Hylidae
e Surinam Golden-eyed Tree Frog
f Rainette coriacée

3984 ***Phrynohyas imitatrix***
Anura - Hylidae
e Rio Golden-eyed Tree Frog

3985 ***Phrynohyas mesophaea***
Anura - Hylidae
e Porto Alegro Golden-eyed Tree Frog

3986 ***Phrynohyas resinifictrix***
Anura - Hylidae
e Mission Golden-eyed Tree Frog; Milk Frog
f Rainette kunawala

3987 ***Phrynohyas venulosa***
Anura - Hylidae
e Veined Tree Frog; Warty Tree Frog; Marbled Tree Frog
d Giftlaubfrosch; Krötenlaubfrosch
f Rainette réticulaire; Rainette flûteuse

3988 ***Phrynomantis***
Anura - Microhylidae
e Snake-neck Frogs; Rubber Frogs (CSA); Red-marked Frogs (CSA); Red-marked short-headed Frogs

3989 ***Phrynomantis affinis***
Anura - Microhylidae
e Pweto Snake-neck Frog; Spotted Rubber Frog (CSA); Red-spotted Frog (CSA); Northern Red-spotted Frog

3990 ***Phrynomantis annectens***
Anura - Microhylidae
e Cape Snake-neck Frog; Marbled Rubber Frog (CSA); Red-spotted Frog (CSA); Red-spotted Namibia Frog (CSA)

3991 ***Phrynomantis bifasciatus***
Anura - Microhylidae
e South African Snake-neck Frog; Red-banded Rubber Frog (CSA); Walking Frog; Banded Rubber Frog (CSA); Red-banded Frog (CSA); Two-banded Frog (CSA)
d Wendehalsfrosch; Südafrikanischer Wendehalsfrosch

3992 ***Phrynomantis microps***
Anura - Microhylidae
e Accra Snake-neck Frog

3993 ***Phrynomantis somalicus***
Anura - Microhylidae
e Somali Snake-neck Frog

3994 ***Phrynomedusa***
Anura - Hylidae
e Coloured Leaf Frogs

3995 ***Phrynomedusa appendiculata***
Anura - Hylidae
e Santa Catarina Leaf Frog

3996 ***Phrynomedusa bokermanni***
Anura - Hylidae
e Bokermann's Leaf Frog

3997 ***Phrynomedusa fimbriata***
Anura - Hylidae
e Spinyknee Leaf Frog

3998 ***Phrynomedusa marginata***
Anura - Hylidae
e Bicoloured Leaf Frog

3999 ***Phrynomedusa vanzolinii***
Anura - Hylidae
e Vanzolini's Leaf Frog

4000 ***Phrynopus***
Anura - Leptodactylidae
e Andes Frogs

4001 ***Phrynopus bagrecitoi***
Anura - Leptodactylidae
e Bagrecito Andes Frog

4002 ***Phrynopus bracki***
Anura - Leptodactylidae
e Brack's Andes Frog

4003 ***Phrynopus brunneus***
Anura - Leptodactylidae
e Carchi Andes Frog

4004 ***Phrynopus columbianus***
Anura - Leptodactylidae
e Colombian Andes Frog

4005 ***Phrynopus cophites***
Anura - Leptodactylidae
e Cuzco Andes Frog

4006 ***Phrynopus flavomaculatus***
Anura - Leptodactylidae
e Yellow-spotted Andes Frog

4007 ***Phrynopus juninensis***
Anura - Leptodactylidae
e Junin Andes Frog

4008 ***Phrynopus kempffi***
Anura - Leptodactylidae
e Kempff's Andes Frog

4009 ***Phrynopus laplacai***
Anura - Leptodactylidae
e La Placa's Andes Frog

4010 ***Phrynopus lucida***
Anura - Leptodactylidae
e Canatella's Andes Frog

4011 ***Phrynopus montium***
Anura - Leptodactylidae
e Cascas Andes Frog

4012 ***Phrynopus nanus***
Anura - Leptodactylidae
e Santander Andes Frog

4013 ***Phrynopus nebulanastes***
Anura - Leptodactylidae
e Canchaque Andes Frog

4014 ***Phrynopus parkeri***
Anura - Leptodactylidae
e Parker's Andes Frog

4015 ***Phrynopus peraccai***
Anura - Leptodactylidae
e Peracca's Andes Frog

4016 ***Phrynopus pereger***
Anura - Leptodactylidae
e Ayuacucho Andes Frog

4017 ***Phrynopus peruanus***
Anura - Leptodactylidae
e Peters's Andes Frog

4018 ***Phrynopus peruvianus***
Anura - Leptodactylidae
e Peru Andes Frog

4019 ***Phrynopus simonsii***
Anura - Leptodactylidae
e Paramo Andes Frog

4020 ***Phrynopus wettsteini***
Anura - Leptodactylidae
e Wettstein's Andes Frog

4021 ***Phyllobates***
Anura - Dendrobatidae
e Golden Poison Frogs; Golden Poison-arrow Frogs; Poison-arrow Frogs; Leaf-climbers
d Blattsteigerfrösche; Pfeilgiftfrösche; Blattsteiger (pl)
f Phyllobates

4022 ***Phyllobates aurotaenia***
Anura - Dendrobatidae
e Kokoe Poison Frog; Kokoe Poison-arrow Frog; Kokoé

4023 ***Phyllobates bicolor***
Anura - Dendrobatidae
e Black-legged Poison Frog ; Black-legged Poison-arrow Frog ; Two-toned Poison-arrow Frog; Neara
d Zweifarben Blattsteiger; Zweifarbiger Blattsteiger

4024 ***Phyllobates lugubris***
Anura - Dendrobatidae
e Lovely Poison Frog; Lovely Poison-arrow Frog; Lovely Poison-dart Frog
d Trauer-Blattsteiger; Düsterer Blattsteiger

4025 ***Phyllobates terribilis***
Anura - Dendrobatidae
e Golden Poison Frog; Golden Poison-arrow Frog
d Goldener Giftfrosch; Gelber Pfeilgiftfrosch; Schrecklicher Giftfrosch; Schrecklicher Blattsteiger

4026 ***Phyllobates vittatus***
Anura - Dendrobatidae
e Golfodulcean Poison Frog; Golfodulcean Poison-arrow Frog; Orange-and-black Poison-dart Frog
d Gestreifter Blattsteigerfrosch; Gestreifter Blattsteiger

4027 ***Phyllodytes***
Anura - Hylidae
e Heart-tongue Frogs; Heart-tongued Frogs
d Blattschlüpfer

4028 ***Phyllodytes acuminatus***
Anura - Hylidae
e Alagoas Heart-tongue Frog

4029 ***Phyllodytes auratus***
Anura - Hylidae
e Trinidad Heart-tongue Frog

4030 ***Phyllodytes brevirostris***
Anura - Hylidae
e Alhandra Heart-tongue Frog

4031 ***Phyllodytes kautskyi***
Anura - Hylidae
e Brazilian Heart-tongue Frog

4032 ***Phyllodytes luteolus***
Anura - Hylidae
e Yellow Heart-tongue Frog
d Gelblicher Laubkleber

4033 ***Phyllodytes melanomystax***
Anura - Hylidae
e Bahia Heart-tongue Frog

4034 ***Phyllodytes tubercolosus***
Anura - Hylidae
e Maracas Heart-tongue Frog

4035 ***Phyllomedusa***
Anura - Hylidae
e Leaf Frogs
d Greiffrösche; Makifrösche
f Rainettes-makis; Phylloméduses

4036 ***Phyllomedusa atelopoides***
Anura - Hylidae
e Tosady Leaf Frog

4037 ***Phyllomedusa ayeaye***
Anura - Hylidae
e Reticulate Leaf Frog

4038 ***Phyllomedusa baltea***
Anura - Hylidae
e Purple-sided Leaf Frog

4039 ***Phyllomedusa bicolor***
Anura - Hylidae
e Giant Monkey Frog; Blue-and-yellow Frog
d Riesengreiffrosch
f Rainette bicolore; Phylloméduse bicolore; Phyllomeduse de Guyane bicolore

4040 ***Phyllomedusa boliviana***
Anura - Hylidae
e Red-rimmed Leaf Frog

4041 ***Phyllomedusa buckleyi***
Anura - Hylidae
e Warty Leaf Frog

4042 ***Phyllomedusa burmeisteri***
Anura - Hylidae
e Common Walking Leaf Frog; Walking Leaf Frog
f Rainette de Burmeister; Phylloméduse de Burmeister

4043 ***Phyllomedusa centralis***
Anura - Hylidae
e Matto Grosso Leaf Frog

4044 ***Phyllomedusa coelestis***
Anura - Hylidae
e Starry Leaf Frog

4045 ***Phyllomedusa danieli***
Anura - Hylidae
e Antioquia Leaf Frog

4046 ***Phyllomedusa distincta***
Anura - Hylidae
e Sao Paulo Walking Leaf Frog; Sao Paolo Walking Leaf Frog

4047 ***Phyllomedusa duellmani***
Anura - Hylidae
e Purple-and-orange Leaf Frog

4048 ***Phyllomedusa ecuatoriana***
Anura - Hylidae
e Agua Rica Leaf Frog

4049 ***Phyllomedusa hypocondrialis***
Anura - Hylidae
e Orange-legged Leaf Frog; Orange-sided eaf Frog
d Affenfrosch; Greiffrosch
f Rainette hypochondriale; Phylloméduse hypochondriale

4050 ***Phyllomedusa iheringii***
Anura - Hylidae
e Southern Walking Leaf Frog

4051 ***Phyllomedusa lemur***
Anura - Hylidae
e Lemur Leaf Frog; Lemur Frog
d Lemurenfrosch

4052 ***Phyllomedusa medianai***
Anura - Hylidae
e Rancho Grande Leaf Frog

4053 ***Phyllomedusa palliata***
Anura - Hylidae
e Jaguar Leaf Frog

4054 ***Phyllomedusa perinesos***
Anura - Hylidae
e Orange-spotted Leaf Frog

4055 ***Phyllomedusa psilopygion***
Anura - Hylidae
e Flecked Leaf Frog

4056 ***Phyllomedusa rohdei***
Anura - Hylidae
e Mertens's Leaf Frog; Rhode's Tree Frog; Rhode's Leaf Frog
d Greiffrosch; Makifrosch
f Rainette de Rhode

4057 ***Phyllomedusa sauvagii***
Anura - Hylidae
e Painted-belly Monkey Frog

4058 ***Phyllomedusa tarsius***
Anura - Hylidae
e Brownbelly Leaf Frog

4059 ***Phyllomedusa tetraploidea***
Anura - Hylidae
e Hidden Walking Frog

4060 ***Phyllomedusa tomptoterna***
Anura - Hylidae
e Tiger-striped Leaf Frog
d Makifrosch
f Phyllomeduse de Cope

4061 ***Phyllomedusa trinitatis***
Anura - Hylidae
e Trinidad Leaf Frog

4062 ***Phyllomedusa vaillanti***
Anura - Hylidae
e White-lined Leaf Frog
f Phyllomeduse de Vaillant

4063 ***Phyllomedusa venusta***
Anura - Hylidae
e Lovely Leaf Frog

4064 ***Phyllonastes***
Anura - Leptodactylidae
e Heyer's Leaf Frogs

4065 ***Phyllonastes heyeri***
Anura - Leptodactylidae
e Heyer's Leaf Frog

4066 ***Phyllonastes lochites***
Anura - Leptodactylidae
e Ecuador Leaf Frog

4067 ***Phyllonastes lynchi***
Anura - Leptodactylidae
e Lynch's Leaf Frog

4068 ***Phyllonastes myrmecoides***
Anura - Leptodactylidae
e Loreto Leaf Frog

4069 ***Physalaemus***
Anura - Leptodactylidae
e Dwarf Frogs; Foam Frogs
d Sumpffrösche
f Physalemes

4070 ***Physalaemus aguirrei***
Anura - Leptodactylidae
e Linhares Dwarf Frog

4071 ***Physalaemus albifrons***
Anura - Leptodactylidae
e Bahia Dwarf Frog

4072 ***Physalaemus albonotatus***
Anura - Leptodactylidae
e Menwig Frog

4073 ***Physalaemus barrioi***
Anura - Leptodactylidae
e Bocaina Dwarf Frog

4074 ***Physalaemus biligonigerus***
Anura - Leptodactylidae
e Weeping Frog

4075 ***Physalaemus bokermanni***
Anura - Leptodactylidae
e Bokerman's Dwarf Frog

4076 ***Physalaemus centralis***
Anura - Leptodactylidae
e Central Dwarf Frog

4077 ***Physalaemus cicada***
Anura - Leptodactylidae
e Maracas Dwarf Frog

4078 ***Physalaemus coloradorum***
Anura - Leptodactylidae
e Colorado Dwarf Frog

4079 ***Physalaemus crombiei***
Anura - Leptodactylidae
e Crombie's Dwarf Frog

4080 ***Physalaemus cuvieri***
Anura - Leptodactylidae
e Barker Frog

4081 ***Physalaemus deimaticus***
Anura - Leptodactylidae
e Jaboticatubas Dwarf Frog

4082 ***Physalaemus enesafae***
Anura - Leptodactylidae
e Heatwole's Dwarf Frog

4083 ***Physalaemus ephippifer***
Anura - Leptodactylidae
e Steindachner's Dwarf Frog

4084 ***Physalaemus evangelistai***
Anura - Leptodactylidae
e Evangelista's Dwarf Frog

4085 ***Physalaemus fernandezae***
Anura - Leptodactylidae
e Whistling Dwarf Frog

4086 ***Physalaemus fischeri***
Anura - Leptodactylidae
e Fischer's Dwarf Frog

4087 ***Physalaemus fuscomaculatus***
Anura - Leptodactylidae
e Brown-spotted Dwarf Frog
d Braunkehliger Frosch; Braunfleckiger Sumpffrosch

4088 ***Physalaemus gracilis***
Anura - Leptodactylidae
e Graceful Dwarf Frog

4089 ***Physalaemus henselii***
Anura - Leptodactylidae
e Hensel's Dwarf Frog

4090 ***Physalaemus jordanensis***
Anura - Leptodactylidae
e Jordan Dwarf Frog

4091 ***Physalaemus kroyeri***
Anura - Leptodactylidae
e Kroyer's Dwarf Frog

4092 ***Physalaemus lisei***
Anura - Leptodactylidae
e Braun's Dwarf Frog

4093 ***Physalaemus maculiventris***
Anura - Leptodactylidae
e Montagnes Dwarf Frog

4094 ***Physalaemus moreirae***
Anura - Leptodactylidae
e Sorocaba Dwarf Frog

4095 ***Physalaemus nanus***
Anura - Leptodactylidae
e Santa Catarina Dwarf Frog

4096 ***Physalaemus nattereri***
Anura - Leptodactylidae
e Cuyaba Dwarf Frog; False-eyed Frog

4097 ***Physalaemus obtectus***
Anura - Leptodactylidae
e Linhares Dwarf Frog

4098 ***Physalaemus olfersii***
Anura - Leptodactylidae
e Atlantic Forest Dwarf Frog

4099 ***Physalaemus petersi***
Anura - Leptodactylidae
e Peters's Dwarf Frog
f Physaleme de Peters

4100 ***Physalaemus pustulatus***
Anura - Leptodactylidae
e Guayaquil Dwarf Frog; Tungara Frog

4101 ***Physalaemus pustulosus***
Anura - Leptodactylidae
e Tungara Frog; Narrow-mouthed Toad

d Schaumnestfrosch; Wimmerfrosch; Tungara-Frosch

4102 ***Physalaemus riograndensis***
Anura - Leptodactylidae
e Rio Grande Dwarf Frog

4103 ***Physalaemus rupestris***
Anura - Leptodactylidae
e Ibitipoca Dwarf Frog

4104 ***Physalaemus sanctafecinus***
Anura - Leptodactylidae
e Helvetia Dwarf Frog

4105 ***Physalaemus signifer***
Anura - Leptodactylidae
e Girard's Dwarf Frog

4106 ***Physalaemus soaresi***
Anura - Leptodactylidae
e Santa Cruz Dwarf Frog

4107 ***Physalaemus spinigerus***
Anura - Leptodactylidae
e Iguape Dwarf Frog

4108 ***Phyzelaphryne***
Anura - Leptodactylidae
e Miriam's Frogs

4109 ***Phyzelaphryne miriamae***
Anura - Leptodactylidae
e Miriam's Frog

4110 ***Pipa***
Anura - Pipidae
e Surinam Toad-Frogs; Surinam Toads
d Wabenkröten; Pipa-Kröten
f Pipas

4111 ***Pipa arrabali***
Anura - Pipidae
e Arrabal's Surinam Toad-Frog; Arrabal's Surinam Toad

4112 ***Pipa aspera***
Anura - Pipidae
e Albina Surinam Toad-Frog; Albina Surinam Toad
f Pipa rugueuse

4113 ***Pipa carvalhoi***
Anura - Pipidae
e Carvalho's Surinam Toad-Frog; Carvalho's Surinam Toad
d Mittlere Wabenkröte; Kleine Wabenkröte

4114 ***Pipa myersi***
Anura - Pipidae
e Myers Surinam Toad-Frog; Myers Surinam Toad

4115 ***Pipa parva***
Anura - Pipidae
e Sabana Surinam Toad-Frog; Sabana Surinam Toad ; Dwarf Toad
d Zwerg Wabenkröte
f Pipa nain

4116 ***Pipa pipa***
Anura - Pipidae
e Surinam Toad-Frog; Surinam Toad
d Wabenkröte; Große Wabenkröte; Surinam-Wabenkröte
f Pipa de Surinam; Crapaud de Surinam; Pipa américaine

4117 ***Pipa snethlageae***
Anura - Pipidae
e Utinga Surinam Toad-Frog; Utinga Surinam Toad
d Zwerg-Pipa

4118 **Pipidae**
Anura
e Tongueless Frogs
d Zungenlose Kröten; Zungenlose; Wabenkröten

f Pipidés; Aglosses

4119 ***Platymantis***
Anura - Ranidae
e Wrinkled Ground Frogs; Platymantis
d Runzelfrösche

4120 ***Platymantis acrochordus***
Anura - Ranidae
e Bougainville Wrinkled Ground Frog

4121 ***Platymantis aculeodactylus***
Anura - Ranidae
e Torokina Wrinkled Ground Frog

4122 ***Platymantis akarithmus***
Anura - Ranidae
e Pomugu Wrinkled Ground Frog

4123 ***Platymantis batantae***
Anura - Ranidae
e Batanta Wrinkled Ground Frog

4124 ***Platymantis boulengeri***
Anura - Ranidae
e Boulenger's Wrinkled Ground Frog; Boulenger's Platymantis
d Boulengers Runzelfrosch

4125 ***Platymantis cheesmanae***
Anura - Ranidae
e Cheesman's Wrinkled Ground Frog

4126 ***Platymantis cornutus***
Anura - Ranidae
e Horned Wrinkled Ground Frog

4127 ***Platymantis corrugatus***
Anura - Ranidae
e Philippine Wrinkled Ground Frog
d Philippinischer Runzelfrosch

4128 ***Platymantis dorsalis***
Anura - Ranidae
e Dumeril's Wrinkled Ground Frog
f Cornufère dorsale

4129 ***Platymantis gilliardi***
Anura - Ranidae
e Gilliard's Wrinkled Ground Frog

4130 ***Platymantis guentheri***
Anura - Ranidae
e Günther's Wrinkled Ground Frog
d Günthers Runzelfrosch

4131 ***Platymantis guppyi***
Anura - Ranidae
e Treasury Wrinkled Ground Frog
d Salomonen-Runzelfrosch

4132 ***Platymantis hazelae***
Anura - Ranidae
e Hazel's Wrinkled Ground Frog

4133 ***Platymantis ingeri***
Anura - Ranidae
e Inger's Wrinkled Ground Frog

4134 ***Platymantis insulatus***
Anura - Ranidae
e Gigante Wrinkled Ground Frog

4135 ***Platymantis lawtoni***
Anura - Ranidae
e Lawton's Wrinkled Ground Frog

4136 ***Platymantis levigatus***
Anura - Ranidae
e Tablas Wrinkled Ground Frog

4137 ***Platymantis macrops***
Anura - Ranidae
e Aresi Wrinkled Ground Frog

4138 ***Platymantis macrosceles***
Anura - Ranidae
e Ti Wrinkled Ground Frog

4139 ***Platymantis magnus***
Anura - Ranidae
e Kavieng Wrinkled Ground Frog

4140 ***Platymantis meyeri***
Anura - Ranidae
e Meyer's Wrinkled Ground Frog

4141 ***Platymantis mimicus***
Anura - Ranidae
e Numundo Wrinkled Ground Frog

4142 ***Platymantis myersi***
Anura - Ranidae
e Myers's Wrinkled Ground Frog

4143 ***Platymantis neckerei***
Anura - Ranidae
e Necker's Wrinkled Ground Frog

4144 ***Platymantis nexipus***
Anura - Ranidae
e Baining Wrinkled Ground Frog

4145 ***Platymantis papuensis***
Anura - Ranidae
e Papua Wrinkled Ground Frog

4146 ***Platymantis parkeri***
Anura - Ranidae
e Parker's Wrinkled Ground Frog

4147 ***Platymantis pelewensis***
Anura - Ranidae
e Palau Wrinkled Ground Frog

4148 ***Platymantis polillensis***
Anura - Ranidae
e Polillo Wrinkled Ground Frog

4149 ***Platymantis punctatus***
Anura - Ranidae
e Dotted Wrinkled Ground Frog

4150 ***Platymantis reticulatus***
Anura - Ranidae
e Reticulate Wrinkled Ground Frog

4151 ***Platymantis rhipiphalcus***
Anura - Ranidae
e Kadrian Wrinkled Ground Frog

4152 ***Platymantis schmidti***
Anura - Ranidae
e Schmidt's Wrinkled Ground Frog

4153 ***Platymantis solomonis***
Anura - Ranidae
e Solomon Wrinkled Ground Frog

4154 ***Platymantis spelaeus***
Anura - Ranidae
e Cave Wrinkled Ground Frog

4155 ***Platymantis subterrestris***
Anura - Ranidae
e Luzon Wrinkled Ground Frog

4156 ***Platymantis vitianus***
Anura - Ranidae
e Viti Wrinkled Ground Frog; Fiji Land Frog
d Fidji-Frosch; Fidschi-Bodenfrosch

4157 ***Platymantis vitiensis***
Anura - Ranidae
e Levuka Wrinkled Ground Frog; Fiji Tree Frog
d Fidji-Frosch; Fidschi-Baumfrosch

4158 ***Platymantis weberi***
Anura - Ranidae
e Weber's Wrinkled Ground Frog

4159 ***Platypelis***
Anura - Microhylidae
e Giant Tree Frogs

4160 ***Platypelis alticola***
Anura - Microhylidae
e Madagascar Giant Tree Frog

4161 ***Platypelis barbouri***
Anura - Microhylidae
e Barbour's Giant Tree Frog

4162 ***Platypelis cowani***
Anura - Microhylidae
e Betsileo Giant Tree Frog

4163 ***Platypelis grandis***
Anura - Microhylidae
e Boulenger's Giant Tree Frog

4164 ***Platypelis milloti***
Anura - Microhylidae
e Nosy Giant Tree Frog

4165 ***Platypelis occultans***
Anura - Microhylidae
e Island Giant Tree Frog

4166 ***Platypelis pollicaris***
Anura - Microhylidae
e Common Giant Tree Frog; Giant Tree Frog

4167 ***Platypelis tsaratananaensis***
Anura - Microhylidae
e Tsaratanana Giant Tree Frog; Tsaratanan Giant Tree Frog

4168 ***Platypelis tuberifera***
Anura - Microhylidae
e Ambatoharanana Giant Tree Frog

4169 ***Plectrohyla***
Anura - Hylidae
e Spikethumb Frogs
f Rainettes-casques

4170 ***Plectrohyla acanthodes***
Anura - Hylidae
e Thorny Spikethumb Frog

4171 ***Plectrohyla avia***
Anura - Hylidae
e Greater Spikethumb Frog

4172 ***Plectrohyla dasypus***
Anura - Hylidae
e Honduras Spikethumb Frog

4173 ***Plectrohyla glandulosa***
Anura - Hylidae
e Forest Spikethumb Frog

4174 ***Plectrohyla guatemalensis***
Anura - Hylidae
e Guatemala Spikethumb Frog; Guatgemalan Spikethumb Frog

4175 ***Plectrohyla hartwegi***
Anura - Hylidae
e Hartweg's Spikethumb Frog

4176 ***Plectrohyla ixil***
Anura - Hylidae
e Ixil Spikethumb Frog

4177 ***Plectrohyla lacerata***
Anura - Hylidae
e Popeyed Spikethumb Frog

4178 ***Plectrohyla matudai***
Anura - Hylidae
e Matuda's Spikethumb Frog

4179 ***Plectrohyla pokomchi***
Anura - Hylidae
e Rio Sanaja Spikethumb Frog

4180 ***Plectrohyla pycnochila***
Anura - Hylidae
e Thick-lipped Spikethumb Frog

4181 ***Plectrohyla queechi***
Anura - Hylidae
e Las Palmas Spikethumb Frog

4182 ***Plectrohyla sagorum***
Anura - Hylidae
e Arcane Spikethumb Frog

4183 ***Plectrohyla tecunumani***
Anura - Hylidae
e Cave Spikethumb Frog

4184 ***Plectrohyla teuchestes***
Anura - Hylidae
e Alta Verapaz Spikethumb Frog

4185 ***Plethodon***
Caudata - Plethodontidae
e Woodland Salamanders
d Waldsalamander (pl)
f Pléthodons; Salamandres de forêt

4186 ***Plethodon ainsworthi***
Caudata - Plethodontidae
e Catahoula Salamander

4187 ***Plethodon albagula***
Caudata - Plethodontidae
e Western Slimy Salamander; White-throated Slimy Salamander

4188 ***Plethodon amplus***
Caudata - Plethodontidae
e Blue Ridge Gray-cheek Salamander (NA)

4189 ***Plethodon aureolus***
Caudata - Plethodontidae
e Tellico Salamander

4190 ***Plethodon caddoensis***
Caudata - Plethodontidae
e Caddo Mountain Salamander

4191 ***Plethodon chattahoochee***
Caudata - Plethodontidae
e Chattahoochee Slimy Salamander

4192 ***Plethodon cheoah***
Caudata - Plethodontidae
e Cheoah Bald Salamander

4193 ***Plethodon chlorobryonis***
Caudata - Plethodontidae
e Atlantic Coast Slimy Salamander; Carolina Slimy Salamander

4194 ***Plethodon cinereus***
Caudata - Plethodontidae
e Eastern Red-backed Salamander; Redback Salamander; Red-backed Salamander (NA); Lead-backed Salamander
d Rotrücken-Waldsalamander; Aschgrauer Erdsalamander
f Salamandre cendrée; Salamandre cendrée à dos rouge; Salamandre cendrée à dos rayé; Salamandre rayé (Qué)

4195 ***Plethodon cylindraceus***
Caudata - Plethodontidae
e White-spotted Slimy Salamander

4196 ***Plethodon dorsalis***
Caudata - Plethodontidae
e Northern Zigzag Salamander; Zigzag Salamander
d Zickzack Salamander
f Salamandre zigzag

4197 ***Plethodon dorsalis angusticlavius***
Caudata - Plethodontidae
e Ozark Zigzag Salamander; Ozark Red-backed Salamander

4198 ***Plethodon dorsalis dorsalis***
Caudata - Plethodontidae
e Eastern Zigzag Salamander

4199 ***Plethodon dunni***
Caudata - Plethodontidae
e Dunn's Salamander
d Dunns Salamander
f Salamandre de Dunn

4200 ***Plethodon electromorphus***
Caudata - Plethodontidae
e Northern Ravine Salamander

4201 ***Plethodon elongatus***
Caudata - Plethodontidae
e Del Norte Salamander
d Del Norte-Salamander

4202 ***Plethodon elongatus croceator***
Caudata - Plethodontidae
e Yellow-blotched Salamander

4203 ***Plethodon elongatus klauberi***
Caudata - Plethodontidae
e Large-blotched Salamander

4204 ***Plethodon elongatus platensis***
Caudata - Plethodontidae
e Sierra Nevada Salamander

4205 ***Plethodon elongatus xanthoptica***
Caudata - Plethodontidae
e Yellow-eyed Salamander

4206 ***Plethodon fourchensis***
Caudata - Plethodontidae
e Fourche Mountain Salamander

4207 ***Plethodon glutinosus***
Caudata - Plethodontidae
e Northern Slimy Salamander; Slimy Salamander (NA)
d Silber-Waldsalamander; Silbersalamander
f Salamandre gluante; Pléthodon visqueux

4208 ***Plethodon gordoni***
Caudata - Plethodontidae
e Mary's Peak Salamander

4209 ***Plethodon grobmani***
Caudata - Plethodontidae
e Southeastern Slimy Salamander; Florida Slimy Salamander

4210 ***Plethodon hoffmani***
Caudata - Plethodontidae
e Valley and Ridge Salamander

4211 ***Plethodon hubrichti***
Caudata - Plethodontidae
e Peaks of Otter Salamander

4212 ***Plethodon idahoensis***
Caudata - Plethodontidae
e Coeur d'Alene Salamander

4213 ***Plethodon jordani***
Caudata - Plethodontidae
e Jordan's Salamander; Jordan's Apalachian Woodland Salamander; Jordan's Red-cheeked Salamander
d Rotwangen-Waldsalamander; Appalachen-Waldsalamander; Jordans Salamander
f Salamandre de Jordan; Salamandre à joues rouges

4214 ***Plethodon kentucki***
Caudata - Plethodontidae
e Cumberland Plateau Salamander; Cumberland Plateau Wood Salamander

4215 ***Plethodon kiamichi***
Caudata - Plethodontidae
e Kiamichi Slimy Salamander

4216 ***Plethodon kisatchie***
Caudata - Plethodontidae
e Louisiana Slimy Salamander

4217 ***Plethodon larselli***
Caudata - Plethodontidae
e Larch Mountain Salamander
d Rotbauch-Waldsalamander

4218 ***Plethodon meridianus***
Caudata - Plethodontidae
e South Moutain Gray-cheeked Salamander (NA); Metcalf's Salamander

4219 ***Plethodon metcalfi***
Caudata - Plethodontidae
e Southern Gray-cheeked Salamanader (NA)

4220 ***Plethodon mississippi***
Caudata - Plethodontidae
e Mississippi Slimy Salamander

4221 ***Plethodon montanus***
Caudata - Plethodontidae
e Northern Gray-cheeked Salamander (NA)

4222 ***Plethodon neomexicanus***
Caudata - Plethodontidae
e Jemez Mountains Salamander
d Neumexikanischer Walldsalamander

4223 ***Plethodon nettingi***
Caudata - Plethodontidae
e Cheat Mountain Salamander; Netting's Salamander

4224 ***Plethodon ocmulgee***
Caudata - Plethodontidae
e Ocmulgee Slimy Salamander

4225 ***Plethodon oconaluftee***
Caudata - Plethodontidae
e Balsam Mountains Salamander

4226 ***Plethodon ouatchitae***
Caudata - Plethodontidae
e Rich Mountain Salamander

4227 ***Plethodon petraeus***
Caudata - Plethodontidae
e Pigeon Mountain Salamander

4228 ***Plethodon punctatus***
Caudata - Plethodontidae
e White-spotted Salamander; Cow Knob Salamander (NA)

4229 ***Plethodon richmondi***
Caudata - Plethodontidae
e Southern Ravine Salamander; Ravine Salamander
f Salamandre des ravins

4230 ***Plethodon savannah***
Caudata - Plethodontidae
e Savannah Slimy Salamander

4231 ***Plethodon sequoyah***
Caudata - Plethodontidae
e Sequoyah Slimy Salamander

4232 ***Plethodon serratus***
Caudata - Plethodontidae
e Southern Red-backed Salamander; Southern Redback Salamander; Ouachita Red-backed Salamander

4233 ***Plethodon shenandoah***
Caudata - Plethodontidae
e Shenandoah Salamander

4234 ***Plethodon shermani***
Caudata - Plethodontidae
e Red-legged Salamander

4235 ***Plethodon stormi***
Caudata - Plethodontidae
e Siskiyou Mountain Salamander

4236 ***Plethodon teyahalee***
Caudata - Plethodontidae

e Southern Appalachian Salamander; Teyahalee Salamander

4237 ***Plethodon vandykei***
Caudata - Plethodontidae
e Van Dyke's Salamander
d Washington-Salamander

4238 ***Plethodon vandykei vandykei***
Caudata - Plethodontidae
e Washington Salamander

4239 ***Plethodon variolatus***
Caudata - Plethodontidae
e South Carolina Slimy Salamander

4240 ***Plethodon vehiculum***
Caudata - Plethodontidae
e Western Red-backed Salamander; Western Redback Salamander
d Oregon-Waldsalamander; Westlicher Rotrückensalamander
f Salamandre à dos rayé

4241 ***Plethodon ventralis***
Caudata - Plethodontidae
e Southern Zigzag Salamander

4242 ***Plethodon virginia***
Caudata - Plethodontidae
e Shenandoah Mountain Salamander

4243 ***Plethodon websteri***
Caudata - Plethodontidae
e Webster's Salamander

4244 ***Plethodon wehrlei***
Caudata - Plethodontidae
e Wehrle's Salamander

4245 ***Plethodon welleri***
Caudata - Plethodontidae
e Weller's Salamander
d Wellers Waldsalamander
f Salamandre de Weller

4246 ***Plethodon yonahlossee***
Caudata - Plethodontidae
e Yonahlossee Salamander
d Yohnalossee-Salamander

4247 **Plethodontidae**
Caudata
e Lungless Salamanders
d Lungenlose Molche; Lungenlose Salamander (pl)
f Salamandres sans poumons; Pléthodontidés

4248 ***Plethodontohyla***
Anura - Microhylidae
e Digging Frogs
d Madagaskar-Engmunfrösche

4249 ***Plethodontohyla alluaudi***
Anura - Microhylidae
e Fort Dauphin's Digging Frog

4250 ***Plethodontohyla angulifera***
Anura - Microhylidae
e Werner's Digging Frog

4251 ***Plethodontohyla bipunctata***
Anura - Microhylidae
e Forest Digging Frog

4252 ***Plethodontohyla brevipes***
Anura - Microhylidae
e Betsileo Digging Frog

4253 ***Plethodontohyla coudreaui***
Anura - Microhylidae
e Betampona Digging Frog

4254 ***Plethodontohyla guentherpetersi***
Anura - Microhylidae
e Tsaratanana Digging Frog; Tsaratanan Digging Frog

4255 ***Plethodontohyla inguinalis***
Anura - Microhylidae

e Boulenger's Digging Frog

4256 ***Plethodontohyla laevipes***
Anura - Microhylidae
e Madagascar Digging Frog

4257 ***Plethodontohyla laevis***
Anura - Microhylidae
e Sakana Digging Frog

4258 ***Plethodontohyla minuta***
Anura - Microhylidae
e Tiny Digging Frog

4259 ***Plethodontohyla notosticta***
Anura - Microhylidae
e Mahanoro Digging Frog
d Schöner Madagaskar-Engmundfrosch

4260 ***Plethodontohyla ocellata***
Anura - Microhylidae
e Ocellated Digging Frog

4261 ***Plethodontohyla serratopalpebrosa***
Anura - Microhylidae
e Guibe's Digging Frog

4262 ***Plethodontohyla tuberata***
Anura - Microhylidae
e Interior Digging Frog

4263 ***Pleurodeles***
Caudata - Salamandridae
e Ribbed Newt
d Rippenmolche
f Pleurodèles

4264 ***Pleurodeles poireti***
Caudata - Salamandridae
e Algerian Ribbed Newt; Algerian Newt; Poiret's Newt
d Poiretscher Rippenmolch; Algerischer Rippenmolch
f Triton de Poiret; Pleurodèle de Poiret; Triton algérien

4265 ***Pleurodeles waltl***
Caudata - Salamandridae
e Spanish Ribbed Newt; Sharp-ribbed Salamander; Waltl Newt; Ribbed Newt; Pleurodele Newt; Sharp-ribbed Newt; Spanish Newt
d Spanischer Rippenmolch; Rippenmolch; Spanischer Rippensalamander
f Pleurodèle de Waltl; Triton à côtes; Triton de Waltl; Triton d'Espagne
i Salamandra di Waltl

4266 ***Pleurodema***
Anura - Leptodactylidae
e Four-eyed Frogs
d Augenkröten

4267 ***Pleurodema bibroni***
Anura - Leptodactylidae
e Four-eyed Frog
d Bibron-Kröte; Chilenischer Südfrosch; Vieraugenkröte

4268 ***Pleurodema borellii***
Anura - Leptodactylidae
e Rufous Four-eyed Frog

4269 ***Pleurodema brachyops***
Anura - Leptodactylidae
e Colombian Four-eyed Frog
d Rotschenkel-Augenkröte

4270 ***Pleurodema bufonina***
Anura - Leptodactylidae
e Grey Four-eyed Frog

4271 ***Pleurodema cinerea***
Anura - Leptodactylidae
e Juliaca Four-eyed Frog

4272 ***Pleurodema diplolistris***
Anura - Leptodactylidae
e Peters's Four-eyed Frog

4273 ***Pleurodema guayapae***
Anura - Leptodactylidae
e Guayapa Four-eyed Frog

4274 ***Pleurodema kriegi***
Anura - Leptodactylidae
e Achala's Four-eyed Frog

4275 ***Pleurodema marmorata***
Anura - Leptodactylidae
e Marbled Four-eyed Frog

4276 ***Pleurodema nebulosa***
Anura - Leptodactylidae
e Mendoza Four-eyed Frog

4277 ***Pleurodema thaul***
Anura - Leptodactylidae
e Chile Four-eyed Frog
f Crapaud thaul

4278 ***Pleurodema tucumana***
Anura - Leptodactylidae
e Spotted Flanks Four-eyed Frog

4279 ***Polypedates***
Anura - Rhacophoridae
e Whipping Frogs
f Polypédates

4280 ***Polypedates chenfui***
Anura - Rhacophoridae
e Chinese Whipping Frog

4281 ***Polypedates colletti***
Anura - Rhacophoridae
e Collet's Whipping Frog; Collet's Tree Frog (SEA)
d Collets Ruderfrosch

4282 ***Polypedates cruciger***
Anura - Rhacophoridae
e Sri Lanka Whipping Frog
d Kreuztragender Ruderfrosch

4283 ***Polypedates dennysii***
Anura - Rhacophoridae
e Blanford's Whipping Frog

4284 ***Polypedates dugritei***
Anura - Rhacophoridae
e Sichuan Whipping Frog

4285 ***Polypedates eques***
Anura - Rhacophoridae
e Günther's Whipping Frog

4286 ***Polypedates feae***
Anura - Rhacophoridae
e Thao Whipping Frog

4287 ***Polypedates hungfuensis***
Anura - Rhacophoridae
e Hongfoshan Whipping Frog

4288 ***Polypedates leucomystax***
Anura - Rhacophoridae
e Java Whipping Frog; East Asian Tree Frog (ISC); Four-lined Tree Frog; Common Tree Frog (SEA)
d Weißbartruderfrosch; Bananenfrosch; Hausfrosch
f Polypédate à moustaches

4289 ***Polypedates longinasus***
Anura - Rhacophoridae
e Southern Whipping Frog

4290 ***Polypedates macrotis***
Anura - Rhacophoridae
e Baram Whipping Frog; Dark-eared Tree Frog; Dark-eared Tree Frog (SEA)
d Schwarzohr-Ruderfrosch

4291 ***Polypedates maculatus***
Anura - Rhacophoridae
e Spotted Whipping Frog; Common Tree Frog (ISC)

4292 ***Polypedates megacephalus***
Anura - Rhacophoridae
e Hong Kong Whipping Frog; Northern Tree Frog (SEA)
d Nördlicher Ruderfrosch

4293 ***Polypedates mutus***
Anura - Rhacophoridae
e Burmese Whipping Frog

4294 ***Polypedates omeimontis***
Anura - Rhacophoridae
e Omei Whipping Frog
d Omei-Ruderfrosch

4295 ***Polypedates otilophus***
Anura - Rhacophoridae
e Bongon Whipping Frog; Bony-headed Rhacophorus; File-eared Tree Frog

4296 ***Polypedates taeniatus***
Anura - Rhacophoridae
e Bengal Whipping Frog

4297 ***Polypedates yaoshanensis***
Anura - Rhacophoridae
e Yaoshan Whipping Frog

4298 ***Polypedates zed***
Anura - Rhacophoridae
e Narayanghat Whipping Frog

4299 ***Potomotyphlus***
Gymnophiona - Typhlonectidae
e Kaup's Caecilians
f Potamotyphle

4300 ***Potomotyphlus kaupi***
Gymnophiona - Typhlonectidae
e Kaup's Caecilian
f Potamotyphle de Kaup

4301 ***Poyntonia***
Anura - Ranidae
e Reserve Frogs

4302 ***Poyntonia paludicola***
Anura - Ranidae
e Reserve Frog; Kogelberg Reserve Frog; Marsh Frog (CSA)

4303 ***Prasilinia***
Gymnophiona - Caeciliaidae
e Cooper's Caecilians

4304 ***Prasilinia cooperi***
Gymnophiona - Caeciliaidae
e Cooper's Caecilian; Cooper's Black Caecilian

4305 ***Probreviceps***
Anura - Microhylidae
e Big-fingered Frogs

4306 ***Probreviceps macrodactylus***
Anura - Microhylidae
e Usambara Big-fingered Frog

4307 ***Probreviceps rhodesianus***
Anura - Microhylidae
e Zimbabwe Big-fingered Frog; Forest Short-headed Frog; Highland Rain Frog (CSA); Forest Rain Frog (CSA)

4308 ***Probreviceps uluguruensis***
Anura - Microhylidae
e Uluguru Big-fingered Frog

4309 ***Proceratophyrs***
Anura - Leptodactylidae
e Smooth Horned Frogs

4310 ***Proceratophyrs appendiculata***
Anura - Leptodactylidae
e Günther's Smooth Horned Frog; Günther's Horned Frog
d Südamerikanischer Zipfelfrosch
f Grenouille cornue de Günther

4311 ***Proceratophyrs avelinoi***
Anura - Leptodactylidae
e Avelino's Smooth Horned Frog

4312 ***Proceratophyrs bigibbosa***
Anura - Leptodactylidae
e Peters's Smooth Horned Frog

4313 ***Proceratophyrs boiei***
Anura - Leptodactylidae
e Bahia Smooth Horned Frog; Boie's Frog
d Buchstabenfrosch
f Grenouille de Boie

4314 ***Proceratophyrs cristiceps***
Anura - Leptodactylidae
e Müller's Smooth Horned Frog

4315 ***Proceratophyrs cristinae***
Anura - Leptodactylidae
e Cristina's Smooth Horned Frog

4316 ***Proceratophyrs fryi***
Anura - Leptodactylidae
e Fry's Smooth Horned Frog

4317 ***Proceratophyrs goyana***
Anura - Leptodactylidae
e Goia's Smooth Horned Frog

4318 ***Proceratophyrs laticeps***
Anura - Leptodactylidae
e Linhares Smooth Horned Frog

4319 ***Proceratophyrs moehringi***
Anura - Leptodactylidae
e Möhring's Smooth Horned Frog

4320 ***Proceratophyrs palustris***
Anura - Leptodactylidae
e Plateau Smooth Horned Frog

4321 ***Proceratophyrs precrenulata***
Anura - Leptodactylidae
e Brazilian Smooth Horned Frog

4322 ***Proceratophyrs schirchi***
Anura - Leptodactylidae
e Santo Smooth Horned Frog

4323 **Proteidae**
Caudata
e Olms and Waterdogs
d Olme
f Protées

4324 ***Proteus***
Caudata - Proteidae
e Olms
d Grottenolme; Europäische Grottenolme
f Protées; Anguillards; Olms; Protées anguillards
i Protei

4325 ***Proteus anguinus***
Caudata - Proteidae
e Olm
d Grottenolm; Europäischer Grottenolm
f Protée; Anguillard d'Europe; Olm; Protée anguillard
i Proteo

4326 ***Pseudacris***
Anura - Hylidae
e Chorus Frogs
d Chorfrösche

4327 ***Pseudacris brachyphona***
Anura - Hylidae
e Mountain Chorus Frog

4328 ***Pseudacris brimleyi***
Anura - Hylidae
e Brimley's Chorus Frog

4329 ***Pseudacris cadaverina***
Anura - Hylidae
e California Tree Frog; California Chorus Frog
d Kalifornischer Laubfrosch

4330 ***Pseudacris clarkii***
Anura - Hylidae
e Spotted Chorus Frog

4331 ***Pseudacris crucifer***
Anura - Hylidae
e Spring Peeper (NA); Spring Piper
d Wasserpfeifer
f Rainette crucifère

4332 ***Pseudacris crucifer bartramiana***
Anura - Hylidae
e Southern Spring Peeper

4333 ***Pseudacris crucifer crucifer***
Anura - Hylidae
e Northern Spring Peeper

4334 ***Pseudacris feriarum***
Anura - Hylidae
e Upland Chorus Frog

4335 ***Pseudacris feriarum feriarium***
Anura - Hylidae
e Upland Chorus Frog

4336 ***Pseudacris feriarum kalmi***
Anura - Hylidae
e New Jersey Chorus Frog

4337 ***Pseudacris maculata***
Anura - Hylidae
e Boreal Chorus Frog
f Rainette faux-grillon boréale (Qué)

4338 ***Pseudacris nigrita***
Anura - Hylidae
e Southern Chorus Frog; Chorus Frog
d Chorfrosch
f Rainette à verrues

4339 ***Pseudacris nigrita nigrita***
Anura - Hylidae
e Southern Chorus Frog; Striped Southern Chorus Frog

4340 ***Pseudacris nigrita verrucosa***
Anura - Hylidae
e Florida Chorus Frog

4341 ***Pseudacris ocularis***
Anura - Hylidae
e Little Grass Frog
d Zwerglaubfrosch

4342 ***Pseudacris ornata***
Anura - Hylidae
e Ornate Chorus Frog
d Sumpflaubfrosch; Schmuckfrosch
f Rainette des marais

4343 ***Pseudacris regilla***
Anura - Hylidae
e Pacific Tree Frog; Pacific Chorus Frog
d Pazifik-Laubfrosch; Königs-Laubfrosch; Pazifik-Baumfrosch; Regilla-Laubfrosch
f Rainette du Pacifique; Rainette arboricole du Pacifique

4344 ***Pseudacris regilla curta***
Anura - Hylidae
e Baja California Chorus Frog

4345 ***Pseudacris regilla hypochondriaca***
Anura - Hylidae
e San Lucan Chorus Frog

4346 ***Pseudacris streckeri***
Anura - Hylidae
e Strecker's Chorus Frog
d Streckers Chorfrosch

4347 ***Pseudacris streckeri illinoensis***
Anura - Hylidae
e Illinois Chorus Frog

4348 ***Pseudacris streckeri streckeri***
Anura - Hylidae
e Strecker's Chorus Frog

4349 ***Pseudacris triseriata***
Anura - Hylidae
e Striped Chorus Frog; Northern Chorus Frog; Chorus Frog (NA)
d Hochland-Chorfrosch; Westlicher Chorfrosch
f Rainette faux-grillon; Rainette faux-grillon de l'Ouest (Qué)

4350 ***Pseudacris triseriata triseriata***
Anura - Hylidae
e Midland Chorus Frog; Western Chorus Frog

4351 ***Pseudhymenochirus***
Anura - Pipidae
e Merlin's Clawed Frogs
d Zwergkrallenfrösche

4352 ***Pseudhymenochirus merlini***
Anura - Pipidae
e Merlin's Clawed Frog
d Zwergkrallenfrosch

4353 **Pseudidae**
Anura
e Harlequin Frogs
d Harlekinfrösche
f Pseudidées

4354 ***Pseudis***
Anura - Pseudidae
e Swimming Frogs
d Trugfrösche
f Pseudis

4355 ***Pseudis minuta***
Anura - Pseudidae
e Lesser Swimming Frog

4356 ***Pseudis paradoxa***
Anura - Pseudidae
e Swimming Frog; Paradox Frog; Paradoxical Frog; Shrinking Frog
d Harlekinfrosch; Großer Harlekinfrosch; Großer Trugfrosch
f Pseudis de Mérian; Grenouille paradoxale

4357 ***Pseudobranchus***
Caudata - Sirenidae
e Dwarf Sirens
d Zwergarmmolche
f Sirènes

4358 ***Pseudobranchus axanthus***
Caudata - Sirenidae
e Payne's Prairie Siren; Southern Dwarf Siren (NA); Narrow-striped Dwarf Siren

4359 ***Pseudobranchus axanthus axanthus***
Caudata - Sirenidae
e Narrow-striped Dwarf Siren

4360 ***Pseudobranchus axanthus belli***
Caudata - Sirenidae
e Everglades Dwarf Siren

4361 ***Pseudobranchus striatus***
Caudata - Sirenidae
e Northern Dwarf Siren; Dwarf Siren
d Zwergarmmolch; Gestreifter Zwergarmmolch
f Sirène naine à larges bandes

4362 ***Pseudobranchus striatus lustricolus***
Caudata - Sirenidae
e Gulf Hammock Dwarf Siren

4363 ***Pseudobranchus striatus spheniscus***
Caudata - Sirenidae

e Slender Dwarf Siren

4364 ***Pseudobranchus striatus striatus***
Caudata - Sirenidae
e Broad-striped Dwarf Siren

4365 ***Pseudobufo***
Anura - Bufonidae
e False Toads; Aquatic Toads
d Falsche Kröten
f Faux-crapauds

4366 ***Pseudobufo subasper***
Anura - Bufonidae
e False Toad
d Falsche Kröte
f Faux-crapaud

4367 ***Pseudoeurycea***
Caudata - Plethodontidae
e False Brook Salamanders
d Mexiko-Salamander (pl)

4368 ***Pseudoeurycea altamontana***
Caudata - Plethodontidae
e Morelos False Brook Salamander
d Morelos-Salamander

4369 ***Pseudoeurycea anitae***
Caudata - Plethodontidae
e Anita's False Brook Salamander
d Mexiko-Höhlensalamander

4370 ***Pseudoeurycea belli***
Caudata - Plethodontidae
e Bell's False Brook Salamander; Tlaconete Salamander
d Gelbflecken-Mexiko-Salamander

4371 ***Pseudoeurycea belli belli***
Caudata - Plethodontidae
e Bell's False Brook Salamander

4372 ***Pseudoeurycea belli boneti***
Caudata - Plethodontidae
e Oaxacan False Brook Salamander

4373 ***Pseudoeurycea belli sierraoccidentalis***
Caudata - Plethodontidae
e Sonoran False Brook Salamander

4374 ***Pseudoeurycea brunnata***
Caudata - Plethodontidae
e Brown False Brook Salamander

4375 ***Pseudoeurycea cephalica***
Caudata - Plethodontidae
e Chunky False Brook Salamander

4376 ***Pseudoeurycea cephalica cephalica***
Caudata - Plethodontidae
e Chunky False Brook Salamander

4377 ***Pseudoeurycea cephalica manni***
Caudata - Plethodontidae
e Mann's False Brook Salamander

4378 ***Pseudoeurycea cephalica rubrimembris***
Caudata - Plethodontidae
e Red-legged False Brook Salamander

4379 ***Pseudoeurycea cochranae***
Caudata - Plethodontidae
e Cochran's False Brook Salamander

4380 ***Pseudoeurycea conanti***
Caudata - Plethodontidae
e Conant's False Brook Salamander

4381 ***Pseudoeurycea expectata***
Caudata - Plethodontidae
e Jalpa False Brook Salamander

4382 ***Pseudoeurycea firscheini***
Caudata - Plethodontidae
e Firschein's False Brook Salamander

d Firscheins Mexiko-Salamander

4383 ***Pseudoeurycea gadovi***
Caudata - Plethodontidae
e Gadow'sFalse Brook Salamander

4384 ***Pseudoeurycea galeanae***
Caudata - Plethodontidae
e Galeana False Brook Salamander
d Galeana-Schleuderzalamander

4385 ***Pseudoeurycea goebeli***
Caudata - Plethodontidae
e Goebel's False Brook Salamander

4386 ***Pseudoeurycea juarezi***
Caudata - Plethodontidae
e Sierra Juarez Salamander; Juarez Salamander

4387 ***Pseudoeurycea leprosa***
Caudata - Plethodontidae
e Leprous False Brook Salamander

4388 ***Pseudoeurycea longicauda***
Caudata - Plethodontidae
e Longtail False Brook Salamander

4389 ***Pseudoeurycea melanomolga***
Caudata - Plethodontidae
e Black False Brook Salamander
d Schwarzer Schleuderzungensalamander

4390 ***Pseudoeurycea mystax***
Caudata - Plethodontidae
e Moustache False Brook Salamander; Mustache False Brook Salamander (NA)
d Schluchten-Mexiko-Salamander

4391 ***Pseudoeurycea nigromaculata***
Caudata - Plethodontidae
e Black-spotted Salamander

4392 ***Pseudoeurycea parva***
Caudata - Plethodontidae
e Dwarf False Brook Salamander

4393 ***Pseudoeurycea praecellens***
Caudata - Plethodontidae
e Admirable False Brook Salamander
d Veracruz-Pygmäensalamander

4394 ***Pseudoeurycea rex***
Caudata - Plethodontidae
e Royal False Brook Salamander

4395 ***Pseudoeurycea robertsi***
Caudata - Plethodontidae
e Roberts's False Brook Salamander

4396 ***Pseudoeurycea saltator***
Caudata - Plethodontidae
e Leaping False Brook Salamander

4397 ***Pseudoeurycea scandens***
Caudata - Plethodontidae
e Tamaulipan False Brook Salamander
d Grauflecken-Mexikosalamander

4398 ***Pseudoeurycea smithi***
Caudata - Plethodontidae
e Smith's False Brook Salamander

4399 ***Pseudoeurycea unguidentis***
Caudata - Plethodontidae
e Claw-toed False Brook Salamander; Clawtooth False Brook Salamander

4400 ***Pseudoeurycea werleri***
Caudata - Plethodontidae
e Werler's False Brook Salamander

4401 ***Pseudopaludicola***
Anura - Leptodactylidae
e Swamp Frogs
f Grenouilles naines du marais

4402 ***Pseudopaludicola boliviana***
Anura - Leptodactylidae
e Bolivian Swamp Frog

4403 ***Pseudopaludicola ceratophryes***
Anura - Leptodactylidae
e Leticia Swamp Frog

4404 ***Pseudopaludicola falcipes***
Anura - Leptodactylidae
e Hensel's Swamp Frog
d Grillenfrosch

4405 ***Pseudopaludicola llanera***
Anura - Leptodactylidae
e Lynch's Swamp Frog

4406 ***Pseudopaludicola mysticalis***
Anura - Leptodactylidae
e Cope's Swamp Frog

4407 ***Pseudopaludicola pusilla***
Anura - Leptodactylidae
e Colombian Swamp Frog

4408 ***Pseudopaludicola saltica***
Anura - Leptodactylidae
e Chupada Swamp Frog

4409 ***Pseudopaludicola ternetzi***
Anura - Leptodactylidae
e Goias Swamp Frog

4410 ***Pseudophilautus***
Anura - Rhacophoridae
e Malabar Coast Frog

4411 ***Pseudophilautus temporalis***
Anura - Rhacophoridae
e Malabar Coast Frog

4412 ***Pseudophryne***
Anura - Myobatrachidae
e Crowned Toadlets
d Scheinkröten; Australische Scheinkröten

4413 ***Pseudophryne australis***
Anura - Myobatrachidae
e Red-crowned Toadlet
d Australische Scheinkröte

4414 ***Pseudophryne bibronii***
Anura - Myobatrachidae
e Brown Toadlet

4415 ***Pseudophryne coriacea***
Anura - Myobatrachidae
e Redback Toadlet; Keferstein's Toadlet; Red-backed Toadlet (ANZ)
d Kefersteins Scheinkröte

4416 ***Pseudophryne corroboree***
Anura - Myobatrachidae
e Corroboree Toadlet; Corroboree Frog (ANZ)
d Corroboree-Scheinkröte

4417 ***Pseudophryne dendyi***
Anura - Myobatrachidae
e Dendy Toadlet; Southern Toadlet

4418 ***Pseudophryne douglasi***
Anura - Myobatrachidae
e Douglas's Toadlet
d Douglas-Scheinkröte

4419 ***Pseudophryne guentheri***
Anura - Myobatrachidae
e Günther's Toadlet
d Günthers Scheinkröte

4420 ***Pseudophryne major***
Anura - Myobatrachidae
e Marbled Toadlet

4421 ***Pseudophryne occidentalis***
Anura - Myobatrachidae
e Orange-crowned Toadlet

4422 ***Pseudophryne semimarmorata***
Anura - Myobatrachidae
e Southern Toadlet

4423 ***Pseudotriton***
Caudata - Plethodontidae
e Red and Mud Salamanders; Red Salamanders; Mud Salamanders
d Rotsalamander
f Pseudotriton

4424 ***Pseudotriton montanus***
Caudata - Plethodontidae
e Mud Salamander; Red Salamander (NA); Baird's Red Salamander (NA)
d Schlammsalamander; Östlicher Schlammsalamander

4425 ***Pseudotriton montanus diastictus***
Caudata - Plethodontidae
e Midland Mud Salamander

4426 ***Pseudotriton montanus flavissimus***
Caudata - Plethodontidae
e Gulf Coast Mud Salamander

4427 ***Pseudotriton montanus floridanus***
Caudata - Plethodontidae
e Rusty Mud Salamander

4428 ***Pseudotriton montanus montanus***
Caudata - Plethodontidae
e Eastern Mud Salamander

4429 ***Pseudotriton ruber***
Caudata - Plethodontidae
e Red Salamander
d Roter Wiesensalamander; Rotsalamander; Nördlicher Rotsalamander; Rotmolch; Roter Quellenmolch
f Salamandre rouge; Salamandre rouge du Nord; Salamandre rousse; Pseudotriton rouge
i Salamandra rossa

4430 ***Pseudotriton ruber nitidus***
Caudata - Plethodontidae
e Blue Ridge Red Salamander

4431 ***Pseudotriton ruber ruber***
Caudata - Plethodontidae
e Northern Red Salamander

4432 ***Pseudotriton ruber schencki***
Caudata - Plethodontidae
e Blackchin Red Salamander; Black-chinned Red Salamander

4433 ***Pseudotriton ruber vioscai***
Caudata - Plethodontidae
e Southern Red Salamander

4434 ***Psyllophryne***
Anura - Brachycephalidae
e Isecksohn's Toads

4435 ***Psyllophryne didactyla***
Anura - Brachycephalidae
e Isecksohn's Toad

4436 ***Pternohyla***
Anura - Hylidae
e Burrowing Frogs; Burrowing Tree Frogs
d Grablaubfrösche

4437 ***Pternohyla dentata***
Anura - Hylidae
e Upland Burrowing Frog; Upland Burrowing Tree Frog

4438 ***Pternohyla fodiens***
Anura - Hylidae
e Lowland Burrowing Tree Frog; Northern Casque-headed Frog (NA)
d Grablaubfrosch

4439 ***Pterorana***
Anura - Ranidae
e Khare's Frogs

4440 ***Pterorana khare***
Anura - Ranidae
e Khare's Frog

4441 ***Ptychadena***
Anura - Ranidae
e Grassland Frogs; Ridged Frogs (CSA); Grass Frogs (CSA)

4442 ***Ptychadena aequiplicata***
Anura - Ranidae
e Victoria Grassland Frog

4443 ***Ptychadena anchietae***
Anura - Ranidae
e Banguella Grassland Frog; Plain Grass Frog (CSA); Anchieta's Ridged Frog; Savanna Ridged Frog (CSA); Northern Rana (CSA); Anchieta's Rocket Frog (CSA); Anchieta's Frog (CSA)

4444 ***Ptychadena ansorgii***
Anura - Ranidae
e Angola Grassland Frog

4445 ***Ptychadena broadleyi***
Anura - Ranidae
e Broadley's Grassland Frog

4446 ***Ptychadena bunoderma***
Anura - Ranidae
e Caconda Grassland Frog

4447 ***Ptychadena christyi***
Anura - Ranidae
e Christy's Grassland Frog; Christy's Frog
f Grenouille de Christy

4448 ***Ptychadena chrysogaster***
Anura - Ranidae
e Rwanda Grassland Frog; Yellow-bellied Ridged Frog

4449 ***Ptychadena cooperi***
Anura - Ranidae
e Cooper's Grassland Frog

4450 ***Ptychadena erlangeri***
Anura - Ranidae
e Erlanger's Grassland Frog

4451 ***Ptychadena grandisonae***
Anura - Ranidae
e Muita Grassland Frog

4452 ***Ptychadena guibei***
Anura - Ranidae
e Guibe's Grassland Frog; Guibe's Grass Frog (CSA)

4453 ***Ptychadena ingeri***
Anura - Ranidae
e Inger's Grassland Frog

4454 ***Ptychadena keilingi***
Anura - Ranidae
e Dala Grassland Frog

4455 ***Ptychadena longirostris***
Anura - Ranidae
e Snouted Grassland Frog; Sharp-nosed Frog (CSA)

4456 ***Ptychadena maccarthyensis***
Anura - Ranidae
e MacCarthy Grassland Frog

4457 ***Ptychadena mascareniensis***
Anura - Ranidae
e Mascarene Grassland Frog; Mascarene Grass Frog (CSA); Mascarene Frog (CSA); Mascarene Ridged Frog (CSA); Common Mascarene Frog; Broad-banded Rana (CSA); Mascarene Rocket Frog (CSA)
d Nilfrosch
f Grenouille des mascareignes

4458 ***Ptychadena mossambica***
Anura - Ranidae
e Mozambique Grassland Frog; Mocambique Ridged Frog (CSA); Broad-backed Grass Frog (CSA); Broad-banded Grass Frog (CSA); Mozambique Ridged Frog; Broad-banded Ridged Frog (CSA)

4459 ***Ptychadena nana***
Anura - Ranidae
e Somali Grassland Frog

4460 ***Ptychadena neumanni***
Anura - Ranidae
e Neumann's Grassland Frog

4461 ***Ptychadena newtoni***
Anura - Ranidae
e Newton's Grassland Frog

4462 ***Ptychadena obscura***
Anura - Ranidae
e Kaziba Grassland Frog

4463 ***Ptychadena oxyrhinchus***
Anura - Ranidae
e Kaffirland Grassland Frog; Ridged Frog; Sharp-nosed Frog (CSA); Sharp-nosed Ridged Frog (CSA); Sharp-nosed Grass Frog (CSA); Sharp-snouted Ridged Frog (CSA)

4464 ***Ptychadena peretti***
Anura - Ranidae
e Perret's Grassland Frog

4465 ***Ptychadena perplicata***
Anura - Ranidae
e Lunda Grassland Frog

4466 ***Ptychadena porosissima***
Anura - Ranidae
e Ethiopia Grassland Frog; Striped Grass Frog (CSA); Grassland Ridged Frog (CSA); Three-striped Rana (CSA); Three Striped Ridged Frog (CSA); Steindachner's Frog (CSA)

4467 ***Ptychadena pumilio***
Anura - Ranidae
e Medine Grassland Frog; Dwarf Grass Frog; Spotted-throated Ridged Frog; Dwarf Rana

4468 ***Ptychadena retropunctata***
Anura - Ranidae
e Cameroon Grassland Frog

4469 ***Ptychadena schillukorum***
Anura - Ranidae
e Sudan Grassland Frog

4470 ***Ptychadena schubotzi***
Anura - Ranidae
e Central Grassland Frog

4471 ***Ptychadena stenocephala***
Anura - Ranidae
e Uganda Grassland Frog; Uganda Striped Grass Frog

4472 ***Ptychadena straeleni***
Anura - Ranidae
e Nagero Grassland Frog

4473 ***Ptychadena submascareniensis***
Anura - Ranidae
e Guinea Grassland Frog

4474 ***Ptychadena subpunctata***
Anura - Ranidae
e Bocage's Grassland Frog; African Spotted Frog; Spot-bellied Ridged Frog; Speckle-bellied Grass Frog (CSA)
f Grenouille tachetée africaine

4475 ***Ptychadena superciliaris***
Anura - Ranidae
e Sierra Leone Grassland Frog; Savanna Ridged Frog (CSA); Red-

backed Grass Frog (CSA)

4476 ***Ptychadena taenioscelis***
Anura - Ranidae
e Lukula Grassland Frog; Dwarf Grass Frog (CSA); Stripe-legged Frog (CSA); Spotted-throated Ridged Frog (CSA); Dwarf Rana (CSA)

4477 ***Ptychadena tournieri***
Anura - Ranidae
e Liberia Grassland Frog

4478 ***Ptychadena trinodis***
Anura - Ranidae
e Dakar Grassland Frog

4479 ***Ptychadena upembae***
Anura - Ranidae
e Upemba Grassland Frog

4480 ***Ptychadena uzungwensis***
Anura - Ranidae
e Uzungwe Grassland Frog; Uzungwe Ridged Frog; Uzungwe Grass Frog (CSA)

4481 ***Ptychohyla***
Anura - Hylidae
e Stream Frogs

4482 ***Ptychohyla erythromma***
Anura - Hylidae
e Guerreran Stream Frog

4483 ***Ptychohyla euthysanota***
Anura - Hylidae
e Cloud-forest Stream Frog

4484 ***Ptychohyla hypomykter***
Anura - Hylidae
e Copan Stream Frog

4485 ***Ptychohyla legleri***
Anura - Hylidae
e Legler's Stream Frog

4486 ***Ptychohyla leonhardschultzi***
Anura - Hylidae
e Schultze's Mountain Stream Frog

4487 ***Ptychohyla macrotympanum***
Anura - Hylidae
e Pine-forest Stream Frog

4488 ***Ptychohyla panchoi***
Anura - Hylidae
e Guatemala Stream Frog

4489 ***Ptychohyla salvadorensis***
Anura - Hylidae
e Salvador Stream Frog

4490 ***Ptychohyla sanctaecrucis***
Anura - Hylidae
e Chinamococh Stream Frog

4491 ***Ptychohyla spinipollex***
Anura - Hylidae
e Ceiba Stream Frog

4492 ***Pyxicephalus***
Anura - Ranidae
e African Bullfrogs; Bullfrogs (CSA)

4493 ***Pyxicephalus adspersus***
Anura - Ranidae
e South African Bullfrog; Tschudi's African Bullfrog; Burrowing Bullfrog; Bullfrog (CSA); Pyxie (CSA); Giant Pyxie (CSA); Speckled Frog; Giant Bullfrog (CSA); African Bullfrog (CSA)
d Grabfrosch; Gesprenkelter Grabfrosch; Südafrikanischer Ochsenfrosch; Afrikanischer Ochsenfrosch
f Grenouille fouisseuse tachetée d'Afrique du Sud

4494 ***Pyxicephalus edulus***
Anura - Ranidae
e African Bullfrog (CSA); Bullfrog (CSA); Lesser Bullfrog (CSA)

R

4495 ***Ramanella***
Anura - Microhylidae
e Dot Frogs; Ramanellas

4496 ***Ramanella anamalaiensis***
Anura - Microhylidae
e Anamallai Dot Frog

4497 ***Ramanella minor***
Anura - Microhylidae
e Hassan Dot Frog

4498 ***Ramanella montana***
Anura - Microhylidae
e Mountain Dot Frog; Jerdon's Ramanella (ISC)

4499 ***Ramanella mormorata***
Anura - Microhylidae
e Indian Dot Frog

4500 ***Ramanella obscura***
Anura - Microhylidae
e Sri Lanka Dot Frog; Ceylonese Narrow-mouthed Frog; Ceylonese Ramanella; Obscure Ramanella

4501 ***Ramanella palmata***
Anura - Microhylidae
e Parker's Dot Frog

4502 ***Ramanella triangularis***
Anura - Microhylidae
e Malabar Dot Frog; Triangle-spotted Ramanella

4503 ***Ramanella variegata***
Anura - Microhylidae
e Eluru Dot Frog; Variable Ramanella

4504 ***Rana***
Anura - Ranidae
e True Frogs; Ranas; River Frogs (CSA); Typical Frog; Stream Frogs (CSA)
d Eigentliche Frösche; Echte Frösche; Schiebebrust-Frösche
f Vraies grenouilles
i Rane

4505 ***Rana adenopleura***
Anura - Ranidae
e Olive Frog; Gland-sided Frog (SEA)
d Seitendrüsenfrosch

4506 ***Rana albolabris***
Anura - Ranidae
e White-lipped Frog
d Silberlippenfrosch
f Grenouille à levres blanches

4507 ***Rana altaica***
Anura - Ranidae
e Siberian Frog

4508 ***Rana alticola***
Anura - Ranidae
e Assam Hills Frog; Hill Frog (SEA)
d Bergfrosch

4509 ***Rana altotuberculata***
Anura - Ranidae
e Cabalian Frog

4510 ***Rana amieti***
Anura - Ranidae
e Pangi Territory Frog

4511 ***Rana amnicola***
Anura - Ranidae
e Ilang Frog

4512 ***Rana amurensis***
Anura - Ranidae
e Khabarovsk Frog; Siberian Wood Frog
d Sibirischer Moorfrosch; Sibirischer Frosch
f Grenouille de Sibérie

4513 ***Rana andersoni***
Anura - Ranidae
e Golden Crossband Frog

4514 ***Rana anlungensis***
Anura - Ranidae
e Luntou Shan Frog

4515 ***Rana aragonensis***
Anura - Ranidae
e Grass Frog

4516 ***Rana areolata***
Anura - Ranidae
e Crawfish Frog
d Krebsfrosch; Gopherfrosch
f Grenouille auréolée

4517 ***Rana areolata areolata***
Anura - Ranidae
e Southern Crawfish Frog

4518 ***Rana areolata circulosa***
Anura - Ranidae
e Northern Crawfish Frog

4519 ***Rana arkfaki***
Anura - Ranidae
e Arfak Mountains Frog; Large River Frog

4520 ***Rana arvalis***
Anura - Ranidae
e Swedish Frog; Moor Frog
d Moorfrosch
f Grenouille oxyrhine; Grenouille des champs; Grenouille champêtre; Grenouille des tourbières; Grenouille des marais; Grenouille à museau pointu
i Rana arvale

4521 ***Rana arvalis arvalis***
Anura - Ranidae
d Ungarischer Moorfrosch

4522 ***Rana asiatica***
Anura - Ranidae
e Asian Frog

4523 ***Rana asperrima***
Anura - Ranidae
e Nkongsamba Frog

4524 ***Rana aurantiaca***
Anura - Ranidae
e Trivandrum Frog; Golden Frog (ISC); Indian Golden Frog

4525 ***Rana aurora***
Anura - Ranidae
e Red-legged Frog
f Grenouille à pattes rouges

4526 ***Rana aurora aurora***
Anura - Ranidae
e Northern Red-legged Frog

4527 ***Rana aurora draytonii***
Anura - Ranidae
e California Red-legged Frog

4528 ***Rana balcanica***
Anura - Ranidae
e Greek Marsh Frog
i Rana verde balcanica

4529 ***Rana baramica***
Anura - Ranidae
e Baram River Frog; Masked Swamp Frog (SEA)
d Maskenfrosch

4530 ***Rana bedriagae***
Anura - Ranidae
e Bedriaga's Syrian Frog; Levant Water Frog; Hadera Frog
f Grenouille de Bedriaga

4531 ***Rana bergeri***
Anura - Ranidae
e Italian Pool Frog; Southern Non-hybrid Frog; Berger's Frog
d Italienischer Wasserfrosch
f Grenouille de Berger; Grenouille verte italienne non-hybride
i Rana di Berger; Rana verde di Berger

4532 ***Rana berlandieri***
Anura - Ranidae
e Rio Grande Leopard Frog (NA); Rio Grande Frog

4533 ***Rana blairi***
Anura - Ranidae
e Plains Leopard Frog

4534 ***Rana bonaspei***
Anura - Ranidae
e Jonkersberg Frog; Mountain Frog; Cape Grass Frog; Banded Stream Frog (CSA)

4535 ***Rana boylei***
Anura - Ranidae
e Foothill Yellow-legged Frog; Yellow-legged Frog

4536 ***Rana boylei boyleyi***
Anura - Ranidae
e California Yellow-legged Frog

4537 ***Rana boylei muscosa***
Anura - Ranidae
e San Bernardino Yellow-legged Frog

4538 ***Rana brownorum***
Anura - Ranidae
e Brown's Leopard Frog

4539 ***Rana bwana***
Anura - Ranidae
e Rio Chipillico

4540 ***Rana caldwelli***
Anura - Ranidae
e Caldwell's Chinese Frog

4541 ***Rana camerani***
Anura - Ranidae
e Caucasus Frog
d Kaukasus-Frosch
f Grenouille de Camerano
i Rana di Camerano

4542 ***Rana capito***
Anura - Ranidae
e Gopher Frog

4543 ***Rana capito aesopus***
Anura - Ranidae
e Florida Gopher Frog

4544 ***Rana capito capito***
Anura - Ranidae
e Carolina Gopher Frog; Florida Gopher Frog

4545 ***Rana capito sevosa***
Anura - Ranidae
e Dusky Gopher Frog

4546 ***Rana cascadae***
Anura - Ranidae
e Cascades Frog

4547 ***Rana catesbeiana***
Anura - Ranidae
e North American Bullfrog; Bullfrog (NA); American Bullfrog
d Ochsenfrosch; Amerikanischer Ochsenfrosch; Nordamerikanischer Ochsenfrosch
f Grenouille taureau; Grenouille-taureau d'Amérique; Grenouille-

boeuf; Ouaouaron (Qué)
i Rana toro

4548 ***Rana celebensis***
Anura - Ranidae
e Celebes Frog

4549 ***Rana cerigensis***
Anura - Ranidae
e Karpathos Water Frog

4550 ***Rana chalconota***
Anura - Ranidae
e Schlegel's Java Frog; White-lipped Frog (SEA)
d Weißlippenfrosch

4551 ***Rana chaochiaoensis***
Anura - Ranidae
e Chaochiao City Frog

4552 ***Rana chapaensis***
Anura - Ranidae
e Chapa Frog

4553 ***Rana chensinensis***
Anura - Ranidae
e Inkiapo Frog; Asiatic Grass Frog; North China Wood Frog

4554 ***Rana chevronta***
Anura - Ranidae
e Toudahoe Frog

4555 ***Rana chiricahuensis***
Anura - Ranidae
e Chiricahua Leopard Frog; Wright's Leopard Frog

4556 ***Rana chosenica***
Anura - Ranidae
e Seoul Frog

4557 ***Rana clamitans***
Anura - Ranidae
e Green Frog; Bronze Frog
d Schreifrosch
f Grenouille verte; Grenouille américaine; Grenouille bronzée; Grenouille criarde

4558 ***Rana clamitans clamitans***
Anura - Ranidae
e Bronze Frog

4559 ***Rana clamitans melanota***
Anura - Ranidae
e Green Frog; Northern Green Frog

4560 ***Rana cordofana***
Anura - Ranidae
e Cordofan Frog

4561 ***Rana crassiovis***
Anura - Ranidae
e Korinchi Frog

4562 ***Rana cretensis***
Anura - Ranidae
e Cretan Water Frog

4563 ***Rana cubitalis***
Anura - Ranidae
e Siam Frog; Dark-eared Frog (SEA)
d Schwarzohrfrosch

4564 ***Rana curtipes***
Anura - Ranidae
e Malabar Frog; Bicoloured Frog (ISC)

4565 ***Rana daemelii***
Anura - Ranidae
e Australian Wood Frog; Wood Frog (ANZ)

4566 ***Rana dalmatina***
Anura - Ranidae
e Spring Frog; Agile Frog; Dalmatian Frog
d Springfrosch

f Grenouille pisseuse; Grenouille agile
i Rana agile; Saltapré

4567 ***Rana danieli***
Anura - Ranidae
e Mawphlang Forest Frog

4568 ***Rana darlingi***
Anura - Ranidae
e Darling's Frog; Darling's Golden-backed Frog (CSA); Golden-backed Frog (CSA)

4569 ***Rana daunchina***
Anura - Ranidae
e Hungchun-ping Frog

4570 ***Rana debussyi***
Anura - Ranidae
e Battak Mountains Frog

4571 ***Rana desaegeri***
Anura - Ranidae
e Byangolo Frog

4572 ***Rana dunni***
Anura - Ranidae
e Patzcuaro Frog

4573 ***Rana dybowskii***
Anura - Ranidae
e Dybowski's Frog

4574 ***Rana elberti***
Anura - Ranidae
e Lesser Sundas Frog

4575 ***Rana emilijanovi***
Anura - Ranidae
e Imienpo Station Frog

4576 ***Rana epeirotica***
Anura - Ranidae
e Epirus Water Frog
d Epirus-Frosch
f Grenouille épirote

4577 ***Rana erythraea***
Anura - Ranidae
e Red-eared Frog; Golden-lined Frog; Green Paddy Frog; Red-eared Frog (SEA)
d Rotohrfrosch; Rotohrfrosch
f Grenouille rouge

4578 ***Rana esculenta***
Anura - Ranidae
e Edible Frog
d Teichfrosch; Wasserfrosch
f Grenouille verte; Grenouille comestible; Grenouille commune; Grande grenouille verte
i Rana verde esculenta; Rana verde; Rana verde minore; Rana verde comune; Rana verdi; Ranòcc

4579 ***Rana everetti***
Anura - Ranidae
e Zamboanga Frog

4580 ***Rana fasciata***
Anura - Ranidae
e Striped Frog; Striped Stream Frog (CSA); Tanganyika Striped Grass Frog; Striped Grass Frog (CSA); Long-toed Frog (CSA); Long-toed Grass Frog; Striped Rana (CSA)
f Grenouille à bandes

4581 ***Rana fisheri***
Anura - Ranidae
e Vegas Valley Frog; Las Vegas Leopard Frog; Vegas Valley Leopard Frog

4582 ***Rana florensis***
Anura - Ranidae
e Flores Frog

4583 ***Rana forreri***
Anura - Ranidae
e Forrer's Grass Frog

4584 ***Rana fukiensis***
Anura - Ranidae
e Futsing Hsien Frog

4585 ***Rana galamensis***
Anura - Ranidae
e Marble-legged Frog; Golden-backed Frog (CSA); Lake Galam Frog; Galam Golden-backed Frog (CSA)
d Goldrückenfrosch
f Grenouille de Galam

4586 ***Rana garoensis***
Anura - Ranidae
e Garo Hills Frog

4587 ***Rana garritor***
Anura - Ranidae
e Eilogo Estate Frog

4588 ***Rana ghoshi***
Anura - Ranidae
e Manipur Frog

4589 ***Rana glandulosa***
Anura - Ranidae
e Sarawak Frog; Rough-sided Frog; Waterdog (SEA)
d Wasserhund

4590 ***Rana gracilis***
Anura - Ranidae
e Gravenhorst's Frog; Slender Frog

4591 ***Rana graeca***
Anura - Ranidae
e Stream Frog; Greek Frog; Grecian Frog; Balkan Stream Frog
d Griechischer Frosch
f Grenouille grecque
i Rana greca

4592 ***Rana grafi***
Anura - Ranidae
e Graf's Hybrid Frog
d Grafscher Hybridfrosch
f Grenouille de Graf

4593 ***Rana grahami***
Anura - Ranidae
e Yunnanfu Frog

4594 ***Rana grandocula***
Anura - Ranidae
e Big-eyed Frog

4595 ***Rana grayii***
Anura - Ranidae
e Gray's Spotted Frog; Gray's Grass Frog; Gray's Frog (CSA); Clicking Stream Frog (CSA); Spotted Rana (CSA); Gray's Grass Frog (CSA)
d Grays Frosch
f Grenouille de Gray

4596 ***Rana grisea***
Anura - Ranidae
e Went Mountains Frog

4597 ***Rana grylio***
Anura - Ranidae
e Pig Frog
d Schweinsfrosch

4598 ***Rana guentheri***
Anura - Ranidae
e Günther's Amoy Frog; Günther's Frog; Red-and-black Frog

4599 ***Rana heckscheri***
Anura - Ranidae
e River Frog

4600 ***Rana hejiangensis***
Anura - Ranidae
e Hejiang Frog

4601 ***Rana hispanica***
Anura - Ranidae
e Italian Hybrid Frog; Southern Hybrid Frog
d Italienischer Hybridfrosch
f Grenouille d'Uzzell
i Rana verde di Uzzell

4602 ***Rana holsti***
Anura - Ranidae
e Okinawa Shima Frog; Dagger Frog
d Dolchfrosch
f Grenouille à poignard

4603 ***Rana holtzi***
Anura - Ranidae
e Turkish Frog; Holtz's Frog
d Holtzs Frosch
f Grenouille de Holtz
i Rana di Holtz

4604 ***Rana hosii***
Anura - Ranidae
e Mount Dulit Frog; Hose's Frog; Poisonous Rock Frog; Hose's Rock Frog (SEA)
d Hoses Frosch; Hoses Felsenfrosch
f Grenouille de Hose

4605 ***Rana huanrenensis***
Anura - Ranidae
e Huanren Frog

4606 ***Rana hubiensis***
Anura - Ranidae
e Hubei Frog

4607 ***Rana humeralis***
Anura - Ranidae
e Bhamo Frog

4608 ***Rana hymenopus***
Anura - Ranidae
e Natal Drakensberg Frog; Berg Stream Frog (CSA); Drakensberg Rana (CSA); Drakensberg Frog (CSA)

4609 ***Rana iberica***
Anura - Ranidae
e Iberian Frog
d Spanischer Frosch
f Grenouille ibérique; Grenouille d'Espagne; Grenouille espagnole

4610 ***Rana ijimae***
Anura - Ranidae
e Ijima Frog

4611 ***Rana ishikawae***
Anura - Ranidae
e Amami Oshima Frog

4612 ***Rana italica***
Anura - Ranidae
e Italian Frog; Italian Stream Frog
d Italienischer Frosch
f Grenouille italienne
i Rana appeninica; Rana rossa appeninica

4613 ***Rana japonica***
Anura - Ranidae
e Agile Frog
d Japanischer Springfrosch

4614 ***Rana jimiensis***
Anura - Ranidae
e Jimi Riuver Frog

4615 ***Rana johni***
Anura - Ranidae
e Moore's Frog

4616 ***Rana juliani***
Anura - Ranidae
e Maya Mountains Frog

4617 ***Rana kampnei***
Anura - Ranidae

e Bander Baru Frog

4618 ***Rana kreffti***
Anura - Ranidae
e San Cristoval Frog

4619 ***Rana kurtmuelleri***
Anura - Ranidae
e Balkan Frog
d Balkan-Frosch
f Grenouille rieuse

4620 ***Rana kwangwuensis***
Anura - Ranidae
e Kuang-wu Shan Frog

4621 ***Rana latastei***
Anura - Ranidae
e Italian Agile Frog
d Italienischer Springfrosch; Italienischer Frosch
f Grenouille italienne; Grenouille de Lataste; Grenouille agile d'Italie
i Rana di Lataste

4622 ***Rana lateralis***
Anura - Ranidae
e Kakarit Frog

4623 ***Rana latouchii***
Anura - Ranidae
e Kuan Frog

4624 ***Rana lemairii***
Anura - Ranidae
e Lofoi Frog

4625 ***Rana leptoglossa***
Anura - Ranidae
e Cope's Assam Frog; Cope's Frog (SEA)
d Copes Frosch

4626 ***Rana lepus***
Anura - Ranidae
e Anderson's Cameroon Frog

4627 ***Rana lessonae***
Anura - Ranidae
e Pool Frog; Little Water Frog; Edible Frog
d Tümpelfrosch; Kleiner Wasserfrosch; Wasserfrosch
f Grenouille de Lessona; Petite grenouille verte
i Rana di Lessona; Rana verde; Rana verde minore

4628 ***Rana livida***
Anura - Ranidae
e Tenasserim Frog; Green Cascade Frog; Burmese Rock Frog (SEA)
d Birmesischer Felsenfrosch

4629 ***Rana longicrus***
Anura - Ranidae
e Taipa Frog

4630 ***Rana longipes***
Anura - Ranidae
e Bamileke Plateau Frog

4631 ***Rana luctuosa***
Anura - Ranidae
e Malaysian Frog; Mahagony Frog (SEA)
d Mahagoni-Frosch

4632 ***Rana lungshengensis***
Anura - Ranidae
e Lung-shen-hsien Frog

4633 ***Rana luteiventris***
Anura - Ranidae
e Columbia Spotted Frog; Great Basin Spotted Frog

4634 ***Rana luzonensis***
Anura - Ranidae
e Luzon Frog

4635 ***Rana macrocnemis***
Anura - Ranidae
e Brusa Frog; Iranian Long-legged Wood Frog; Caucasian Frog
d Kleinasiatischer Frosch; Kaukasischer Frosch
f Grenouille du Caucase
i Rana gambalunga

4636 ***Rana macrodactyla***
Anura - Ranidae
e Guangdong Frog; Marbled Slender Frog; Long-toed Frog

4637 ***Rana macroglossa***
Anura - Ranidae
e Guatemala Plateau Frog

4638 ***Rana macrops***
Anura - Ranidae
e Masarang Mountains Frog

4639 ***Rana maculata***
Anura - Ranidae
e Highlands Frog; Highland Frog

4640 ***Rana magnaocularis***
Anura - Ranidae
e Northwest Mexico Leopard Frog

4641 ***Rana malabarica***
Anura - Ranidae
e Malabar Hills Frog; Fungoid Frog (ISC)
f Grenouille de Malabar

4642 ***Rana maosonensis***
Anura - Ranidae
e Mao-Son Frog

4643 ***Rana margaretae***
Anura - Ranidae
e Margareta's Frog

4644 ***Rana margariana***
Anura - Ranidae
e Irawady Frog

4645 ***Rana maslini***
Anura - Ranidae
e Maslin's Wood Frog

4646 ***Rana melanomenta***
Anura - Ranidae
e Papahag Island Frog

4647 ***Rana miadis***
Anura - Ranidae
e Little Corn Island Frog

4648 ***Rana milleti***
Anura - Ranidae
e Dalat Frog

4649 ***Rana minima***
Anura - Ranidae
e Fujian Frog

4650 ***Rana miopus***
Anura - Ranidae
e Khao Wang Hip Frog; Three-striped Frog (SEA)
d Dreistreifenfrosch

4651 ***Rana moellendorffi***
Anura - Ranidae
e Culion Frog

4652 ***Rana moluccana***
Anura - Ranidae
e Moluccas Frog

4653 ***Rana montezumae***
Anura - Ranidae
e Montezuma Leopard Frog

4654 ***Rana montivaga***
Anura - Ranidae

e Langbian Plateau Frog; Stream Frog (SEA)
d Bachfrosch

4655 ***Rana mortensii***
Anura - Ranidae
e Koh Chang Island Frog

4656 ***Rana muscosa***
Anura - Ranidae
e Mountain Yellow-legged Frog

4657 ***Rana narina***
Anura - Ranidae
e Ryukyu Island Frog

4658 ***Rana neovolcanica***
Anura - Ranidae
e Transverse Volcanic Leopard Frog

4659 ***Rana nicobariensis***
Anura - Ranidae
e Nicobar Island Frog; Nicobar Frog; Nicobar Cricket Frog (SEA)
d Nikobaren-Frosch

4660 ***Rana nigrolineata***
Anura - Ranidae
e Meng-yang Frog

4661 ***Rana nigromaculata***
Anura - Ranidae
e Black-spotted Frog; Dark-spotted Frog; Chinese Edible Frog; Common Pond Frog
d Chinesischer Wasserfrosch

4662 ***Rana nigrotympanica***
Anura - Ranidae
e Black-eared Frog

4663 ***Rana nigrovittata***
Anura - Ranidae
e Black-striped Frog; Dark-sided Frog (SEA)
d Schwarzseitenfrosch

4664 ***Rana novaeguineae***
Anura - Ranidae
e New Guinea Frog

4665 ***Rana oatesii***
Anura - Ranidae
e Toungoo Frog

4666 ***Rana obbianus***
Anura - Ranidae
e Calabresi's African Bullfrog

4667 ***Rana occidentalis***
Anura - Ranidae
e Ivory Coast Frog

4668 ***Rana okaloossae***
Anura - Ranidae
e Florida Bog Frog

4669 ***Rana okinavana***
Anura - Ranidae
e Okinawa Frog

4670 ***Rana omiltemana***
Anura - Ranidae
e Guerreran Leopard Frog

4671 ***Rana onca***
Anura - Ranidae
e Relict Leopard Frog
d Vegas-Pantherfrosch

4672 ***Rana ornativentris***
Anura - Ranidae
e Nikko Frog

4673 ***Rana palmipes***
Anura - Ranidae
e Amazon River Frog; Springchicken
d Südamerikanischer Wasserfrosch
f Grenouille de Spix

4674 ***Rana palustris***
Anura - Ranidae
e Pickerel Frog
d Sumpffrosch; Nordamerikanischer Sumpffrosch
f Grenouille des marais

4675 ***Rana papua***
Anura - Ranidae
e Papua Frog
d Papua-Fröschchen
f Grenouille papoue

4676 ***Rana parkeriana***
Anura - Ranidae
e Congolo Frog

4677 ***Rana perezi***
Anura - Ranidae
e Coruna Frog; Perez's Frog; Iberian Green Frog; Iberian Water Frog
d Iberischer Seefrosch; Iberischer Wasserfrosch
f Grenouille de Perez
i Rana verde spagnola

4678 ***Rana persimilis***
Anura - Ranidae
e Sumatra Frog

4679 ***Rana pipiens***
Anura - Ranidae
e Northern Leopard Frog (NA); Leopard Frog; Meadow Frog (NA)
d Leopardfrosch
f Grenouille léopard; Grenouille mugissante

4680 ***Rana pirica***
Anura - Ranidae
e Hokkaido Frog

4681 ***Rana plancyi***
Anura - Ranidae
e Peking Frog

4682 ***Rana pleuraden***
Anura - Ranidae
e Yunkwei Plateau Frog

4683 ***Rana porosa***
Anura - Ranidae
e Kanagawa Frog

4684 ***Rana pretiosa***
Anura - Ranidae
e Spotted Frog; Oregon Spotted Frog
f Grenouille maculée; Rana luteiventris

4685 ***Rana psaltes***
Anura - Ranidae
e Kampir Falls Frog

4686 ***Rana pueblae***
Anura - Ranidae
e Puebla Frog

4687 ***Rana pustulosus***
Anura - Ranidae
e Mexican Cascades Frog; Rough-skinned Frog; Cascade Frog

4688 ***Rana pyrenaica***
Anura - Ranidae
e Pyrenees Frog; Pyrenean Frog
d Pyrenäenfrosch
f Grenouille des Pyrénés

4689 ***Rana raniceps***
Anura - Ranidae
e Peters's Malaysian Frog

4690 ***Rana ridibunda***
Anura - Ranidae
e Marsh Frog; Lake Frog; Laughing Frog
d Seefrosch; Märzfrosch
f Grenouille rieuse; Grenouille des marais; Grenouille des étangs
i Rana verde maggiore

4691 ***Rana rugosa***
Anura - Ranidae
e Wrinkled Frog; Japanese Wrinkled Frog
d Runzelfrosch
f Grenouille ridée; Grenouille rugeuse

4692 ***Rana ruwenzorica***
Anura - Ranidae
e Ruwenzori Range Frog

4693 ***Rana saharica***
Anura - Ranidae
e Sahara Frog; North African Green Frog
f Grenouille verte d'Afrique du Nord

4694 ***Rana sakuraii***
Anura - Ranidae
e Napparagawa River Frog

4695 ***Rana sanguinea***
Anura - Ranidae
e Calamianes Frog

4696 ***Rana sangzhiensis***
Anura - Ranidae
e Sangzhi Frog

4697 ***Rana sauteri***
Anura - Ranidae
e Kanshirei Village Frog

4698 ***Rana schmackeri***
Anura - Ranidae
e Schmacker's Frog
d Schmackers Frosch

4699 ***Rana scutigera***
Anura - Ranidae
e Haut Sanuk Frog

4700 ***Rana senchalensis***
Anura - Ranidae
e Senchal Lake Frog

4701 ***Rana septentrionalis***
Anura - Ranidae
e Mink Frog
d Minkfrosch
f Grenouille du Nord (Qué)

4702 ***Rana shgiperica***
Anura - Ranidae
e Virpazar Frog; Albanian Pool Frog; Lake Skutari Water Frog
d Skutari-Wasserfrosch
f Grenouille d'Albanie

4703 ***Rana shuchinae***
Anura - Ranidae
e Sichuan Frog

4704 ***Rana siberu***
Anura - Ranidae
e Siberut Island Frog

4705 ***Rana sierramadrensis***
Anura - Ranidae
e Sierra Madre Occidental Frog

4706 ***Rana signata***
Anura - Ranidae
e Matang Frog; Spotted Stream Frog (SEA)
d Geflecktersumpffrosch

4707 ***Rana similis***
Anura - Ranidae
e Laguna del Bay Frog

4708 ***Rana spectabilis***
Anura - Ranidae
e Showy Leopard Frog

4709 ***Rana sphenocephala***
Anura - Ranidae
e Southern Leopard Frog

d Südlicher Leopardfrosch

4710 ***Rana sphenocephala sphenocephala***
Anura - Ranidae
e Florida Leopard Frog

4711 ***Rana sphenocephala utricularia***
Anura - Ranidae
e Southern Leopard Frog

4712 ***Rana spinidactyla***
Anura - Ranidae
e Spine-fingered Frog

4713 ***Rana spinulosa***
Anura - Ranidae
e Spiny Frog

4714 ***Rana springbokensis***
Anura - Ranidae
e Springbok Frog; Namaqua Stream Frog (CSA)

4715 ***Rana subaquavocalis***
Anura - Ranidae
e Ramsey Canyon Leopard Frog

4716 ***Rana subaspera***
Anura - Ranidae
e Amami Oshima Frog

4717 ***Rana supragrisea***
Anura - Ranidae
e Papua Grey Frog

4718 ***Rana swinhoana***
Anura - Ranidae
e Bangkimtsing Frog

4719 ***Rana sylvatica***
Anura - Ranidae
e Wood Frog
d Waldfrosch
f Grenouille des bois

4720 ***Rana sylvatica cantabrigensis***
Anura - Ranidae
e Northern Wood Frog

4721 ***Rana sylvatica sylvatica***
Anura - Ranidae
e Eastern Wood Frog (NA)

4722 ***Rana tagoe***
Anura - Ranidae
e Tagoe Frog

4723 ***Rana taipehensis***
Anura - Ranidae
e Taipei Frog; Striped Slender Frog

4724 ***Rana tarahumarae***
Anura - Ranidae
e Tarahumara Frog

4725 ***Rana taylori***
Anura - Ranidae
e Peralta Frog

4726 ***Rana temporalis***
Anura - Ranidae
e Kerala Frog; Bronze Frog (ISC)

4727 ***Rana temporaria***
Anura - Ranidae
e European Common Frog; European Frog; Common Frog; Brown Frog; Grass Frog
d Brauner Grasfrosch; Breitschnauziger Grasfrosch; Landfrosch; Grasfrosch
f Grenouille rousse; Grenouille muette
i Rana rossa montana; Rana rossa di montagna; Rana montana; Rana remporaria

4728 ***Rana temporaria honnorati***
Anura - Ranidae
e Heron's Frog
f Grenouille à longues pattes

4729 ***Rana tenggerensis***
Anura - Ranidae
e Yellow River Frog

4730 ***Rana tiannanensis***
Anura - Ranidae
e Dawei Hill Frog

4731 ***Rana tientaiensis***
Anura - Ranidae
e Tiantai Frog

4732 ***Rana tlaloci***
Anura - Ranidae
e Tlaloc's Leopard Frog

4733 ***Rana trilobata***
Anura - Ranidae
e Bigfoot Leopard Frog

4734 ***Rana tsushimensis***
Anura - Ranidae
e Tsushima Frog

4735 ***Rana umbraculata***
Anura - Ranidae
e Umzimkulu River Frog; Umbraculate Frog (CSA)

4736 ***Rana vaillanti***
Anura - Ranidae
e Vaillant's Frog

4737 ***Rana varians***
Anura - Ranidae
e Palawan Frog

4738 ***Rana versabilis***
Anura - Ranidae
e Kwangsi Frog

4739 ***Rana vibicaria***
Anura - Ranidae
e Rancho Redondo Frog

4740 ***Rana virgatipes***
Anura - Ranidae
e Carpenter Frog; Sphagnum Frog (NA)
d Zimmermann-Frosch
f Grenouille charpentier

4741 ***Rana wageri***
Anura - Ranidae
e Natal Uplands Frog; Plain Stream Frog (CSA); Plain Rana (CSA); Plain Stream Rana; Wager's Frog (CSA); Wager's Rana (CSA)

4742 ***Rana warszewitschii***
Anura - Ranidae
e Warszewitsch's Frog

4743 ***Rana weinigensis***
Anura - Ranidae
e Weining Frog

4744 ***Rana wittei***
Anura - Ranidae
e Molo Frog; Barred Rana (CSA)

4745 ***Rana wuchuanensis***
Anura - Ranidae
e Wuchuang Frog

4746 ***Rana yavapaiensis***
Anura - Ranidae
e Yavapai Leopard Frog; Lowland Leopard Frog (NA)

4747 ***Rana zweifeli***
Anura - Ranidae
e Zweifel's Frog

4748 **Ranidae**
Anura
e True Frogs; Typical Frogs
d Echte Frösche; Frösche; Raniden
f Grenouilles; Vraies grenouilles ; Ranidés

4749 ***Relictivomer***
Anura - Microhylidae
e Plump Frogs

4750 ***Relictivomer pearsei***
Anura - Microhylidae
e Plump Frog; Colombian Plump Frog

4751 **Rhacophoridae**
Anura
e Flying Frogs; Tree Frogs
d Ruderfrösche; Flugfrösche; Rhacophoriden
f Rhacophoridés

4752 ***Rhacophorus***
Anura - Rhacophoridae
e Flying Frogs
d Flugfrösche
f Rhacophores

4753 ***Rhacophorus angulirostris***
Anura - Rhacophoridae
e Kina Balu Flying Frog

4754 ***Rhacophorus annamensis***
Anura - Rhacophoridae
e Annam Flying Frog

4755 ***Rhacophorus appendiculatus***
Anura - Rhacophoridae
e Philippine Flying Frog; Frilled Tree Frog (SEA)
d Frilled Tree Frog (SEA)

4756 ***Rhacophorus arboreus***
Anura - Rhacophoridae
e Kinugassa Flying Frog

4757 ***Rhacophorus baluensis***
Anura - Rhacophoridae
e Balu Flying Frog; Spotted Tree Frog (SEA)
d Gefleckter Flugfrosch

4758 ***Rhacophorus bimaculatus***
Anura - Rhacophoridae
e Mindanao Flying Frog

4759 ***Rhacophorus bipunctatus***
Anura - Rhacophoridae
e Himalaya Flying Frog; Twin-spotted Tree Frog (SEA)
d Zweiflecken-Flugfrosch

4760 ***Rhacophorus bisacculus***
Anura - Rhacophoridae
e Loei Flying Frog; Phu Kradung Tree Frog (SEA)
d Phu-Kradung-Flugfrosch

4761 ***Rhacophorus calcadensis***
Anura - Rhacophoridae
e Langbian Flying Frog

4762 ***Rhacophorus calcaneus***
Anura - Rhacophoridae
e Vietnam Flying Frog

4763 ***Rhacophorus depressus***
Anura - Rhacophoridae
e Ahl's Flying Frog

4764 ***Rhacophorus dulitensis***
Anura - Rhacophoridae
e Dulit Flying Frog

4765 ***Rhacophorus edentulus***
Anura - Rhacophoridae
e Celebes Flying Frog

4766 ***Rhacophorus everetti***
Anura - Rhacophoridae
e Everett's Flying Frog

4767 ***Rhacophorus fasciatus***
Anura - Rhacophoridae
e Banded Flying Frog

4768 ***Rhacophorus gauni***
Anura - Rhacophoridae
e Inger's Flying Frog

4769 ***Rhacophorus georgii***
Anura - Rhacophoridae
e Tuwa Flying Frog

4770 ***Rhacophorus gongshanensis***
Anura - Rhacophoridae
e Yunnan Flying Frog

4771 ***Rhacophorus harrissoni***
Anura - Rhacophoridae
e Harrisson's Flying Frog

4772 ***Rhacophorus hecticus***
Anura - Rhacophoridae
e Samar Flying Frog

4773 ***Rhacophorus javanus***
Anura - Rhacophoridae
e Java Flying Frog

4774 ***Rhacophorus kajau***
Anura - Rhacophoridae
e Dring's Flying Frog

4775 ***Rhacophorus malabaricus***
Anura - Rhacophoridae
e Malabar Flying Frog; Malabar Gliding Frog (ISC)

4776 ***Rhacophorus maximus***
Anura - Rhacophoridae
e Nepal Flying Frog; Giant Tree Frog (ISC)

4777 ***Rhacophorus modestus***
Anura - Rhacophoridae
e Boulenger's Flying Frog

4778 ***Rhacophorus moltrechti***
Anura - Rhacophoridae
e Nantou Flying Frog

4779 ***Rhacophorus monticola***
Anura - Rhacophoridae
e Loka Flying Frog

4780 ***Rhacophorus namdaphaensis***
Anura - Rhacophoridae
e Namdapha Flying Frog

4781 ***Rhacophorus nigropalmatus***
Anura - Rhacophoridae
e Abah River Flying Frog; Wallace's Flying Frog
d Wallaces Baumfrosch

4782 ***Rhacophorus nigropunctatus***
Anura - Rhacophoridae
e Weining Flying Frog

4783 ***Rhacophorus notater***
Anura - Rhacophoridae
e Daban Flying Frog

4784 ***Rhacophorus owstoni***
Anura - Rhacophoridae
e Japanese Flying Frog

4785 ***Rhacophorus oxycephalus***
Anura - Rhacophoridae
e Red-headed Flying Frog
d Spitzkopfruderfrosch

4786 ***Rhacophorus pardalis***
Anura - Rhacophoridae
e Panther Flying Frog; Harlequin Tree Frog (SEA)
d Borneo-Ruderfrosch; Gefleckter Flugfrosch

4787 ***Rhacophorus poecilonotus***
Anura - Rhacophoridae
e Sumatra Flying Frog

4788 ***Rhacophorus prasinatus***
Anura - Rhacophoridae
e Tributary Flying Frog

4789 ***Rhacophorus promianus***
Anura - Rhacophoridae
e Jor Flying Frog; Malayan Tree Frog (SEA)
d Malaiischer Flugfrosch

4790 ***Rhacophorus reinwardtii***
Anura - Rhacophoridae
e Reinwardt's Flying Frog; Reinwardt's Gliding Frog; Javan Flying Frog
d Flugfrosch; Java-Flugfrosch; Reinwardts Flugfrosch
f Grenouille volante; Grenouille-parachute; Rhacophore de Reinwardt

4791 ***Rhacophorus rhodopus***
Anura - Rhacophoridae
e Hunan Flying Frog

4792 ***Rhacophorus robinsoni***
Anura - Rhacophoridae
e Robinson's Flying Frog; Robinson's Tree Frog (SEA)
d Robinsons Flugfrosch

4793 ***Rhacophorus rufipes***
Anura - Rhacophoridae
e Malaysian Flying Frog

4794 ***Rhacophorus schlegelii***
Anura - Rhacophoridae
e Schlegel's Flying Frog; Japanese Gliding Frog; Japanese Tree Frog; Schlegel's Gliding Frog
d Japanischer Ruderfrosch
f Rhacophore de Schlegel; Grenouille chantante

4795 ***Rhacophorus taipeianbus***
Anura - Rhacophoridae
e Taipei Flying Frog

4796 ***Rhacophorus taroensis***
Anura - Rhacophoridae
e Patsarlamdan Flying Frog

4797 ***Rhacophorus translineatus***
Anura - Rhacophoridae
e Medog Flying Frog

4798 ***Rhacophorus tuberculatus***
Anura - Rhacophoridae
e Warty Flying Frog

4799 ***Rhacophorus tunkui***
Anura - Rhacophoridae
e Johore Flying Frog; Tunku's Tree Frog (SEA)
d Tunkus Flugfrosch

4800 ***Rhacophorus turpes***
Anura - Rhacophoridae
e Htingnan Flying Frog

4801 ***Rhacophorus verrucopus***
Anura - Rhacophoridae
e Beibeng Flying Frog; Boulenger's Tree Frog (SEA)
d Boulengers Flugfrosch

4802 ***Rhacophorus viridis***
Anura - Rhacophoridae
e Green Flying Frog

4803 ***Rhamphophryne***
Anura - Bufonidae
e Beaked Toads

4804 ***Rhamphophryne acrolopha***
Anura - Bufonidae
e Cerro Mali Beaked Toad

4805 ***Rhamphophryne festae***
Anura - Bufonidae
e Valle Santiago Beaked Toad

4806 ***Rhamphophryne lindae***
Anura - Bufonidae
e Murri Beaked Toad

4807 ***Rhamphophryne macrorhina***
Anura - Bufonidae
e Santa Rita Beaked Toad

4808 ***Rhamphophryne nicefori***
Anura - Bufonidae
e Colombian Beaked Toad

4809 ***Rhamphophryne proboscidae***
Anura - Bufonidae
e Beaked Toad

4810 ***Rhamphophryne rostrata***
Anura - Bufonidae
e Mesopotamia Beaked Toad

4811 ***Rhamphophryne tenrec***
Anura - Bufonidae
e Antioquia Beaked Toad

4812 ***Rhamphophryne truebae***
Anura - Bufonidae
e Trueb's Beaked Toad

4813 ***Rhanodon***
Caudata - Hynobiidae
e Siberian Salamanders
d Froschzahnmolche

4814 ***Rhanodon sibiricus***
Caudata - Hynobiidae
e Siberian Salamander
d Sibirischer Bachrandmolch; Sibirischer Froschzahnmolch
f Salamandre du pays Sept Fleuves; Salamandre du Semiretchensk

4815 ***Rhanodon tsinpaensis***
Caudata - Hynobiidae
e Shanxi Salamander

4816 ***Rheobatrachus***
Anura - Myobatrachidae
e Gastric-brooding Frogs
d Conondale-Magenbrüter (pl)

4817 ***Rheobatrachus silus***
Anura - Myobatrachidae
e Gastric-brooding Frog; Southern Gastric--brooding Frog; Platypus Frog
d Conondale-Magenbrüter

4818 ***Rhinatrema***
Gymnophiona - Rhinatrematidae
e Two-lined Caecilians; Two-banded Caecilians
d Nasewühlen
f Rhinatrèmes

4819 ***Rhinatrema bivittatum***
Gymnophiona - Rhinatrematidae
e Two-lined Caecilian; Two-banded Caecilian
d Nasewühle
f Rhinatrème à deux bandes

4820 **Rhinatrematidae**
Gymnophiona
e Beaked Caecilians; Tailed Caecilians
d Wühlen
f Rhinatrématidés; Cécilies

4821 ***Rhinoderma***
Anura - Rhinodermatidae
e Darwin's Frogs; Mouth-breeding Frogs; Darwin's Toads; Vaqueros
d Darwin-Nasenfrösche
f Rhinodermes

4822 ***Rhinoderma darwinii***
Anura - Rhinodermatidae
e Darwin's Toad; Darwin's Frog
d Darwin-Nasenfrösch; Nasenfrosch; Darwin-Frosch; Chilenischer Nasenfrosch
f Rhinoderme de Darwin

4823 ***Rhinoderma rufum***
Anura - Rhinodermatidae
e Chile Darwin's Frog

4824 **Rhinodermatidae**
Anura
e Mouth-breeding Frogs; Darwin's Frogs
d Nasenfrösche
f Rhinodermes

4825 **Rhinophrynidae**
Caudata
e Burrowing Toads
d Grabkröten; Nasenkröten; Nasenkrötchen
f Rhinophrynidés

4826 ***Rhinophrynus***
Anura - Rhinodermatidae
e Mexican Burrowing Toads; Burrowing Toads; Burrowing Frogs; Mexican Burrowing Frogs
d Nasenfrösche
f Crapauds fournisseurs mexicains; Rhinophrynes

4827 ***Rhinophrynus dorsalis***
Anura - Rhinodermatidae
e Mexican Burrowing Toad; Mexican Burrowing Frog
d Mexikanische Grabkröte; Nasenkröte; Nasenfrosch; Mexikanische Nasenkröte
f Crapaud fournisseur mexicain; Rhinophryne à raie dorsale

4828 ***Rhombophryne***
Anura - Microhylidae
e Burrowing Frogs

4829 ***Rhombophryne testudo***
Anura - Microhylidae
e Burrowing Frog; Nosy Be Burrowing Frog

4830 ***Rhyacosiredon***
Caudata - Ambystomatidae
e Siredons; Stream Salamanders
d Mexikanische Hochland-Querzahnsalamander; Mexikanische Hochland-Querzahnmolche

4831 ***Rhyacosiredon altamirani***
Caudata - Ambystomatidae
e Mountain Stream Salamander; Mountain Stream Siredon
d Ajusko-Gebirgsbachsalamander

4832 ***Rhyacosiredon leorae***
Caudata - Ambystomatidae
e Leora's Stream Salamander; Leora's Stream Siredon

4833 ***Rhyacosiredon rivularis***
Caudata - Ambystomatidae
e Michoacan Stream Salamander; Michoacan Stream Siredon
d Viktoria-Gebirgsbachmolch; Viktoria-Gebirgsbachsalamander

4834 ***Rhyacosiredon zempoalaensis***
Caudata - Ambystomatidae
e Lake Zempola Salamander; Lake Zempola Siredon
d Zempoala-Gebirgsbachmolch; Zempoala-Gebirgsbachsalamander

4835 ***Rhyacotriton***
Caudata - Rhyacotritonidae
e Olympic Salamanders; Torrent Salamanders
d Olymp-Querzahnmolche; Olymp-Querzahnsalamander (pl)
f Salamandres olympiques

4836 ***Rhyacotriton cascadae***
Caudata - Rhyacotritonidae
e Cascades Salamander; Cascade Torrent Salamander (NA)

4837 ***Rhyacotriton kezeri***
Caudata - Rhyacotritonidae
e Oregon Olympic Salamander; Columbia Torrent Salamander

4838 ***Rhyacotriton olympicus***
Caudata - Rhyacotritonidae
e Olympic Salamander; Olympic Torrent Salamander
d Olymp-Querzahnmolch; Olymp-Querzahnsalamander
f Salamandre olympique

4839 ***Rhyacotriton olympicus olympicus***
Caudata - Rhyacotritonidae
e Northern Olympic Salamander

4840 ***Rhyacotriton olympicus variegatus***
Caudata - Rhyacotritonidae
e Southern Olympic Salamander; Southern Torrent Salamander

4841 **Rhyacotritonidae**
Caudata
e Olympic Salamanders
d Olympysche Salamander (pl)
f Salamandres olympiques

S

4842 ***Salamandra***
Caudata - Salamandridae
e Fire Salamanders; Salamanders
d Salamander (pl); Echte Salamander (pl); Eigentliche Salamander (pl)
f Salamandres
i Salamandre

4843 ***Salamandra algira***
Caudata - Salamandridae
e Algerian Salamander
d Algerischer Feuersalamander
f Salamandre algire

4844 ***Salamandra atra***
Caudata - Salamandridae
e Alpine Salamander; European Black Salamander
d Alpensalamander; Tattermandl; Mohrensalamander
f Salamandre noire; Salamandre des Alpes; Salamandre alpestre
i Salamandra atra; Salamandra nera; Salamandra nera alpina; Salamandra alpina

4845 ***Salamandra corsica***
Caudata - Salamandridae
e Corsican Fire Salamander
d Korsischer Feuersalamander
f Salamandre de Corse

4846 ***Salamandra lanzai***
Caudata - Salamandridae
e Lanza's Fire Salamander; Lanza's Alpine Salamander
d Lanzas Alpensalamander
f Salamandre de Lanza; Grande salamandre noire
i Salamandra nera di Lanza

4847 ***Salamandra salamandra***
Caudata - Salamandridae
e Fire Salamander; European Fire Salamander; European Salamander; Spotted Salamander
d Feuersalamander
f Salamandre; Salamandre commune; Salamandre maculée; Salamandre terrestre; Salamandre de feu; Salamandre tachetée
i Salamandra pezzata; Salamandra gialla e nera; Salamandra comune; Sacaramàndla; Scaràmandla

4848 ***Salamandra salamandra algira***
Caudata - Salamandridae
e Fire Salamander
f Salamandre tachetée

4849 ***Salamandra salamandra terrestris***
Caudata - Salamandridae
d Gebänderter Feuersalamander
f Salamandre tachetée

4850 ***Salamandrella***
Caudata - Hynobiidae
e Dybowski's Salamanders; Siberian Salamanders
d Sibirische Winkelzahnmolche
f Salamandres de Sibérie; Salamandrelles de Sibérie

4851 ***Salamandrella keyserlingii***
Caudata - Hynobiidae
e Dybowski's Salamander; Siberian Salamander; Manchurian Salamander
d Sibirischer Winkelzahnmolch
f Salamandre de Sibérie; Salamandrelle de Sibérie

4852 **Salamandridae**
Caudata
e Salamanders and Newts; Fire Salamanders and Newts

d Molche und Salamander; Salamandriden; Salamander und Molche; Echte Salamander und Molche
f Tritons et Salamandres vulgaires; Salamandres; Salamandres vraies; Salamandridés

4853 ***Salamandrina***
Caudata - Salamandridae
e Spectacled Salamanders
d Brillensalamander (pl); Italienische Brillensalamander (pl)
f Salamandrines
i Salamandrine

4854 ***Salamandrina terdigitata***
Caudata - Salamandridae
e Spectacled Salamander
d Brillensalamander; Italienischer Brillensalamander
f Salamandrine à lunettes; Salamandrine à trois orteils; Salamandre à lunettes d'Italie
i Salamandrina dagli occhiali

4855 ***Scaphiophryne***
Anura - Microhylidae
e Rain Frogs

4856 ***Scaphiophryne brevis***
Anura - Microhylidae
e Brown Rain Frog

4857 ***Scaphiophryne calcarata***
Anura - Microhylidae
e Mocquard's Rain Frog

4858 ***Scaphiophryne gottlebei***
Anura - Microhylidae
e Red Rain Frog

4859 ***Scaphiophryne madagascariensis***
Anura - Microhylidae
e Madagascar Rain Frog

4860 ***Scaphiophryne marmorts***
Anura - Microhylidae
e Marbled Rain Frog

4861 ***Scaphiophryne obscura***
Anura - Microhylidae
e Grandidier's Rain Frog

4862 ***Scaphiophryne pustulosa***
Anura - Microhylidae
e Green Rain Frog

4863 ***Scaphiophryne verrucosa***
Anura - Microhylidae
e Angel's Rain Frog

4864 ***Scaphiopus***
Anura - Pelobatidae
e Eastern Spadefoot Toads; Spadefoot Toads; Southern Spadefoot Toads; Southern Spadefoots
d Schaufelfüße; Schaufelfußfrösche; Schaufelfrösche; Schaufelfußkröten

4865 ***Scaphiopus couchii***
Anura - Pelobatidae
e Couch's Spadefoot; Couch's Spadefoot Toad
d Südlicher Schaufelfuß; Couchs Schaufelfuß
f Pélobate de Couch

4866 ***Scaphiopus holbrookii***
Anura - Pelobatidae
e Eastern Spadefoot; Eastern Spadefoot Toad; American Spadefoot; American Spadefoot Toad; Spadefoot Toad (NA); Hermit Toad (NA); Garlic Toad (NA)
d Östlicher Schaufelfuß
f Pélobate de l'Est

4867 ***Scaphiopus holbrookii holbrookii***
Anura - Pelobatidae
e Eastern Spadefoot; Eastern Spadefoot Toad

4868 ***Scaphiopus hurterii***
Anura - Pelobatidae
e Hurter's Spadefoot Toad; Spadefoot Toad; Southern Spadefoot Toad; Hurter's Spadefoot

4869 ***Scarthyala***
Anura - Hylidae
e Madre de Dios Tree Frogs

4870 ***Scarthyla ostinodactylus***
Anura - Hylidae
e Madre de Dios Tree Frog

4871 ***Schismaderma***
Anura - Bufonidae
e African Split-skin Toads; Red Toads

4872 ***Schismaderma carens***
Anura - Bufonidae
e African Split-skin Toad; Red Toad (CSA)
f Crapaud rouge

4873 ***Schistometopum***
Gymnophiona - Caeciliaidae
e Guinea Caecilians

4874 ***Schistometopum brevirostre***
Gymnophiona - Caeciliaidae
e Sao Tome Caecilian

4875 ***Schistometopum ephele***
Gymnophiona - Caeciliaidae
e Aqua Ize Caecilian

4876 ***Schistometopum garzonheydti***
Gymnophiona - Caeciliaidae
e Insel Fernando Po Caecilian; Fernando Po Caecilian

4877 ***Schistometopum gregori***
Gymnophiona - Caeciliaidae
e Witu Caecilian; Mud-dwelling Caecilian

4878 ***Schistometopum thomense***
Gymnophiona - Caeciliaidae
e Island Caecilian; Sao Tome Caecilian
d Buntwühle
f Cécilie de Sao Tome

4879 ***Scinax***
Anura - Hylidae
e Snouted Tree Frogs; Mexican Treefrogs
f Rainettes

4880 ***Scinax acuminata***
Anura - Hylidae
e Mato Grosso Snouted Tree Frog

4881 ***Scinax agilis***
Anura - Hylidae
e Agile Snouted Tree Frog

4882 ***Scinax albicans***
Anura - Hylidae
e Teresopolis Snouted Tree Frog

4883 ***Scinax alcatraz***
Anura - Hylidae
e Alcatraz Snouted Tree Frog

4884 ***Scinax alleni***
Anura - Hylidae
e Allen's Snouted Tree Frog

4885 ***Scinax altera***
Anura - Hylidae
e Crubixa Snouted Tree Frog

4886 ***Scinax argyreornata***
Anura - Hylidae
e Rio Mutum Snouted Tree Frog

4887 ***Scinax ariadne***
Anura - Hylidae
e Serra da Bocaina Snouted Tree Frog

4888 ***Scinax atrata***
Anura - Hylidae
e Peixoto's Snouted Tree Frog

4889 ***Scinax aurata***
Anura - Hylidae
e Santa Innes Snouted Tree Frog
d Goldstreifiger Laubkleber

4890 ***Scinax baerthae***
Anura - Hylidae
e Dwarf Snouted Tree Frog

4891 ***Scinax baumgardneri***
Anura - Hylidae
e Baumgardner's Snouted Tree Frog

4892 ***Scinax blairi***
Anura - Hylidae
e Blair's Snouted Tree Frog

4893 ***Scinax boesemani***
Anura - Hylidae
e Boeseman's Snouted Tree Frog
f Rainette de Boeseman

4894 ***Scinax boulengeri***
Anura - Hylidae
e Boulenger's Snouted Tree Frog

4895 ***Scinax brieni***
Anura - Hylidae
e Brien's Snouted Tree Frog

4896 ***Scinax caldarum***
Anura - Hylidae
e Caldas Snouted Tree Frog

4897 ***Scinax canastrensis***
Anura - Hylidae
e Canastra Snouted Tree Frog

4898 ***Scinax cardosa***
Anura - Hylidae
e Ravolta Snouted Tree Frog

4899 ***Scinax carnevalli***
Anura - Hylidae
e Rio Dolce Snouted Tree Frog

4900 ***Scinax castroviejoi***
Anura - Hylidae
e Temperate Snouted Tree Frog

4901 ***Scinax catharinae***
Anura - Hylidae
e Catharina Snouted Tree Frog

4902 ***Scinax chiquitana***
Anura - Hylidae
e Rio Negro Snouted Tree Frog

4903 ***Scinax crospedospila***
Anura - Hylidae
e Campo Belo Snouted Tree Frog

4904 ***Scinax cruentomma***
Anura - Hylidae
e Manaus Snouted Tree Frog
f Rainette de Duellman

4905 ***Scinax cuspidata***
Anura - Hylidae
e Coastal Lowland Snouted Tree Frog
d Bromelienfrosch

4906 ***Scinax cyanocephala***
Anura - Hylidae
e Blue-headed Snouted Tree Frog
f Rainette cynocéphale

4907 ***Scinax danae***
Anura - Hylidae
e Bolivar Snouted Tree Frog

4908 ***Scinax duartei***
Anura - Hylidae
e Duarte's Snouted Tree Frog

4909 ***Scinax ehrhardti***
Anura - Hylidae
e Ehrhardt's Snouted Tree Frog

4910 ***Scinax elaeochroa***
Anura - Hylidae
e Sipurio Snouted Tree Frog

4911 ***Scinax eurydice***
Anura - Hylidae
e Maracas Snouted Tree Frog

4912 ***Scinax exigua***
Anura - Hylidae
e Gran Sabana Snouted Tree Frog

4913 ***Scinax flavoguttata***
Anura - Hylidae
e Yellowbelly Snouted Tree Frog

4914 ***Scinax funerea***
Anura - Hylidae
e Moyobamba Snouted Tree Frog

4915 ***Scinax fuscomarginata***
Anura - Hylidae
e Brown-bordered Snouted Tree Frog

4916 ***Scinax fuscovaria***
Anura - Hylidae
e Snouted Tree Frog

4917 ***Scinax garbei***
Anura - Hylidae
e Einrunepe Snouted Tree Frog

4918 ***Scinax goinorum***
Anura - Hylidae
e Tarauaca Snouted Tree Frog

4919 ***Scinax hayii***
Anura - Hylidae
e Hay's Snouted Tree Frog

4920 ***Scinax heyeri***
Anura - Hylidae
e Heyer's Snouted Tree Frog

4921 ***Scinax hiemalis***
Anura - Hylidae
e Sousas Snouted Tree Frog

4922 ***Scinax humilis***
Anura - Hylidae
e Rio Babi Snouted Tree Frog

4923 ***Scinax jolyi***
Anura - Hylidae
f Rainette de Joly

4924 ***Scinax jureia***
Anura - Hylidae
e Jureia Snouted Tree Frog

4925 ***Scinax kautskyi***
Anura - Hylidae
e Kautsky's Snouted Tree Frog

4926 ***Scinax kennedyi***
Anura - Hylidae
e Kennedy's Snouted Tree Frog

4927 ***Scinax lindsayi***
Anura - Hylidae
e Lindsay's Snouted Tree Frog

4928 ***Scinax littoralis***
Anura - Hylidae
e Rio Verde Snouted Tree Frog

4929 ***Scinax littorea***
Anura - Hylidae
e Coastal Snouted Tree Frog

4930 ***Scinax longilinea***
Anura - Hylidae
e Lutz's Snouted Tree Frog

4931 ***Scinax louisotavioi***
Anura - Hylidae
e Santa Barbara Snouted Tree Frog

4932 ***Scinax machadoi***
Anura - Hylidae
e Machado's Snouted Tree Frog

4933 ***Scinax maracaya***
Anura - Hylidae
e Alpinopolis Snouted Tree Frog

4934 ***Scinax megapodia***
Anura - Hylidae
e Miranda's Snouted Tree Frog

4935 ***Scinax melloi***
Anura - Hylidae
e Southeastern Brazil Snouted Tree Frog

4936 ***Scinax nasica***
Anura - Hylidae
e Lesser Snouted Tree Frog

4937 ***Scinax nebulosa***
Anura - Hylidae
e Spix's Snouted Tree Frog
f Rainette nébuleuse

4938 ***Scinax obtriangulata***
Anura - Hylidae
e Sao Paulo Snouted Tree Frog; Sao Paolo Snouted Tree Frog

4939 ***Scinax opalina***
Anura - Hylidae
e Opal Snouted Tree Frog

4940 ***Scinax oreites***
Anura - Hylidae
e Balzapata Snouted Tree Frog

4941 ***Scinax pachychrus***
Anura - Hylidae
e Pocao Snouted Tree Frog

4942 ***Scinax pedromedinai***
Anura - Hylidae
e Hele's Snouted Tree Frog

4943 ***Scinax perpusilla***
Anura - Hylidae
e Bandeirantes Snouted Tree Frog
d Zwerglaubfrosch

4944 ***Scinax proboscidea***
Anura - Hylidae
e Gran Rio Snouted Tree Frog
f Rainette proboscidienne

4945 ***Scinax quinquefasciata***
Anura - Hylidae
e Fowler's Snouted Tree Frog

4946 ***Scinax ranki***
Anura - Hylidae
e Rank's Snouted Tree Frog

4947 ***Scinax rizibilis***
Anura - Hylidae
e Santo Andre Snouted Tree Frog

4948 ***Scinax rostrata***
Anura - Hylidae
e Caracas Snouted Tree Frog

4949 ***Scinax rubra***
Anura - Hylidae
e Red Snouted Tree Frog; Daudin's Tree Frog
d Daudins Baumfrosch
f Rainette rouge

4950 ***Scinax rugillata***
Anura - Hylidae
e Quevedo Snouted Tree Frog

4951 ***Scinax similis***
Anura - Hylidae
e Cochran's Snouted Tree Frog

4952 ***Scinax squalirostris***
Anura - Hylidae
e Striped Snouted Tree Frog

4953 ***Scinax staufferi***
Anura - Hylidae
e Stauffer's Snouted Tree Frog; Stauffer's Tree Frog

4954 ***Scinax staufferi staufferi***
Anura - Hylidae
e Northern Stauffer's Tree Frog

4955 ***Scinax strigilata***
Anura - Hylidae
e Bahia Snouted Tree Frog

4956 ***Scinax trachythorax***
Anura - Hylidae
e San Luis Snouted Tree Frog

4957 ***Scinax trapicheiroi***
Anura - Hylidae
e Three-lined Snouted Tree Frog

4958 ***Scinax vautieri***
Anura - Hylidae
e Maldonado Snouted Tree Frog

4959 ***Scinax v-signata***
Anura - Hylidae
e Forest Snouted Tree Frog

4960 ***Scinax wandae***
Anura - Hylidae
e Villavicenzio Snouted Tree Frog

4961 ***Scinax x-signata***
Anura - Hylidae
e Venezuela
f Rainette à taches oranges

4962 **Scolecomorphidae**
Gymnophiona
e Tropical Caecilians
d Scolemorphiden

4963 ***Scolecomorphus***
Gymnophiona - Scolecomorphidae
e Tropical Caecilians

4964 ***Scolecomorphus kirkii***
Gymnophiona - Scolecomorphidae
e Lake Tanganyika Caecilian; Kirk's Caecilian

4965 ***Scolecomorphus uluguruensis***
Gymnophiona - Scolecomorphidae
e Nyingwa Caecilian

4966 ***Scolecomorphus vittatus***
Gymnophiona - Scolecomorphidae
e Banded Caecilian; African Caecilian

4967 ***Scotobleps***
Anura - Arthroleptidae
e Gabon Forest Frogs; Gaboon Forest Frogs

4968 ***Scotobleps gabonicus***
Anura - Arthroleptidae
e Gabon Forest Frog; Gaboon Forest Frogs

4969 ***Scutiger***
Anura - Megophryidae
e Lazy Toads
d Schildträger

4970 ***Scutiger adungensis***
Anura - Megophryidae
e Adung Lazy Toad

4971 ***Scutiger boulengeri***
Anura - Megophryidae

e Boulenger's Lazy Toad

4972 ***Scutiger chintingensis***
Anura - Megophryidae
e Chinting Lazy Toad

4973 ***Scutiger chuanbeiensis***
Anura - Megophryidae
e Sichuan Lazy Toad

4974 ***Scutiger glandulatus***
Anura - Megophryidae
e Hopachai Lazy Toad

4975 ***Scutiger gongshanensis***
Anura - Megophryidae
e Gongshan Lazy Toad

4976 ***Scutiger granulosus***
Anura - Megophryidae
e Yunnan Lazy Toad

4977 ***Scutiger jingdongensis***
Anura - Megophryidae
e Jingdong Lazy Toad

4978 ***Scutiger lichuanensis***
Anura - Megophryidae
e Lichuan Lazy Toad

4979 ***Scutiger liupanensis***
Anura - Megophryidae
e Liupan Lazy Toad

4980 ***Scutiger maculatus***
Anura - Megophryidae
e Spotted Lazy Toad

4981 ***Scutiger major***
Anura - Megophryidae
e Common Lazy Toad; Lazy Toad

4982 ***Scutiger mammatus***
Anura - Megophryidae
e Tungsolo Lazy Toad

4983 ***Scutiger multipunctatus***
Anura - Megophryidae
e Dotted Lazy Toad

4984 ***Scutiger nepalensis***
Anura - Megophryidae
e Nepal Lazy Toad

4985 ***Scutiger ningshanensis***
Anura - Megophryidae
e Ningshan Lazy Toad

4986 ***Scutiger nyingchiensis***
Anura - Megophryidae
e Nyingchi Lazy Toad

4987 ***Scutiger occidentalis***
Anura - Megophryidae
e Asian Lazy Toad; Kashmir Snow Toad (ISC)

4988 ***Scutiger omeimontis***
Anura - Megophryidae
e Omei Lazy Toad

4989 ***Scutiger pingii***
Anura - Megophryidae
e Chinese Lazy Toad

4990 ***Scutiger pingwuensis***
Anura - Megophryidae
e Pingwu Lazy Toad

4991 ***Scutiger popei***
Anura - Megophryidae
e Pope's Lazy Toad

4992 ***Scutiger rhodostigmatus***
Anura - Megophryidae
e Guizhou Lazy Toad

4993 ***Scutiger ruginosus***
Anura - Megophryidae
e Chaochiao Lazy Toad

4994 ***Scutiger rugosus***
Anura - Megophryidae
e Kanding Lazy Toad

4995 ***Scutiger schimidti***
Anura - Megophryidae
e Schmidt's Lazy Toad
d Schmidtscher Krötenfrosch

4996 ***Scutiger sikkimensis***
Anura - Megophryidae
e Sikkim Lazy Toad; Sikkim Snow Toad (ISC)

4997 ***Scutiger tuberculatus***
Anura - Megophryidae
e Bumpy Lazy Toad

4998 ***Scutiger weigoldi***
Anura - Megophryidae
e Weigold's Lazy Toad

4999 ***Scutiger xiangchengensis***
Anura - Megophryidae
e Xiangcheng Lazy Toad

5000 ***Scythrophrys***
Anura - Leptodactylidae
e Banhado Frogs

5001 ***Scythrophrys sawayae***
Anura - Leptodactylidae
e Banhado Frog

5002 ***Semnodactylus***
Anura - Hyperoliidae
e Weale's Frogs

5003 ***Semnodactylus wealei***
Anura - Hyperoliidae
e Weale's Frog; Rattling Frog (CSA); Rattling Kassina (CSA); Long-toed Running Frog (CSA); Long-toed Kassina (CSA)

5004 ***Silurana***
Anura - Pipidae
e Tropical Clawed Frogs

5005 ***Silurana epitropicalis***
Anura - Pipidae
e Cameroon Clawed Frog

5006 ***Silurana tropicalis***
Anura - Pipidae
e Tropical Clawed Frog
d Gespornter Krallenfrosch

5007 ***Siphonops***
Gymnophiona - Caeciliaidae
e Ringed Caecilians
d Ringelwühlen
f Siphonops

5008 ***Siphonops annulatus***
Gymnophiona - Caeciliaidae
e Ringed Caecilian; Ringed South American Caecilian; Blue-ringed Caecilian; South American Caecilian
d Ringelwühle
f Cécilie annelée; Siphnops annelé

5009 ***Siphonops hardyi***
Gymnophiona - Caeciliaidae
e Hardy's Caecilian

5010 ***Siphonops insularis***
Gymnophiona - Caeciliaidae
e Insular Caecilian

5011 ***Siphonops leucoderus***
Gymnophiona - Caeciliaidae
e Salvador Caecilian

5012 ***Siphonops paulensis***
Gymnophiona - Caeciliaidae
e Boettger's Caecilian

5013 ***Siren***
Caudata - Sirenidae
e Sirens
d Armmolche
f Sirènes

5014 ***Siren intermedia***
Caudata - Sirenidae
e Lesser Siren
d Kleiner Armmolch

5015 ***Siren intermedia intermedia***
Caudata - Sirenidae
e Eastern Lesser Siren

5016 ***Siren intermedia nettingi***
Caudata - Sirenidae
e Western Lesser Siren

5017 ***Siren intermedia texana***
Caudata - Sirenidae
e Rio Grande Lesser Siren

5018 ***Siren lacertina***
Caudata - Sirenidae
e Siren; Greater Siren; Mud Eel(NA)
d Großer Armmolch
f Sirène lacertine
i Grande sirena

5019 **Sirenidae**
Caudata
e Siren and Dwarf Sirens; Sirens
d Armmolche; Sireniden
f Sirènes; Sirénidés

5020 ***Smilisca***
Anura - Hylidae
e Cross-banded Tree Frogs; Mexican Tree Frogs; Smiliscas
d Baumfrösche

5021 ***Smilisca baudinii***
Anura - Hylidae
e Mexican Tree Frog; Common Mexican Tree Frog; Mexican Smilisca

5022 ***Smilisca cyanosticta***
Anura - Hylidae
e Blue-spotted Mexican Tree Frog; Blue-spotted Smilisca

5023 ***Smilisca phaeota***
Anura - Hylidae
e New Granada Cross-banded Tree Frog; Central American Smilisca

5024 ***Smilisca puma***
Anura - Hylidae
e Nicaragua Cross-banded Tree Frog; Tawny Smilisca

5025 ***Smilisca sila***
Anura - Hylidae
e Panama Cross-banded Tree Frog; Pug-nosed Smilisca

5026 ***Smilisca sordida***
Anura - Hylidae
e Veragua Cross-banded Tree Frog; Costa Rica Smilisca

5027 ***Somuncuria***
Anura - Leptodactylidae
e El Rincon Stream Frogs

5028 ***Somuncuria somuncurensis***
Anura - Leptodactylidae
e El Rincon Stream Frog

5029 **Sooglossidae**
Anura
e Seychelles Frogs; Seychelles Islands Frogs
d Seychellen-Frösche; Sooglossiden

f Grenouilles des Seychelles

5030 ***Sooglossus***
Anura - Sooglossidae
e Seychelles Frogs
d Seychellen-Frösche
f Grenouilles des Seychelles

5031 ***Sooglossus gardineri***
Anura - Sooglossidae
e Gardiner's Seychelles Frog; Gardiners Frog; Seychelles Frog
d Gardiners Seychellen-Frosch

5032 ***Sooglossus sechellensis***
Anura - Sooglossidae
e Seychelles Frog
d Seychellen-Frosch
f Grenouille des Seychelles

5033 ***Spea***
Anura - Pelobatidae
e Western Spadefoot Toads; Western Spadefoots

5034 ***Spea bombifrons***
Anura - Pelobatidae
e Plains Spadefoot Toad; Plains Spadefoot
d Flachland-Schaufelfuß
f Crapaud des plaines

5035 ***Spea hammondii***
Anura - Pelobatidae
e Western Spadefoot; Western Spadefoot Toad; Hammond's Spadefoot
d Westlicher Schaufelfuß

5036 ***Spea intermontana***
Anura - Pelobatidae
e Great Basin Spadefoot; Great Basin Spadefoot Toad
f Crapaud du Grand Bassin

5037 ***Spea multiplicata***
Anura - Pelobatidae
e New Mexico Spadefoot; New Mexico Spadefoot Toad (NA); Southern Spadefoot (NA)

5038 ***Spea multiplicata multiplicata***
Anura - Pelobatidae
e New Mexico Spadefoot

5039 ***Spea multiplicata stagnalis***
Anura - Pelobatidae
e Mexican Spadefoot; Desert Spadefoot

5040 ***Spelaeophryne***
Anura - Microhylidae
e African Striped Frogs

5041 ***Spelaeophryne methneri***
Anura - Microhylidae
e African Striped Frog

5042 ***Speleomantes***
Caudata - Plethodontidae
e European Salamanders; Cave Salamanders
d Höhlensalamander (pl); Schleuderzungensalamander (pl); Europäische Höhlensalamander (pl)
f Spélerpès; Hydromantes; Salamandres cavernicoles; Spéléomantes
i Geotritoni

5043 ***Speleomantes ambrosii***
Caudata - Plethodontidae
e Ambrosi's Cave Salamander; French Cave Salamander
d Französischer Höhlensalamander; Ambrosis Höhlensalamander
f Spélerpès brun; Spélerpès italien; Hydromante brun; Hydromante italien; Spéléomante de Strinati; Hydromante d'talie; Hydromantes de Ligurie
i Geotritone di Ambrosi

5044 ***Speleomantes flavus***
Caudata - Plethodontidae
e Stefani's Salamander; Monte Albo Cave Salamander; Yellowish Cave Salamander
d Gelblicher Höhlensalamander; Monte Albo-Höhlensalamander
f Spélerpès du Mont Albo
i Geotritone giallo

5045 ***Speleomantes genei***
Caudata - Plethodontidae
e Sardinia Salamander; Sardinian Cave Salamander; Sardinian Brown Cave Salamander; Gene's Cave Salamander
d Sardischer Schleuderzungensalamander; Sardinischer Schleuderzungensalamander; Sardischer Höhlensalamander; Sardinischer Höhlensalamander; Genés Höhlensalamander
f Hydromante sarde; Salamandre cavernicole de Sardaigne; Salamandre cavernicole sarde; Spélerpès de Génie; Geotritone sardo; Geotritone
i Geotritone sardo

5046 ***Speleomantes imperialis***
Caudata - Plethodontidae
e Imperial Salamander; Scented Cave Salamander
d Duftender Höhlensalamander
f Spélerpès impérial
i Geotritone imperiale; Geotritone maggiore

5047 ***Speleomantes italicus***
Caudata - Plethodontidae
e Apennine Salamander; Italian Cave Salamander
d Italienischer Höhlensalamander
f Salamandre cavernicole d'Italie; Salamandre cavernicole italienne; Spélerpès italien
i Geotritone italiano; Geotritone

5048 ***Speleomantes supramontis***
Caudata - Plethodontidae
e Nuoro Salamander; Supramontane Cave Salamander
d Supramontes-Höhlensalamander
f Spélerpès de Supramonte
i Geotritone del Supramonte

5049 ***Spenophryne***
Anura - Microhylidae
e Land Frogs

5050 ***Spenophryne adelphe***
Anura - Microhylidae
e Peeping Land Frog

5051 ***Spenophryne brevicrus***
Anura - Microhylidae
e Hellwing Land Frog

5052 ***Spenophryne brevipes***
Anura - Microhylidae
e Victoria Land Frog

5053 ***Spenophryne cornuta***
Anura - Microhylidae
e Horned Land Frog

5054 ***Spenophryne crassa***
Anura - Microhylidae
e Papua Land Frog

5055 ***Spenophryne dentata***
Anura - Microhylidae
e Alotau Land Frog

5056 ***Spenophryne fryi***
Anura - Microhylidae
e Whistling Land Frog

5057 ***Spenophryne gracilipes***
Anura - Microhylidae
e Golden Land Frog

5058 ***Spenophryne hooglandi***
Anura - Microhylidae
e New Guinea Land Frog

5059 ***Spenophryne macrorrhyncha***
Anura - Microhylidae
e Manikion Land Frog

5060 ***Spenophryne mehelyi***
Anura - Microhylidae
e Mehely's Land Frog

5061 ***Spenophryne palmipes***
Anura - Microhylidae
e Dauman Land Frog

5062 ***Spenophryne pluvialis***
Anura - Microhylidae
e Flecked Land Frog

5063 ***Spenophryne polysticta***
Anura - Microhylidae
e Morobe Land Frog

5064 ***Spenophryne pusilla***
Anura - Microhylidae
e Aru Land Frog

5065 ***Spenophryne rhodactyla***
Anura - Microhylidae
e Owen Stanley Land Frog

5066 ***Spenophryne robusta***
Anura - Microhylidae
e Chirping Land Frog

5067 ***Spenophryne schlaginhaufeni***
Anura - Microhylidae
e Wandolleck's Land Frog

5068 ***Sphaenorhynchus***
Anura - Hylidae
e Lime Tree Frogs

5069 ***Sphaenorhynchus bromelicola***
Anura - Hylidae
e Bahia Lime Tree Frog

5070 ***Sphaenorhynchus carneus***
Anura - Hylidae
e Napo Lime Tree Frog

5071 ***Sphaenorhynchus dorisae***
Anura - Hylidae
e Doris's Lime Treefrog

5072 ***Sphaenorhynchus lacteus***
Anura - Hylidae
e Orinoco Lime Tree Frog
f Rainette orangée; Rainette lactée

5073 ***Sphaenorhynchus orophilus***
Anura - Hylidae
e Lutz's Lime Tree Frog

5074 ***Sphaenorhynchus palustris***
Anura - Hylidae
e Linhare Lime Tree Frog

5075 ***Sphaenorhynchus pauloalvina***
Anura - Hylidae
e Paulo's Lime Tree Frog

5076 ***Sphaenorhynchus planicola***
Anura - Hylidae
e Rio Lime Tree Frog

5077 ***Sphaenorhynchus platycephalus***
Anura - Hylidae
e South American Lime Tree Frog

5078 ***Sphaenorhynchus prasinus***
Anura - Hylidae
e Bokermann's Lime Tree Frog

5079 ***Sphaenorhynchus surdus***
Anura - Hylidae
e Cochran's Lime Tree Frog

5080 ***Spinophrynoides***
Anura - Bufonidae
e Ethiopian Toads

5081 ***Spinophrynoides osgoodi***
Anura - Bufonidae
e Ethiopian Toad; Osgood's Ethiopian Toad

5082 ***Staurois***
Anura - Ranidae
e Splash Frog

5083 ***Staurois latopalmatus***
Anura - Ranidae
e Sabah Splash Frog

5084 ***Staurois natator***
Anura - Ranidae
e Mindanao Splash Frog

5085 ***Staurois tuberlinguis***
Anura - Ranidae
e Borneo Splash Frog

5086 ***Stefania***
Anura - Hylidae
e Stefania Tree Frogs

5087 ***Stefania evansi***
Anura - Hylidae
e Groete Creek Tree Frog; Evan's Essequibo Tree Frog

5088 ***Stefania ginesi***
Anura - Hylidae
e Tepuis Tree Frog

5089 ***Stefania goini***
Anura - Hylidae
e Vegas Falls Tree Frog

5090 ***Stefania marahuaquensis***
Anura - Hylidae
e Marahuaca Tree Frog

5091 ***Stefania riae***
Anura - Hylidae
e Stefania Tree Frog

5092 ***Stefania roraimae***
Anura - Hylidae
e Roraima Tree Frog

5093 ***Stefania woodleyi***
Anura - Hylidae
e Woodley's Tree Frog

5094 ***Stephopaedes***
Anura - Bufonidae
e Mahenge Toads; Chirinda Forest Toads

5095 ***Stephopaedes anotis***
Anura - Bufonidae
e Mashonaland Toad; Chirinda Forest Toad; Chirinda Toad (CSA)

5096 ***Stephopaedes loveridgei***
Anura - Bufonidae
e Mahenge Toad

5097 ***Stereochilus***
Caudata - Plethodontidae
e Many-lined Salamanders
d Streifensalamander (pl)
f Salamandres rayées

5098 ***Stereochilus marginatus***
Caudata - Plethodontidae
e Many-lined Salamander
d Streifensalamander
f Salamandre rayée

5099 ***Stereocyclops***
Anura - Microhylidae
e Brazilian Dumpy Frogs

5100 ***Stereocyclops incrassatus***
Anura - Microhylidae
e Brazilian Dumpy Frog

5101 ***Stumpffia***
Anura - Microhylidae
e Stump-toed Frogs

5102 ***Stumpffia gimmeli***
Anura - Microhylidae
e Benavony Stump-toed Frog

5103 ***Stumpffia grandis***
Anura - Microhylidae
e Giant Stump-toed Frog

5104 ***Stumpffia psologlossa***
Anura - Microhylidae
e Madagascar Stump-toed Frog

5105 ***Stumpffia pygmaea***
Anura - Microhylidae
e Andoany Stump-toed Frog

5106 ***Stumpffia tetradactyla***
Anura - Microhylidae
e Maromandia Stump-toed Frog

5107 ***Stumpffia tridactyla***
Anura - Microhylidae
e Guibe's Stump-toed Frog

5108 ***Sylvacaecilia***
Gymnophiona - Caeciliaidae
e Aleku Caecilians

5109 ***Sylvacaecilia grandisonae***
Gymnophiona - Caeciliaidae
e Aleku Caecilian

5110 ***Synapturanus***
Anura - Microhylidae
e Disc Frogs

5111 ***Synapturanus mirandariberoi***
Anura - Microhylidae
e Miranda's Disc Frog
f Sinapturan de Miranda-Ribeiro

5112 ***Synapturanus rabus***
Anura - Microhylidae
e Vaupes Disc Frog

5113 ***Synapturanus salseri***
Anura - Microhylidae
e Timbo Disc Frog

5114 ***Syncope***
Anura - Microhylidae
e Silent Frogs

5115 ***Syncope antenori***
Anura - Microhylidae
e Ecuador Silent Frog

5116 ***Syncope carvalhoi***
Anura - Microhylidae
e Carvalho's Silent Frog

T

5117 ***Tachycnemis***
Anura - Hyperoliidae
e Seychelles Islands Frogs; Seychelles Tree Frogs
d Seychellen-Ruderfrösche

5118 ***Tachycnemis seychellensis***
Anura - Hyperoliidae
e Seychelles Islands Frog; Seychelles Tree Frog
d Seychellen-Ruderfrosch

5119 ***Taricha***
Caudata - Salamandridae
e Roughskin Newts; Pacific Newts
d Westamerikanische Wassermolche; Rauhautmolch; Gelbbauchmolche
f Tarichas

5120 ***Taricha granulosa***
Caudata - Salamandridae
e Rough-skinned Newt; Roughskin Newt
d Rauhäutiger Wassermolch; Rauhäutiger Molch; Rauhäutiger Gelbbauchmolch
f Triton rugueux; Taricha granuleuse

5121 ***Taricha granulosa granulosa***
Caudata - Salamandridae
e Northern Roughskin Newt; Rough-skinned Newt

5122 ***Taricha granulosa mazamae***
Caudata - Salamandridae
e Southern Roughskin Newt; Crater Lake Newt; Crater Lake Rough-skinned Newt

5123 ***Taricha rivularis***
Caudata - Salamandridae
e Red-bellied Newt; Redbelly Newt; Western Red-bellied Newt
d Rotbauchmolch

5124 ***Taricha torosa***
Caudata - Salamandridae
e California Newt
d Kalifornischer Molch; Kalifornischer Wassermolch; Gelbbauchmolch; Kalifornischer Gelbbauchmolch
f Triton à ventre rouge de Californie; Triton à ventre rouge

5125 ***Taricha torosa sierrae***
Caudata - Salamandridae
e Sierra Newt

5126 ***Taricha torosa torosa***
Caudata - Salamandridae
e Coast Range Newt; California Newt

5127 ***Taudactylus***
Anura - Myobatrachidae
e Torrent Frogs

5128 ***Taudactylus acutirostris***
Anura - Myobatrachidae
e Sharpsnout; Sharp-snouted Torrent Frog (ANZ); Tinker Frog

5129 ***Taudactylus diurnus***
Anura - Myobatrachidae
e Mount Glorious Torrent Frog (ANZ); Day Frog

5130 ***Taudactylus eungellensis***
Anura - Myobatrachidae
e Eungella Torrent Frog

5131 ***Taudactylus liemi***
Anura - Myobatrachidae
e Palm Torrent Frog

5132 ***Taudactylus pleione***
Anura - Myobatrachidae
e Kroombit Tops Torrent Frog

5133 ***Taudactylus rheophilus***
Anura - Myobatrachidae
e Mountain Torrent Frog

5134 ***Taylorana***
Anura - Ranidae
e Taylors Frogs

5135 ***Taylorana hascheana***
Anura - Ranidae
e Penang Taylors Frog; Hill Forest Frog; Hasche's Frog (SEA)
d Hasches Frosch

5136 ***Taylorana limborgi***
Anura - Ranidae
e Taylors Frog

5137 ***Telmatobius***
Anura - Leptodactylidae
e Water Frogs
d Anden-Pfeiffrösche

5138 ***Telmatobius albiventris***
Anura - Leptodactylidae
e White Water Frog

5139 ***Telmatobius arequipensis***
Anura - Leptodactylidae
e Chili Water Frog

5140 ***Telmatobius atacamensis***
Anura - Leptodactylidae
e Salta Water Frog

5141 ***Telmatobius atahualpai***
Anura - Leptodactylidae
e Amazonas Water Frog

5142 ***Telmatobius brevipes***
Anura - Leptodactylidae
e Huahachuco Water Frog

5143 ***Telmatobius brevirostris***
Anura - Leptodactylidae
e Shortsnout Water Frog

5144 ***Telmatobius carillae***
Anura - Leptodactylidae
e Ancash Water Frog

5145 ***Telmatobius ceiorum***
Anura - Leptodactylidae
e Catamarca Water Frog

5146 ***Telmatobius cirrhacelis***
Anura - Leptodactylidae
e Loja Water Frog

5147 ***Telmatobius colanensis***
Anura - Leptodactylidae
e Colan Water Frog

5148 ***Telmatobius contrerasi***
Anura - Leptodactylidae
e Contreras Water Frog

5149 ***Telmatobius crawfordi***
Anura - Leptodactylidae
e Crawford's Water Frog

5150 ***Telmatobius culeus***
Anura - Leptodactylidae
e Titicaca Water Frog
d Titicaca-Kröte

5151 ***Telmatobius halli***
Anura - Leptodactylidae
e Hall's Water Frog

5152 ***Telmatobius hauthali***
Anura - Leptodactylidae
e Calientes Water Frog

5153 ***Telmatobius hypselocephalus***
Anura - Leptodactylidae
e Jujuy Water Frog

5154 ***Telmatobius ignavus***
Anura - Leptodactylidae
e Piura Water Frog

5155 ***Telmatobius intermedius***
Anura - Leptodactylidae
e Allipacca Water Frog

5156 ***Telmatobius jelskii***
Anura - Leptodactylidae
e Acancocha Water Frog

5157 ***Telmatobius laticeps***
Anura - Leptodactylidae
e Tacuman Water Frog

5158 ***Telmatobius latirostris***
Anura - Leptodactylidae
e Cajamarca Water Frog

5159 ***Telmatobius marmoratus***
Anura - Leptodactylidae
e Marbled Water Frog
d Marmorierter Anden-Pfeiffrosch

5160 ***Telmatobius necopinus***
Anura - Leptodactylidae
e Wiens's Water Frog

5161 ***Telmatobius niger***
Anura - Leptodactylidae
e Black Water Frog

5162 ***Telmatobius oxycephalus***
Anura - Leptodactylidae
e Red-headed Water Frog

5163 ***Telmatobius pefauri***
Anura - Leptodactylidae
e Arico Water Frog

5164 ***Telmatobius peruvianus***
Anura - Leptodactylidae
e Peru Water Frog

5165 ***Telmatobius pinguiculus***
Anura - Leptodactylidae
e Medanitos Water Frog

5166 ***Telmatobius pisanoi***
Anura - Leptodactylidae
e Argentine Water Frog

5167 ***Telmatobius platycephalus***
Anura - Leptodactylidae
e Flatheaded Water Frog

5168 ***Telmatobius rimac***
Anura - Leptodactylidae
e Rimac

5169 ***Telmatobius schreiteri***
Anura - Leptodactylidae
e La Rioja Water Frog

5170 ***Telmatobius scrocchii***
Anura - Leptodactylidae
e Andalgala Water Frog

5171 ***Telmatobius simonsi***
Anura - Leptodactylidae
e Sucre Water Frog

5172 ***Telmatobius stephani***
Anura - Leptodactylidae
e Stephan's Water Frog

5173 ***Telmatobius thompsoni***
Anura - Leptodactylidae
e Thompson's Water Frog

5174 ***Telmatobius truebae***
Anura - Leptodactylidae
e Trueb's Water Frog

5175 ***Telmatobius vellardi***
Anura - Leptodactylidae
e Vellard's Water Frog

5176 ***Telmatobius verrucosus***
Anura - Leptodactylidae
e Chaco Water Frog

5177 ***Telmatobius yuracare***
Anura - Leptodactylidae
e Sehuencas Water Frog

5178 ***Telmatobius zapahuirensis***
Anura - Leptodactylidae
e Zapahuira Water Frog

5179 ***Telmatobufo***
Anura - Leptodactylidae
e False Toads

5180 ***Telmatobufo australis***
Anura - Leptodactylidae
e Pelado Mountains False Toad

5181 ***Telmatobufo bullocki***
Anura - Leptodactylidae
e Bullock's False Toad

5182 ***Telmatobufo venustus***
Anura - Leptodactylidae
e Chile False Toad

5183 ***Tepuihyla***
Anura - Hylidae
e Amazon Tree Frogs

5184 ***Tepuihyla aecii***
Anura - Hylidae
e Monte Duida Tree Frog

5185 ***Tepuihyla edelcae***
Anura - Hylidae
e Ayarzaguena's Tree Frog

5186 ***Tepuihyla galani***
Anura - Hylidae
e Senaris's Tree Frog

5187 ***Tepuihyla rimarum***
Anura - Hylidae
e Gorzula Amazon Tree Frog

5188 ***Tepuihyla rodriguezi***
Anura - Hylidae
e Rodriguez's Amazon Tree Frog

5189 ***Theloderma***
Anura - Rhacophoridae
e Bug-eyed Frogs; Wart Frogs (SEA)
d Warzenfrösche

5190 ***Theloderma asperum***
Anura - Rhacophoridae
e Hill Garden Bug-eyed Frog; Pied Wart Frog (SEA)
d Bunter Warzenfrosch

5191 ***Theloderma bicolor***
Anura - Rhacophoridae
e Chapa Bug-eyed Frog

5192 ***Theloderma corticale***
Anura - Rhacophoridae
e Tonkin Bug-eyed Frog

5193 ***Theloderma gordoni***
Anura - Rhacophoridae
e Gordon's Bug-eyed Frog

5194 ***Theloderma horridum***
Anura - Rhacophoridae
e Thai Bug-eyed Frog; Thorny Wart Frog (SEA)
d Stachliger Warzenfrosch

5195 ***Theloderma leporosa***
Anura - Rhacophoridae
e Malaya Bug-eyed Frog; Malayan Wart Frog (SEA)

d Malaiischer Warzenfrosch

5196 ***Theloderma phrynoderma***
Anura - Rhacophoridae
e Burmese Bug-eyed Frog

5197 ***Theloderma schmardanum***
Anura - Rhacophoridae
e Sri Lanka Bug-eyed Frog

5198 ***Theloderma stellatum***
Anura - Rhacophoridae
e Taylor's Bug-eyed Frog; Taylor's Wart Frog (SEA)
d Taylors Warzenfrosch

5199 ***Thorius***
Caudata - Plethodontidae
e Pygmy Salamanders; Mexican Pygmy Salamanders
d Mexikanischer Pygmäensalamander (pl); Zwergsalamander (pl)
f Salamandres pygmées

5200 ***Thorius dubitus***
Caudata - Plethodontidae
e Acultzingo Pygmy Salamander
d Gebirgs-Zwergsalamander

5201 ***Thorius macdougalli***
Caudata - Plethodontidae
e MacDougall's Pygmy Salamander

5202 ***Thorius maxillabrochus***
Caudata - Plethodontidae
e Zoquitlan Pygmy Salamander

5203 ***Thorius minutissimus***
Caudata - Plethodontidae
e Oaxacan Pygmy Salamander

5204 ***Thorius narisovalis***
Caudata - Plethodontidae
e Upper Cerro Pygmy Salamander

5205 ***Thorius pennatulus***
Caudata - Plethodontidae
e Veracruz Pygmy Salamander
d Borken-Zwergsalamander

5206 ***Thorius pennatulus narismagnus***
Caudata - Plethodontidae
e San Martin Pygmy Salamander

5207 ***Thorius pennatulus pennatulus***
Caudata - Plethodontidae
e Vera Cruz Pygmy Salamander

5208 ***Thorius pulmonaris***
Caudata - Plethodontidae
e Lower Cerro Pygmy Salamander

5209 ***Thorius schmidti***
Caudata - Plethodontidae
e Schmidt's Pygmy Salamander

5210 ***Thorius troglodytes***
Caudata - Plethodontidae
e Taylor's Pygmy Salamander
d Hellbrauner Zwergsalamander

5211 ***Thoropa***
Anura - Leptodactylidae
e River Frogs

5212 ***Thoropa lutzi***
Anura - Leptodactylidae
e Lutz's River Frog

5213 ***Thoropa megatympanum***
Anura - Leptodactylidae
e Big-eared River Frog

5214 ***Thoropa miliaris***
Anura - Leptodactylidae
e Military River Frog

5215 ***Thoropa petropolitana***
Anura - Leptodactylidae

e Petropolis River Frog

5216 ***Thoropa saxatilis***
Anura - Leptodactylidae
e Brazilian River Frog

5217 ***Tomopterna***
Anura - Ranidae
e Old World Bullfrogs; Pyxies (CSA); Burrowing Frogs (CSA); Sand Frogs (CSA); Dwarf Bullfrogs (CSA)

5218 ***Tomopterna breviceps***
Anura - Ranidae
e Punjab Bullfrog; Indian Burrowing Frog (ISC); Burrowing Frog

5219 ***Tomopterna cryptotis***
Anura - Ranidae
e Catequoro Bullfrog; Tremolo Sand Frog (CSA); Striped Burrowing Frog; Striped Pyxie (CSA); Kalahari Pyxie (CSA); Striped Sand Frog (CSA)

5220 ***Tomopterna delalandii***
Anura - Ranidae
e African Bullfrog; Cape Sand Frog (CSA); Common Burrowing Frog (CSA); Pyxie (CSA); Delalande's Pyxie (CSA); Delalande's Frog (CSA)
f Grenouille de Delalande

5221 ***Tomopterna dobsoni***
Anura - Ranidae
e Mangalore Bullfrog; Dobson's Burrowing Frog

5222 ***Tomopterna krugerensis***
Anura - Ranidae
e Kruger Bullfrog; Knocking Sand Frog (CSA); Sandveld Pyxie (CSA); Kruger Burrowing Frog (CSA)

5223 ***Tomopterna labrosa***
Anura - Ranidae
e Madagascar Bullfrog

5224 ***Tomopterna leucorhynchus***
Anura - Ranidae
e Wattakole Bullfrog

5225 ***Tomopterna marmorata***
Anura - Ranidae
e Marbled Bullfrog; Russet-backed Sand Frog (CSA); Marmorate Pyxie (CSA); Mozambique Burrowing Frog; Mozambique Pyxie (CSA); Mocambique Burrowing Frog (CSA); Red-backed Pyxie (CSA)

5226 ***Tomopterna natalensis***
Anura - Ranidae
e Natal Bullfrog; Puddle Frog; Natal Sand Frog (CSA); Natal Burrowing Frog (CSA); Natal Pyxie (CSA)

5227 ***Tomopterna rolandae***
Anura - Ranidae
e Sri Lanka Bullfrog

5228 ***Tomopterna strachani***
Anura - Ranidae
e Pakistan Bullfrog

5229 ***Tomopterna swani***
Anura - Ranidae
e Dharan Bullfrog

5230 ***Tomopterna tuberculosa***
Anura - Ranidae
e Dharan Bullfrog; Warty Frog; Angolan Burrowing Frog; Beaded Pyxie; Beaded Sand Frog (CSA); Beaded Burrowing Frog
f Grenouille verruqueuse

5231 ***Trachycephalus***
Anura - Hylidae
e Casque-headed Tree Frogs; Bone-headed Tree Frogs; Bonyheads
d Panzerlaubfrösche

5232 ***Trachycephalus atlas***
Anura - Hylidae
e Bokermann's Casque-headed Tree Frog

5233 ***Trachycephalus jordani***
Anura - Hylidae
e Jordan's Casque-headed Tree Frog

5234 ***Trachycephalus nigromaculatus***
Anura - Hylidae
e Black-spotted Casque-headed Tree Frog; Black-spotted Bonyhead

5235 ***Trichobatrachus***
Anura - Arthroleptidae
e Hairy Frogs
d Haarfrösche
f Grenouilles poilues

5236 ***Trichobatrachus robustus***
Anura - Arthroleptidae
e Hairy Frog
d Haarfrosch
f Grenouille poilue

5237 ***Triprion***
Anura - Hylidae
e Shovel-headed Tree Frogs; Casquehead Treefrogs

5238 ***Triprion petasutus***
Anura - Hylidae
e Yucatecan Shovel-headed Tree Frog; Yucatan Casquehead Tree Frog

5239 ***Triprion spatulatus***
Anura - Hylidae
e Shovel-headed Tree Frog; Shovelhead Tree Frog

5240 ***Triprion spatulatus reticulatus***
Anura - Hylidae
e Mexican Shovelhead Tree Frog

5241 ***Triprion spatulatus spatulatus***
Anura - Hylidae
e Shovelhead Tree Frog

5242 ***Triturus***
Caudata - Salamandridae
e Alpine Newts; Newts; Pond Newts
d Wassermolche; Echte Wassermolche
f Tritons
i Tritone

5243 ***Triturus alpestris***
Caudata - Salamandridae
e Laurenti's Alpine Newt; Alpine Newt
d Bergmolch; Alpenmolch; Gebirgsmolch
f Triton alpestre; Triton des Alpes; Triton ceinturé
i Tritone alpino; Tritone alpestre; Tritone montano; Salamandra acquaiola; Scaramàndla; Sacaramàndla

5244 ***Triturus alpestris inexpectatus***
Caudata - Salamandridae
e Calabrian Newt
f Triton calabrien
i Tritone alpino della Calabria

5245 ***Triturus boscai***
Caudata - Salamandridae
e Bosca's Newt
d Boscas Wassermolch; Spanischer Wassermolch
f Triton de Bosca; Triton boscai

5246 ***Triturus carnifex***
Caudata - Salamandridae
e Vienna Newt; Italian Crested Newt; Alpine Warty Newt; Italian Great Crested Newt
d Alpen-Kammolch; Italienischer Kammolch
f Triton crêté italien; Triton crêté d'talie
i Tritone crestato meridionale; Tritone crestato; Scaramàndla;

Sacaramàndla; Salamandra acquaiola

5247 ***Triturus cristatus***
Caudata - Salamandridae
e Northern Crested Newt; Great Crested Newt; Great Warty Newt; Crested Newt; Warty Newt
d Kammolch; Großer Teichmolch; Sumpfmolch; Nördlicher Kammolch
f Triton verruqueux; Triton crêté; Triton à crête; Triton à grande crête
i Tritone crestato

5248 ***Triturus cristatus x T.marmoratus***
Caudata - Salamandridae
f Triton de Blasius

5249 ***Triturus dobrogicus***
Caudata - Salamandridae
e Danube Newt; Danube Crested Newt; Danube Warty Newt
d Donau-Kammolch
f Triton crêté du Danube; Triton du Danube

5250 ***Triturus helveticus***
Caudata - Salamandridae
e Palmate Newt
d Schweizer Molch; Fadenmolch; Leistenmolch
f Triton helvétique; Triton palmé; Triton à pattes palmées

5251 ***Triturus italicus***
Caudata - Salamandridae
e Italian Newt
d Italienischer Wasssermolch; Italienischer Teichmolch
f Triton italien; Triton d'Italie
i Tritone italiano

5252 ***Triturus karelini***
Caudata - Salamandridae
e Karelin's Newt; Balkan Crested Newt; Balkan Warty Newt
d Persischer Kammolch; Balkan-Molch; Balkan-Kammolch; Südlicher Kammolch
f Triton de Karelin; Triton d'Asie mineure

5253 ***Triturus marmoratus***
Caudata - Salamandridae
e Marbled Newt
d Marmormolch; Nördlicher Marmormolch
f Triton marbré

5254 ***Triturus montadoni***
Caudata - Salamandridae
e Carpathian Newt; Montadon's Newt
d Karpaten-Molch
f Triton des Carpates; Triton de Montadon

5255 ***Triturus pygmaeus***
Caudata - Salamandridae
e Northern Marbled Newt

5256 ***Triturus vittatus***
Caudata - Salamandridae
e Banded Newt
d Bandmolch; Ophrytischer Bandmolch
f Triton à bandes; Triton du Caucase; Triton rubané

5257 ***Triturus vulgaris***
Caudata - Salamandridae
e Smooth Newt; Common Smooth Newt; Common Newt
d Streifenmolch; Teichmolch
f Triton vulgaire; Triton commun; Triton ponctué; Triton lobé
i Tritone punteggiato; Tritone comune; Scaramàndla; Sacaramàndla

5258 ***Triturus vulgaris meridionalis***
Caudata - Salamandridae
d Südlicher Teichmolch

5259 ***Tylototriton***
Caudata - Salamandridae

e Crocodile Newts
d Krokodilmolche

5260 ***Tylototriton kweichowensis***
Caudata - Salamandridae
e Kweichow Crocodile Newt
d Kweichow-Molch

5261 ***Tylototriton taliangensis***
Caudata - Salamandridae
e Pusakang Crocodile Newt
d Sichuan-Krokodilmolch; Schwarzer Krokodilmolch

5262 ***Tylototriton verrucosus***
Caudata - Salamandridae
e Crocodile Newt; Burmese Newt; Indian Newt; Himalayan Newt (ISC); Mandarin Salamander; Crocodile Salamander (SEA)
d Geknöpfter Krokodilmolch; Geknöpfter Birma-Krokodilmolch; Birma-Krokodilmolch; Gefleckter Krokodilmolch; Geknöpfter Burma-Krokodilmolch; Burma-Krokodilmolch; Krokodilsalamander

5263 ***Typhlomolge***
Caudata - Plethodontidae
e Blind Salamanders; Texas Blind Salamanders
d Brunnenmolche

5264 ***Typhlomolge robusta***
Caudata - Plethodontidae
e Blanco Blind Salamander

5265 ***Typhlonectes***
Gymnophiona - Typhlonectidae
e Aquatic Caecilians
d Schwimmwühlen
f Cécilies

5266 ***Typhlonectes compressicauda***
Gymnophiona - Typhlonectidae
e Cayenne Caecilian
d Schwimmwühle
f Typhlonecte queue-comprimée; Cécilie queue-comprimée

5267 ***Typhlonectes cunhai***
Gymnophiona - Typhlonectidae
e Cunha's Caecilian

5268 ***Typhlonectes eiselti***
Gymnophiona - Typhlonectidae
e Eiselt's Caecilian

5269 ***Typhlonectes natans***
Gymnophiona - Typhlonectidae
e Rio Cauca Caecilian; Rubber Eel

5270 ***Typhlonectes venezuelensis***
Gymnophiona - Typhlonectidae
e Caracas Caecilian

5271 **Typhlonectidae**
Gymnophionia
e Aquatic Caecilians
d Schwimmwühlenverwandten
f Typhlonectidés

5272 ***Typhlotriton***
Caudata - Plethodontidae
e Grotto Salamanders
d Grottensalamander (pl)
f Salamandres des grottes

5273 ***Typhlotriton spelaeus***
Caudata - Plethodontidae
e Grotto Salamander
d Grottensalamander
f Salamandre des grottes

U

5274 ***Uperodon***
Anura - Microhylidae
e Globular Frogs; Balloon Frogs (ISC)
f Upérodontes

5275 ***Uperodon globulosus***
Anura - Microhylidae
e Indian Globular Frog; Balloon Frog (ISC)

5276 ***Uperodon systoma***
Anura - Microhylidae
e Globular Frog; Marbled Balloon Frog (ISC); Indian Marbled Burrowing Frog
f Upérodonte marbrée

5277 ***Uperoleia***
Anura - Myobatrachidae
e Australian Toadlets

5278 ***Uperoleia altissima***
Anura - Myobatrachidae
e Davies's Toadlet

5279 ***Uperoleia arenicola***
Anura - Myobatrachidae
e Alligator Rivers Toadlet

5280 ***Uperoleia aspera***
Anura - Myobatrachidae
e Clicking Toadlet

5281 ***Uperoleia borealis***
Anura - Myobatrachidae
e Rasping Toadlet

5282 ***Uperoleia capitulata***
Anura - Myobatrachidae
e Grass Toadlet

5283 ***Uperoleia crassa***
Anura - Myobatrachidae
e Forest Toadlet

5284 ***Uperoleia fusca***
Anura - Myobatrachidae
e Brown Forest Toadlet

5285 ***Uperoleia glandulosa***
Anura - Myobatrachidae
e Blotched Toadlet

5286 ***Uperoleia inundata***
Anura - Myobatrachidae
e Mottled Toadlet

5287 ***Uperoleia laevigata***
Anura - Myobatrachidae
e Orange-knee Toadlet

5288 ***Uperoleia lithomoda***
Anura - Myobatrachidae
e Ticking Toadlet; Stonemason Toadlet

5289 ***Uperoleia littlejohni***
Anura - Myobatrachidae
e Littlejohn's Toadlet

5290 ***Uperoleia marmorata***
Anura - Myobatrachidae
e Gray's Marbled Toadlet; Yellow-spotted Toadlet (ANZ)

5291 ***Uperoleia martini***
Anura - Myobatrachidae
e Coastal Toadlet

5292 ***Uperoleia micromeles***
Anura - Myobatrachidae
e Metallic Toadlet

5293 ***Uperoleia mimula***
Anura - Myobatrachidae
e Monsoon Toadlet

5294 ***Uperoleia minima***
Anura - Myobatrachidae
e Mitchell Plateau Toadlet

5295 ***Uperoleia mjobergii***
Anura - Myobatrachidae
e Mjoberg Toadlet

5296 ***Uperoleia orientalis***
Anura - Myobatrachidae
e Groote Eyland Toadlet

5297 ***Uperoleia rugosa***
Anura - Myobatrachidae
e Mud Toadlet; Red-groined Toadlet (ANZ); Andersson's Toadlet

5298 ***Uperoleia russelli***
Anura - Myobatrachidae
e Russell's Toadlet

5299 ***Uperoleia talpa***
Anura - Myobatrachidae
e Plains Toadlet

5300 ***Uperoleia trachyderma***
Anura - Myobatrachidae
e Yellow Clay Toadlet

5301 ***Uperoleia tyleri***
Anura - Myobatrachidae
e Tyler's Toadlet

5302 ***Uraeotyphlus***
Gymnophiona - Uraeotyphlidae
e Indian Caecilians

5303 ***Uraeotyphlus malabaricus***
Gymnophiona - Uraeotyphlidae
e Malabar Caecilian

5304 ***Uraeotyphlus menoni***
Gymnophiona - Uraeotyphlidae
e Kerala Caecilian

5305 ***Uraeotyphlus narayani***
Gymnophiona - Uraeotyphlidae
e Kannan Caecilian

5306 ***Uraeotyphlus oxyurus***
Gymnophiona - Uraeotyphlidae
e Red Caecilian

5307 **Uraetoyphlidae**
Gymnophionia
e Indian Caecilians

V

5308 ***Vanzolinus***
Anura - Leptodactylidae
e Amazon Frogs

5309 ***Vanzolinus discodactylus***
Anura - Leptodactylidae
e Amazon Frog; Vanzolini's Amazon Frog

5310 ***Vibrissaphora***
Anura - Megophryidae
e Spiny Toads

5311 ***Vibrissaphora ailaonica***
Anura - Megophryidae
e Ailao Spiny Toad

5312 ***Vibrissaphora boringii***
Anura - Megophryidae
e Taosze Spiny Toad

5313 ***Vibrissaphora leishanensis***
Anura - Megophryidae
e Leishan Spiny Toad

5314 ***Vibrissaphora liui***
Anura - Megophryidae
e Pope's Spiny Toad

W

5315 ***Werneria***
Anura - Bufonidae
e Small-tongue Toads

5316 ***Werneria bambutensis***
Anura - Bufonidae
e Bamboutos Small-tongue Toad

5317 ***Werneria mertensiana***
Anura - Bufonidae
e Mertens's Small-tongue Toad

5318 ***Werneria preussi***
Anura - Bufonidae
e Buea Small-tongue Toad
d Bergbachkröte

5319 ***Werneria tandyi***
Anura - Bufonidae
e Tandy's Small-tongue Toad

5320 ***Wolterstorffina***
Anura - Bufonidae
e Wolterstorff Toads

5321 ***Wolterstorffina mirei***
Anura - Bufonidae
e Mount Okou Wolterstorff Toad

5322 ***Wolterstorffina parvipalmata***
Anura - Bufonidae
e Cameroon Wolterstorff Toad

X

5323 ***Xenobatrachus***
Anura - Microhylidae
e Fanged Frogs

5324 ***Xenobatrachus anorbis***
Anura - Microhylidae
e Fly River Fanged Frog

5325 ***Xenobatrachus arfakianus***
Anura - Microhylidae
e Arfak Fanged Frog

5326 ***Xenobatrachus bidens***
Anura - Microhylidae
e Digul River Fanged Frog

5327 ***Xenobatrachus giganteus***
Anura - Microhylidae
e Giant Fanged Frog

5328 ***Xenobatrachus huon***
Anura - Microhylidae
e Morobe Fanged Frog

5329 ***Xenobatrachus macrops***
Anura - Microhylidae
e Hellwig Fanged Frog

5330 ***Xenobatrachus mehelyi***
Anura - Microhylidae
e Mehely's Fanged Frog

5331 ***Xenobatrachus obesus***
Anura - Microhylidae
e Fat Fanged Frog

5332 ***Xenobatrachus ocellatus***
Anura - Microhylidae
e Ocellated Fanged Frog

5333 ***Xenobatrachus ophiodon***
Anura - Microhylidae
e Hatam Fanged Frog

5334 ***Xenobatrachus rostratus***
Anura - Microhylidae
e Irian Jaya Fanged Frog

5335 ***Xenobatrachus scheepstrai***
Anura - Microhylidae
e Indonesian Fanged Frog

5336 ***Xenobatrachus schiefenhoeveli***
Anura - Microhylidae
e Munggona Fanged Frog

5337 ***Xenobatrachus subcroceus***
Anura - Microhylidae
e Lae Fanged Frog

5338 ***Xenobatrachus tumulus***
Anura - Microhylidae
e Mandang Fanged Frog

5339 ***Xenobatrachusfuscigula***
Anura - Microhylidae
e Kaironk Valley Fanged Frog

5340 ***Xenopus***
Anura - Pipidae
e Common Clawed Frogs; Clawed Frogs; Clawed Toads (CSA); Platannas (CSA)
d Krallenfrösche
f Xénopes; Dactylèthres; Grenouilles à griffes; Crapauds à griffes; Crapauds du Cap
i Xenopi

5341 ***Xenopus amieti***
Anura - Pipidae

e Volcano Clawed Frog

5342 ***Xenopus andrei***
Anura - Pipidae
e Andre's Clawed Frog

5343 ***Xenopus borealis***
Anura - Pipidae
e Marsabit Clawed Frog; Kenya Clawed Frog

5344 ***Xenopus boubaensis***
Anura - Pipidae
e Mawa Clawed Frog

5345 ***Xenopus clivii***
Anura - Pipidae
e Eritrea Clawed Frog; Eritrean Smooth Clawed Frog

5346 ***Xenopus fraseri***
Anura - Pipidae
e Fraser's Clawed Frog
d Fraserscher Spornfrosch

5347 ***Xenopus gilli***
Anura - Pipidae
e Cape Clawed Frog; Cape Smooth Clawed Frog; Cape Platanna (CSA); Gill's Clawed Toad (CSA); Gill's Platanna (CSA); Gill's Clawed Frog (CSA)
d Gills Krallenfrosch; Zwergkrallenfrosch; Südafrikanischer Krallenfrosch; Kleiner Kap-Krallenfrosch
f Xénope du Cap

5348 ***Xenopus laevis***
Anura - Pipidae
e African Clawed Frog; Smooth Clawed Frog; Upland Clawed Frog; African Clawed Toad; Clawed Toad (CSA); Platanna (CSA); Common Platanna (CSA)
d Glatter Krallenfrosch; Großer Krallenfrosch; Platanna; Krallenfrosch; Apothekerfrosch
f Dactylèthre du Cap; Crapaud à griffes; Platanna; Xénope commun
i Xenopo liscio

5349 ***Xenopus laevis laevis***
Anura - Pipidae
e Common Platanna (CSA); Platanna (CSA); Common Clawed Toad (CSA); Common Clawed Frog (CSA)

5350 ***Xenopus laevis poweri***
Anura - Pipidae
e Power's Platanna (CSA); Power's Clawed Toad (CSA); Power's Clawed Frog

5351 ***Xenopus longipes***
Anura - Pipidae
e Savannah Clawed Frog

5352 ***Xenopus muelleri***
Anura - Pipidae
e Müller's Clawed Frog; Müller's Smooth Clawed Frog; Savanna Clawed Frog; Northern Platanna; Tropical Platanna (CSA); Mueller's Platanna (CSA); Northern Clawed Frog (CSA)
d Müllers Krallenfrosch

5353 ***Xenopus pygmaeus***
Anura - Pipidae
e Bouchia Clawed Frog

5354 ***Xenopus ruwenzoriensis***
Anura - Pipidae
e Uganda Clawed Frog

5355 ***Xenopus vestitus***
Anura - Pipidae
e Kivu Clawed Frog

5356 ***Xenopus wittei***
Anura - Pipidae

e De Witte's Clawed Frog

5357 ***Xenorhina***
Anura - Microhylidae
e Snouted Frogs

5358 ***Xenorhina bouwensi***
Anura - Microhylidae
e Arafak Snouted Frog

5359 ***Xenorhina eiponis***
Anura - Microhylidae
e Eipomek Snouted Frog

5360 ***Xenorhina minima***
Anura - Microhylidae
e Parker's Snouted Frog

5361 ***Xenorhina oxycephala***
Anura - Microhylidae
e Schlegel's Snouted Frog

5362 ***Xenorhina parkerorum***
Anura - Microhylidae
e Imigabip Snouted Frog

5363 ***Xenorhina similis***
Anura - Microhylidae
e Zweifel's Snouted Frog

Z

5364 ***Zachaenus***
Anura - Leptodactylidae
e Bug-eyed Frogs

5365 ***Zachaenus carvalhoi***
Anura - Leptodactylidae
e Carvalho's Bug-eyed Frog

5366 ***Zachaenus parvulus***
Anura - Leptodactylidae
e Rio Bug-eyed Frog; Girard's Frog
d Girards Frosch
f Grenouille de Girard

5367 ***Zachaenus roseus***
Anura - Leptodactylidae
e Rosy Bug-eyed Frog

English Index

Deutches Register

D

Index Français

F

Indice Italiano

ISBN 0-444-51374-4